HUMAN DEVELOPMENT 92/93

Twentieth Edition

Editor

Larry Fenson
San Diego State University

Larry Fenson is a professor of psychology at San Diego State University. He received his Ph.D. in child psychology from the Institute of Child Behavior and Development at the University of Iowa in 1968. Dr. Fenson is a member of the MacArthur Foundation Research Network on Infancy and Early Childhood. His research focuses on early conceptual development, and he has authored articles on infant attention, symbolic play, concept development, and language acquisition.

Editor

Judith Fenson
Children's Hospital

Judith Fenson received a B.A. from Ohio University in 1963 and an M.A. from the University of New Mexico in 1965. She is the Language Data Coordinator at the Language Research Center at Children's Hospital. She has contributed to a variety of research projects in medicine, psychology, and linguistics, and has authored a variety of materials and study aids for students in psychology and child development.

Cover illustration by Mike Eagle

Annual Editions
A Library of Information from the Public Press

The Dushkin Publishing Group, Inc.
Sluice Dock, Guilford, Connecticut 06437

The Annual Editions Series

Annual Editions is a series of over 55 volumes designed to provide the reader with convenient, low-cost access to a wide range of current, carefully selected articles from some of the most important magazines, newspapers, and journals published today. Annual Editions are updated on an annual basis through a continuous monitoring of over 300 periodical sources. All Annual Editions have a number of features designed to make them particularly useful, including topic guides, annotated tables of contents, unit overviews, and indexes. For the teacher using Annual Editions in the classroom, an Instructor's Resource Guide with test questions is available for each volume.

VOLUMES AVAILABLE

- Africa
- Aging
- American Government
- American History, Pre-Civil War
- American History, Post-Civil War
- Anthropology
- Biology
- Business and Management
- Business Ethics
- Canadian Politics
- China
- Comparative Politics
- Computers in Education
- Computers in Business
- Computers in Society
- Criminal Justice
- Drugs, Society, and Behavior
- Early Childhood Education
- Economics
- Educating Exceptional Children
- Education
- Educational Psychology
- Environment
- Geography
- Global Issues
- Health
- Human Development
- Human Resources
- Human Sexuality
- International Business
- Japan
- Latin America
- Life Management
- Macroeconomics
- Management
- Marketing
- Marriage and Family
- Microeconomics
- Middle East and the Islamic World
- Money and Banking
- Nutrition
- Personal Growth and Behavior
- Physical Anthropology
- Psychology
- Public Administration
- Race and Ethnic Relations
- Social Problems
- Sociology
- Soviet Union (Commonwealth of Independent States and Central Europe)
- State and Local Government
- Third World
- Urban Society
- Violence and Terrorism
- Western Civilization, Pre-Reformation
- Western Civilization, Post-Reformation
- Western Europe
- World History, Pre-Modern
- World History, Modern
- World Politics

Library of Congress Cataloging in Publication Data
Main entry under title: Annual Editions: Human development. 1992/93.
    1. Child study—Periodicals. 2. Socialization—Periodicals. 3. Old age—Periodicals.
I. Fenson, Larry, *comp.*; Fenson, Judith, *comp.* II. Title: Human development.
ISBN 1-56134-092-8   155'.05   72-91973
HQ768.A55

© 1992 by The Dushkin Publishing Group, Inc. Annual Editions ® is a Registered Trademark of The Dushkin Publishing Group, Inc.

Copyright © 1992 by The Dushkin Publishing Group, Inc., Guilford, Connecticut 06437

All rights reserved. No part of this book may be reproduced, stored, or transmitted by any means—mechanical, electronic, or otherwise—without written permission from the publisher.

Twentieth Edition

Manufactured by The Banta Company, Harrisonburg, Virginia 22801

Editors/Advisory Board

EDITORS

Larry Fenson
San Diego State University

Judith Fenson
Children's Hospital

ADVISORY BOARD

H. Wade Bedwell
Harding University

Judith E. Blakemore
Indiana University-Purdue University

Mary Anne Christenberry
College of Charleston

Karen Duffy
State University College Geneseo

Bonnie Duguid-Siegal
University of Western Sydney

Judy Gray
Colorado Northwestern Community College

Mark Greenberg
University of Washington

Don Hamacheck
Michigan State University

Gregory F. Harper
SUNY College, Fredonia

Alice S. Honig
Syracuse University

Angela J. C. LaSala
Community College of Southern Nevada

Lynda G. MacCulloch
Acadia University

David S. McKell
Northern Arizona University

Carroll Mitchell
Cecil Community College

Martin Murphy
University of Akron

Harriett Ritchie
American River College

Gary M. Schumacher
Ohio University

Diana T. Slaughter
Northwestern University

William H. Strader
Fitchburg State College

James R. Wallace
St. Lawrence University

Karen Zabrucky
Georgia State University

Members of the Advisory Board are instrumental in the final selection of articles for each edition of Annual Editions. Their review of articles for content, level, currentness, and appropriateness provides critical direction to the editor and staff. We think you'll find their careful consideration well reflected in this volume.

STAFF

Ian A. Nielsen, Publisher
Brenda S. Filley, Production Manager
Roberta Monaco, Editor
Addie Raucci, Administrative Editor
Cheryl Greenleaf, Permissions Editor
Diane Barker, Editorial Assistant
Lisa Holmes-Doebrick, Administrative Coordinator
Charles Vitelli, Designer
Shawn Callahan, Graphics
Meredith Scheld, Graphics
Steve Shumaker, Graphics
Libra A. Cusack, Typesetting Supervisor
Juliana Arbo, Typesetter

To the Reader

In publishing ANNUAL EDITIONS we recognize the enormous role played by the magazines, newspapers, and journals of the *public press* in providing current, first-rate educational information in a broad spectrum of interest areas. Within the articles, the best scientists, practitioners, researchers, and commentators draw issues into new perspective as accepted theories and viewpoints are called into account by new events, recent discoveries change old facts, and fresh debate breaks out over important controversies.

Many of the articles resulting from this enormous editorial effort are appropriate for students, researchers, and professionals seeking accurate, current material to help bridge the gap between principles and theories and the real world. These articles, however, become more useful for study when those of lasting value are carefully *collected, organized, indexed,* and *reproduced* in a *low-cost format*, which provides easy and permanent access when the material is needed. That is the role played by *Annual Editions*. Under the direction of each volume's *Editor*, who is an expert in the subject area, and with the guidance of an *Advisory Board*, we seek each year to provide in each ANNUAL EDITION a current, well-balanced, carefully selected collection of the best of the public press for your study and enjoyment. We think you'll find this volume useful, and we hope you'll take a moment to let us know what you think.

Any history of the field of human development will reflect the contributions of the many individuals who helped craft the topical content of the discipline. For example, Binet launched the intelligence test movement, Freud focused attention on personality development, and Watson and Thorndike paved the way for the emergence of social learning theory. However, the philosophical principles that give definition to the field of human development have their direct ancestral roots in the evolutionary biology of Darwin, Wallace, and Spencer, and in the embryology of Preyer. Each of the two most influential developmental psychologists of the early twentieth century, James Mark Baldwin and G. Stanley Hall, was markedly influenced by questions about phylogeny (species' adaptation) and ontogeny (individual adaptation or fittingness). Baldwin's persuasive arguments challenged the assertion that changes in species precede changes in individual organisms. Instead, Baldwin argued, ontogeny not only precedes phylogeny but is the process that shapes phylogeny. Thus, as Robert Cairns points out, developmental psychology has always been concerned with the study of the forces that guide and direct development. Early theories stressed that development was the unfolding of already formed or predetermined characteristics. Many contemporary students of human development embrace the epigenetic principle that asserts that development is an emergent process of active, dynamic, reciprocal, and systemic change. This systems perspective forces one to think about the historical, social, cultural, interpersonal, and intrapersonal forces that shape the developmental process.

The study of human development involves all fields of inquiry comprising the social, natural, and life sciences and professions. The need for depth and breadth of knowledge creates a paradox: While students are being advised to acquire a broad-based education, each discipline is becoming more highly specialized. One way to combat specialization is to integrate the theories and findings from a variety of disciplines with those of the parent discipline. This, in effect, is the approach of *Annual Editions: Human Development 92/93*. This anthology includes articles that discuss the problems, issues, theories, and research findings from many fields of study. In most instances, the articles were written specifically to communicate information about recent scientific findings or controversial issues to the general public. As a result, the articles tend to blend the history of a topic with the latest available information. In many instances, the reader is challenged to consider the personal and social implications of the topic. The articles included in this anthology were selected by the editors with valued advice and recommendations from an advisory board consisting of faculty from community colleges, small liberal arts colleges, and large universities. Evaluations obtained from students, instructors, and advisory board members influenced the decision to retain or replace specific articles. Throughout the year we screen many articles for accuracy, interest value, writing style, and recency of information. Readers can have input into the next edition by completing and returning the article rating form in the back of the book.

Human Development 92/93 is organized into six major units. Unit 1 focuses on the origins of life, including genetic influences on development, and Unit 2 focuses on development during infancy and early childhood. Unit 3 is divided into subsections addressing social, emotional, and cognitive development. Unit 4 addresses issues related to family, school, and cultural influences on development. Units 5 and 6 cover human development from adolescence to old age. In our experience, this organization provides great flexibility for those using the anthology with any standard textbook. The units can be assigned sequentially, or instructors can devise any number of arrangements of individual articles to fit their specific needs. In large lecture classes, this anthology seems to work best as assigned reading to supplement the basic text. In smaller sections, articles can stimulate instructor-student discussions. Regardless of the instructional style used, we hope that our excitement for the study and teaching of human development is evident and catching as you read the articles in this twentieth edition of *Human Development*.

Larry Fenson

Judith Fenson

Editors

Contents

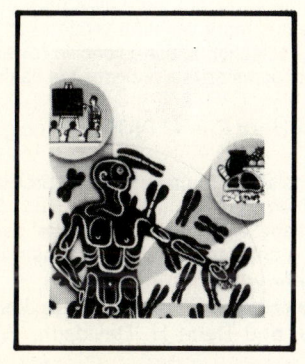

To the Reader ... iv
Topic Guide ... 1
Overview ... 4

Unit 1

Genetic and Prenatal Influences on Development

Seven selections discuss genetic and societal influences on development, focusing on reproductive technology, genetic influences, and chemical effects on development.

1. **Suffer the Little Children: Shameful Bequests to the Next Generation,** *Time,* October 8, 1990. ... 6
   These selections are a forceful indictment of **America's treatment of its children**. From prenatal care to sex education and social services, funding is grossly inadequate, argue a wide spectrum of academics, politicians, and business executives. Given that prevention is far less costly than later "repairs," they urge major reordering of our national priorities.

2. **The Gene Dream,** Natalie Angier, *American Health,* March 1989. ... 12
   An ambitious $3 billion project has been launched to map **the human genome**, the complete set of instructions for making a human being. This work promises to revolutionize medicine but will also create some vexing ethical problems.

3. **Made to Order Babies,** Geoffrey Cowley, *Newsweek,* Special Issue, Winter/Spring 1990. ... 16
   As knowledge from the Genome Project increases, so do dilemmas about how to use such knowledge. Will **future parents pick the child they want** on the basis of characteristics such as hair color, personality, or body type?

4. **Sperm Under Siege,** Anne Merewood, *Health,* April 1991. ... 19
   Both parents, not just the mother, share responsibility for bearing a healthy baby. New findings suggest that the **sperm may be more vulnerable** than previously thought, and may be affected by such factors as smoking, alcohol, drugs, radiation, and chemical exposure.

5. **Clipped Wings,** Lucile F. Newman and Stephen L. Buka, *American Educator,* Spring 1991. ... 23
   This article examines how **prenatal exposure to alcohol, drugs, and smoking** affects children's learning. While many of the children exposed to these conditions are at risk, some promising postnatal treatments are described.

6. **What Crack Does to Babies,** Janice Hutchinson, *American Educator,* Spring 1991. ... 29
   The author describes **what happens when crack enters the body** and how it can damage the developing fetus. It can affect the child's temperament, affective development, cognitive development, and motor skills, producing permanent disabilities. As a consequence, these children are very hard to reach, presenting major challenges to educators and other professionals responsible for the welfare of these children.

7. **Motherhood on Trial,** David Ruben, *Parenting Magazine,* June/July 1990. ... 31
   Should a woman be arrested because she used illegal drugs during pregnancy? The pros and cons of **fetal abuse** prosecution are discussed in this article.

Unit 2

Development During Infancy and Early Childhood

Seven selections discuss development of the brain and development of communication, emotions, and cognitive systems during the first years of life.

Overview ... 36

8. **How Infants See the World,** *U.S. News & World Report,* August 20, 1990. ... 38
   This article profiles new research insights into **affective and emotional development** in infancy.

9. **Diary of a Baby,** Daniel N. Stern, *U.S. News & World Report,* August 20, 1990. ... 40
   Daniel N. Stern gives a sensitive glimpse into what **the thoughts and feelings of a baby** may be like. In the process, he documents and dramatizes the rapid advances in development that occur in the early years of life.

The concepts in bold italics are developed in the article. For further expansion please refer to the Topic Guide and the Index.

10. **Guns and Dolls,** Laura Shapiro, *Newsweek,* May 28, 1990. 44
Are boys and girls born different or are *gender differences* the result of cultural influences? Scientists representing many different perspectives agree that the sexes are more alike than different, and they argue that there is more variation within each sex than between the sexes. This selection offers a concise review of the evidence in a wide array of domains that provides the basis for this consensus.

11. **Preschool: Head Start or Hard Push?** Philip R. Piccigallo, *Social Policy,* Fall 1988. 49
Many parents select preschools with formal education programs for their children with the idea that this will help ensure later success in life. However, many experts warn that the consequences of pressuring young children can actually impair long-term *educational and psychological development.*

12. **How Three Key Countries Shape Their Children,** Joseph J. Tobin, David Y. H. Wu, and Dana H. Davidson, *World Monitor,* April 1989. 53
A typical day in the life of 4-year-olds varies greatly in preschools in China, Japan, and the United States. The contrasts reflect widely *differing values and aspirations for children* in these three cultures.

13. **The Day Care Generation,** Pat Wingert and Barbara Kantrowitz, *Newsweek,* Special Issue, Winter/Spring 1990. 60
Three of the biggest *concerns involving day care* are discussed in this article. Are infants in day care most likely to grow up maladjusted? Is the high turnover of caregivers damaging? Does day care put children at greater risk for contracting minor and major ailments?

14. **Where Pelicans Kiss Seals,** Ellen Winner, *Psychology Today,* August 1986. 63
The development of *drawing is a complex process.* Children's drawings move from scribbles to tadpole people to more technically skilled pictures. This article captures these charming stages and examines the underlying changes in their thinking that account for them.

Overview 70

A. SOCIAL AND EMOTIONAL DEVELOPMENT

15. **Building Confidence,** Bruce A. Baldwin, *PACE Magazine,* October 1988. 72
The author presents a "Confidence Builder's Checklist" to help parents respond to their children in *ways that promote healthy achievement* motivation.

16. **Dealing With Difficult Young Children,** Anne K. Soderman, *Young Children,* July 1985. 75
Children seem to differ in temperamental styles right from the beginning, and these differences become more pronounced with age. This article identifies the major dimensions along which *children vary temperamentally,* and it considers strategies for molding such differences into strengths rather than problems.

17. **The Miracle of Resiliency,** David Gelman, *Newsweek,* Special Issue, Summer 1991. 79
Some children seem especially resilient and able to thrive even under thoroughly negative conditions. A positive self-image and the existence of a good relationship with an adult have been found to be two of the most vital features of these *children's success in dealing with adversity.*

B. COGNITIVE AND LANGUAGE DEVELOPMENT

18. **Three Heads Are Better Than One,** Robert J. Trotter, *Psychology Today,* August 1986. 82
The author reviews Robert J. Sternberg's *triarchic theory of intelligence* and contrasts it with the traditional IQ approach. The triarchic theory de-emphasizes speed and accuracy, focusing instead on *executive processes* related to planning and evaluating. The emphasis on *practical intelligence* and *multidimensional intelligence* has revitalized the study of individual differences in mental abilities.

Unit 3

Development During Childhood

Nine selections examine human development during childhood, paying specific attention to social and emotional development, cognitive and language development, and developmental problems.

The concepts in bold italics are developed in the article. For further expansion please refer to the Topic Guide and the Index.

19. **How Kids Learn,** Barbara Kantrowitz and Pat Wingert, *Newsweek,* April 17, 1989.   88

Young children between the ages of five and eight learn best through **hands-on teaching methods**. They learn at different rates, and are at an age where acquiring social skills is also important. New school programs are being developed that incorporate these findings in the curriculum.

20. **Now We're Talking!** Bernard Ohanian and Greta Vollmer, *Parenting Magazine,* October 1989.   94

Sometime between 10 and 14 months, most kids begin talking. By 5 years of age they have an average vocabulary of 8,000 words. Experts cited in this article encourage parents to make language fun for their children by incorporating rich communication into one-on-one social activities such as playing and reading. However, they also caution against going to extremes and assuming the role of teacher rather than parent.

C. DEVELOPMENTAL PROBLEMS

21. **Suffer the Restless Children,** Alfie Kohn, *The Atlantic,* November 1989.   98

A child may be labeled **hyperactive** and as a result be given a drug, such as Ritalin, to control behavior. However, teachers, parents, and clinicians often do not agree on this label and, in fact, it may be environment-specific. This article raises the question of whether hyperactivity is really a disorder at all.

22. **Tykes and Bytes,** Bettijane Levine, *Los Angeles Times,* July 1, 1990.   105

Children with a range of disabilities that make language and motor activities difficult, if not impossible, can now gain a new lease on life thanks to **computer technology**. Specially designed computers allow these children to communicate and control aspects of their environment. Moreover, many of these children, some as young as 18 months, get to play for the first time in their lives.

23. **Facts About Dyslexia,** *Children Today,* November/December 1985.   108

This informative article **lists symptoms of dyslexia**, examines possible causes, describes three treatment programs, and discusses the prognosis of this language-related disability.

Overview   112

A. PARENTING

24. **Dr. Spock Had It Right,** *U.S. News & World Report,* August 7, 1989.   114

While there are no magic formulas for raising children, Dr. Benjamin Spock has maintained all along that the most **effective parental discipline** combines control with warmth and nurturance.

25. **Positive Parenting,** Bruce A. Baldwin, *PACE Magazine,* November 1988.   116

The author uses the term "Cornucopia Kids" to describe kids today who want it all, now, without working for it. He suggests some guidelines for parents to use at home to help **provide their children with sound values**.

26. **Can Your Career Hurt Your Kids?** Kenneth Labich, *Fortune,* May 20, 1991.   119

Day care is a necessary fact of life for an ever-growing number of single-parent and two-parent families as the proportion of stay-at-home mothers continues to decrease in U.S. society. This article reviews some of the pluses and minuses of day care, and examines the emerging roles being played by companies facilitating and, in part, underwriting day care for their employees.

Unit 4

Family, School, and Cultural Influences on Development

Thirteen selections discuss the impact of home, school, and culture on child rearing and child development. The topics include parenting styles, family structure, stress, cultural influences, and education.

The concepts in bold italics are developed in the article. For further expansion please refer to the Topic Guide and the Index.

B. STRESS AND MALTREATMENT

27. **The Lasting Effects of Child Maltreatment,** Raymond H. Starr, Jr., *The World & I,* June 1990.   123
Do abused children grow up to be abusive adults? Is there a *relationship between maltreatment and criminality*? These and other issues of great contemporary significance are discussed in this informative article.

28. **Children After Divorce,** Judith S. Wallerstein, *The New York Times Magazine,* January 22, 1989.   128
The impact of divorce on the child was always thought to be somewhat passing in nature; the shock of a family broken by divorce was considered an unfortunate experience, but one with only a short-term effect on the child. As this article points out, this has been found to be false, and the longitudinal study discussed finds that the effects of a divorce on the child are much more significant than was originally thought.

29. **Children of Violence,** Lois Timnick, *Los Angeles Times Magazine,* September 3, 1989.   134
How does *growing up in a violent home*, school, or neighborhood affect children? Research indicates that these children are subject to depression or anxiety, and they tend to be more aggressive than peers not exposed to such trauma. These children may become desensitized as their exposure to violence increases. Experts stress the need for early intervention.

30. **Children Under Stress,** *U.S. News & World Report,* October 27, 1986.   140
Children growing up in today's culture are more likely to suffer stress for many specific reasons. The long-term effects of stress have not yet been determined, but some of the clues are very disturbing.

C. CULTURAL INFLUENCES

31. **Rumors of Inferiority,** Jeff Howard and Ray Hammond, *The New Republic,* September 9, 1985.   146
Studies cited in this article link deficiencies in the process of *cognitive development* to continued *social-economic underdevelopment* in America's black population. The authors argue that differences in performance between blacks and whites are related to self-doubt and fear of intellectual abilities on the part of blacks. Internalized negative expectations affect competitive behaviors and cause blacks to avoid situations that might reinforce their feelings of inferiority.

32. **Biology, Destiny, and All That,** Paul Chance, *Across the Board,* July/August 1988.   151
The author reviews evidence for sex differences in *aggression, self-confidence, rational thinking*, and *emotionality*, and finds that while there are some differences, they tend to be small.

33. **Alienation and the Four Worlds of Childhood,** Urie Bronfenbrenner, *Phi Delta Kappan,* February 1986.   156
Disorganized *families* and disorganized *environments* contribute to the adolescent's sense of *alienation* from family, friends, school, and work. This article links dramatic changes in the structure of the family to social upheaval in the United States, which ranks first among the industrialized countries of the world in poverty, teen pregnancy, and teen drug use.

D. EDUCATION

34. **Tracked To Fail,** Sheila Tobias, *Psychology Today,* September 1989.   161
More and more children are tested at an early age and labeled as "fast" or "slow." Unfortunately, this label tends to stick all through school, often producing highly negative effects on *self-esteem and academic success*.

The concepts in bold italics are developed in the article. For further expansion please refer to the Topic Guide and the Index.

Unit 5

Development During Adolescence and Early Adulthood

Seven selections examine some of the effects of social environment, sibling relationships, sex differences, and jealousy on human development during adolescence and early adulthood.

35. **Master of Mastery,** Paul Chance, *Psychology Today,* April 1987. — 165

Benjamin Bloom believes there is too much drill and rote learning in our schools and not enough active participation. He and his colleagues have developed a system called ***mastery learning***. Studies have shown that mastery learning students have done better than 85 percent of students taught in the traditional way.

36. **Not Just for Nerds,** *Newsweek,* April 9, 1990. — 168

American students' knowledge of science and mathematics, as is often reported, lags far behind students in most other industrialized countries. Many regard the gap as a legitimate crisis in the making. This article describes some of the programs that have been proposed or implemented in hopes of reversing the flight from biology, chemistry, and other disciplines that now play such large roles in our lives and in the nation's competitiveness.

Overview — 170

37. **The Myth About Teen-Agers,** Richard Flaste, *The New York Times Magazine,* October 9, 1988. — 172

Adolescence has its ups and downs, but past accounts have generally stressed the negative. The view of adolescence as a period filled with turmoil is gradually being replaced by one picturing the adolescent as complicated and changeable, but basically well-adapted and happy.

38. **Those Gangly Years,** Anne C. Petersen, *Psychology Today,* September 1987. — 175

The ***differing effects of early and late maturation on girls and boys*** are highlighted in this summary of a three-year study of 335 adolescents. The experiences affected by the timing of puberty include satisfaction with appearance, school achievement, moods, and interaction with members of the opposite sex.

39. **A Much Riskier Passage,** David Gelman, *Newsweek,* Special Issue, Summer/Fall 1990. — 180

Today's generation of teenagers differs from its predecessors in a number of ways. According to this article, they are on their own more; they have greater access to drugs, sex, and fast cars; and many have unrewarding jobs in the service industry. In spite of what appears to be major differences, experts say teens reflect the values and life-styles of their parents.

40. **Puberty and Parents: Understanding Your Early Adolescent,** Bruce A. Baldwin, *PACE Magazine,* October 1986. — 184

The striving of adolescents for ***independence*** is linked to parental apprehensions focused on ***risk taking***, ***societal dangers***, and ***lack of control***. The author proposes three ***stages*** of adolescence spanning the years from 10 to 30, lists 15 adolescent ***attitudes***, links them to 5 ***aroused emotions*** in parents, and tops off the article by listing 3 teen needs.

41. **Therapists Find Last Outpost of Adolescence in Adulthood,** Daniel Goleman, *The New York Times,* November 8, 1988. — 189

Emotional dependence on parents does not end with completion of adolescence, but continues well into the 20s, say researchers, who cite the mid- to late 20s as a more realistic end point for psychological dependence.

42. **Proceeding With Caution,** David M. Gross and Sophfronia Scott, *Time,* July 16, 1990. — 191

This article examines today's generation of young Americans between 18 and 29 years of age—***the twentysomething generation***. Their views on such issues as family, dating, education, careers, and role models are presented.

43. **Jealousy and Envy: The Demons Within Us,** Jon Queijo, *Bostonia,* May/June 1988. — 195

Social psychologists view the interrelated ***emotions*** of jealousy and envy in terms of ***motivation and self-esteem***. Hostility and privacy signal jealousy and envy. According to this article, three strategies—***self-reliance, selective ignoring***, and ***self-bolstering***—help individuals to cope with negative emotions.

The concepts in bold italics are developed in the article. For further expansion please refer to the Topic Guide and the Index.

Unit 6

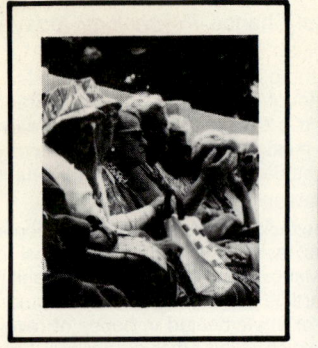

Development During Middle and Late Adulthood

Seven selections explore how family life-styles, loneliness, and depression relate to development during adulthood and consider how physical, cognitive, and social changes affect the aged.

Overview   200

44. **Family Ties: The Real Reason People Are Living Longer,** Leonard A. Sagan, *The Sciences,* March/April 1988.   202

    The author contends that it is impossible to trace the hardiness of modern people directly to improvements in medicine, sanitation, or diet. Personal relationships and the dynamics of a strong family, on the other hand, do have a direct impact. *Good health* is as much a *social and psychological achievement* as a physical one.

45. **The Vintage Years,** Jack C. Horn and Jeff Meer, *Psychology Today,* May 1987.   208

    As a group, America's 28 million citizens over 65 years of age are healthy, vigorous, and economically independent—contrary to many *stereotyped* views of the aged. Most elderly people live in their own homes or apartments, usually with a partner. Many of the elderly suffer from disorders such as *Alzheimer's disease,* and providing care for them can be stressful.

46. **Why Do We Age?** Ken Flieger, *FDA Consumer,* October 1988.   216

    Do our genes govern the length of our life or is aging controlled predominantly by environmental influences? Despite the bold advances made in biology and medicine in the past several decades, these questions remain unanswered. Yet, as this selection explains, the *complexities of the aging process* are being unraveled little by little.

47. **A Vital Long Life: New Treatments for Common Aging Ailments,** Evelyn B. Kelly, *The World & I,* April 1988.   220

    The number of persons 65 and older has increased from 4 percent of the population in 1900 to 12 percent in 1985. In this article, the author reviews some of the new treatments and attitudes that have contributed to this trend.

48. **The Myths of Menopause,** Lisa Davis, *In Health,* May/June 1989.   225

    Menopause is an often feared, frequently misunderstood influence on women's lives. The author provides a great deal of information on this topic and relates steps that can be taken before, during, and after menopause to minimize problems.

49. **Don't Act Your Age!** Carol Tavris, *American Health,* July/August 1989.   229

    This article throws down the gauntlet to theorists, such as Erik Erikson, who argue that life progresses through a series of more or less fixed psychological stages. Author Carol Tavris argues that *life's transitions* are far less fixed and their effects seldom as predictable as the stage theorists maintain. She suggests we celebrate the variety of possible life-styles and take advantage of opportunities, regardless of age.

50. **The Prime of Our Lives,** Anne Rosenfeld and Elizabeth Stark, *Psychology Today,* May 1987.   233

    The authors examine the essence of change, reviewing *Sheehy's mid-life crises*, *Erikson's eight stages of human development*, and *Levinson's ladder*. They suggest that *cohort effects* may provide a better explanation for changes during adulthood. The ages-and-stages approach to adult development, popular in the 1970s, clearly is under attack.

Index   241
Article Review Form   244
Article Rating Form   245

The concepts in bold italics are developed in the article. For further expansion please refer to the Topic Guide and the Index.

Topic Guide

This topic guide suggests how the selections in this book relate to topics of traditional concern to students and professionals involved with the study of human development. It is useful for locating articles that relate to each other for reading and research. The guide is arranged alphabetically according to topic. Articles may, of course, treat topics that do not appear in the topic guide. In turn, entries in the topic guide do not necessarily constitute a comprehensive listing of all the contents of each selection.

TOPIC AREA	TREATED IN:	TOPIC AREA	TREATED IN:
Abortion	3. Made to Order Babies	**Cognitive Development**	9. Diary of a Baby 14. Where Pelicans Kiss Seals 17. Miracle of Resiliency 18. Three Heads Are Better Than One 19. How Kids Learn 22. Tykes and Bytes 23. Facts About Dyslexia 26. Can Your Career Hurt Your Kids? 31. Rumors of Inferiority 32. Biology, Destiny, and All That 45. Vintage Years
Adolescence/ Adolescent Development	33. Alienation and the Four Worlds of Childhood 37. Myth About Teen-Agers 38. Those Gangly Years 39. Much Riskier Passage 41. Therapists Find Last Outpost of Adolescence in Adulthood 43. Jealousy and Envy: The Demons Within Us		
		Competence	17. Miracle of Resiliency 29. Children of Violence
Adulthood	32. Biology, Destiny, and All That 41. Therapists Find Last Outpost of Adolescence in Adulthood 42. Proceeding With Caution 43. Jealousy and Envy: The Demons Within Us 49. Don't Act Your Age! 50. Prime of Our Lives	**Creativity**	14. Where Pelicans Kiss Seals 18. Three Heads Are Better Than One 22. Tykes and Bytes
		Day Care	1. Suffer the Little Children: Shameful Bequests to the Next Generation 13. Day Care Generation 26. Can Your Career Hurt Your Kids?
Aggression/ Violence	10. Guns and Dolls 29. Children of Violence 32. Biology, Destiny, and All That	**Depression/Despair**	28. Children After Divorce 29. Children of Violence 34. Tracked to Fail
Aging	45. Vintage Years 47. Vital Long Life: New Treatments for Common Aging Ailments 48. Myths of Menopause 49. Don't Act Your Age!	**Developmental Disabilities**	5. Clipped Wings 6. What Crack Does to Babies 17. Miracle of Resiliency 22. Tykes and Bytes 23. Facts About Dyslexia
Alienation	1. Suffer the Little Children: Shameful Bequests to the Next Generation 29. Children of Violence 33. Alienation and the Four Worlds of Childhood 43. Jealousy and Envy: The Demons Within Us	**Divorce**	1. Suffer the Little Children: Shameful Bequests to the Next Generation 28. Children After Divorce 30. Children Under Stress
Attachment	8. How Infants See the World 17. Miracle of Resiliency 26. Can Your Career Hurt Your Kids? 44. Family Ties: The Real Reason People Are Living Longer	**Drug Abuse**	1. Suffer the Little Children: Shameful Bequests to the Next Generation 5. Clipped Wings 6. What Crack Does to Babies 7. Motherhood on Trial 39. Much Riskier Passage
Brain Organization/ Function	23. Facts About Dyslexia	**Education/ Educators**	1. Suffer the Little Children: Shameful Bequests to the Next Generation 11. Preschool: Head Start or Hard Push? 12. How Three Key Countries Shape Their Children 18. Three Heads Are Better Than One 19. How Kids Learn
Child Abuse	1. Suffer the Little Children: Shameful Bequests to the Next Generation 17. Miracle of Resiliency 27. Lasting Effects of Child Maltreatment		

TOPIC AREA	TREATED IN:	TOPIC AREA	TREATED IN:
Education/ Educators (cont'd)	21. Suffer the Restless Children 22. Tykes and Bytes 23. Facts About Dyslexia 31. Rumors of Inferiority 33. Alienation and the Four Worlds of Childhood 34. Tracked to Fail 35. Master of Mastery 36. Not Just for Nerds	Giftedness	18. Three Heads Are Better Than One
		High Risk Infants	3. Made to Order Babies 5. Clipped Wings 6. What Crack Does to Babies 7. Motherhood On Trial
		Hyperactivity	21. Suffer the Restless Children
Emotional Development	8. How Infants See the World 9. Diary of a Baby 10. Guns and Dolls 13. Day Care Generation 15. Building Confidence 23. Facts About Dyslexia 24. Dr. Spock Had It Right 26. Can Your Career Hurt Your Kids? 27. Lasting Effects of Child Maltreatment 28. Children After Divorce 29. Children of Violence 30. Children Under Stress 32. Biology, Destiny, and All That 33. Alienation and the Four Worlds of Childhood 34. Tracked to Fail 37. Myth About Teen-Agers 39. Much Riskier Passage 41. Therapists Find Last Outpost of Adolescence in Adulthood 43. Jealousy and Envy: The Demons Within Us 44. Family Ties: The Real Reason People Are Living Longer	Infant Development	8. How Infants See the World 9. Diary of a Baby 13. Day Care Generation
		Intelligence	18. Three Heads Are Better Than One 31. Rumors of Inferiority 34. Tracked to Fail
		Language Development	20. Now We're Talking!
		Learning	11. Preschool: Head Start or Hard Push? 19. How Kids Learn 20. Now We're Talking! 22. Tykes and Bytes 23. Facts About Dyslexia 31. Rumors of Inferiority 35. Master of Mastery 45. Vintage Years
		Loneliness	43. Jealousy and Envy: The Demons Within Us
Erikson's Theory	49. Don't Act Your Age! 50. Prime of Our Lives	Love/Marriage	43. Jealousy and Envy: The Demons Within Us
Exceptionality	14. Where Pelicans Kiss Seals		
Family Development/ Siblings	13. Day Care Generation 21. Suffer the Restless Children 24. Dr. Spock Had It Right 26. Can Your Career Hurt Your Kids? 28. Children After Divorce 30. Children Under Stress 33. Alienation and the Four Worlds of Childhood 37. Myth About Teen-Agers 41. Therapists Find Last Outpost of Adolescence in Adulthood 42. Proceeding With Caution 44. Family Ties: The Real Reason People Are Living Longer	Maternal Employment	13. Day Care Generation 26. Can Your Career Hurt Your Kids? 30. Children Under Stress 32. Biology, Destiny, and All That 33. Alienation and the Four Worlds of Childhood
		Mid-Life Crisis	49. Don't Act Your Age! 50. Prime of Our Lives
		Parenting	1. Suffer the Little Children: Shameful Bequests to the Next Generation 8. How Infants See the World 13. Day Care Generation 15. Building Confidence 16. Dealing With Difficult Children 17. Miracle of Resiliency 24. Dr. Spock Had It Right 25. Positive Parenting 26. Can Your Career Hurt Your Kids? 33. Alienation and the Four Worlds of Childhood 40. Puberty and Parents: Understanding Your Early Adolescent
Fertilization/ Infertility	3. Made to Order Babies 4. Sperm Under Siege		
Genetics	2. Gene Dream 3. Made to Order Babies 4. Sperm Under Siege 46. Why Do We Age?		

TOPIC AREA	TREATED IN:	TOPIC AREA	TREATED IN:
Peers	33. Alienation and the Four Worlds of Childhood 40. Puberty and Parents: Understanding Your Early Adolescent	**Self-Esteem/ Self-Control**	1. Suffer the Little Children: Shameful Bequests to the Next Generation 15. Building Conference 17. Miracle of Resiliency 32. Biology, Destiny, and All That 38. Those Gangly Years 43. Jealousy and Envy: The Demons Within Us 44. Family Ties: The Real Reason People Are Living Longer
Personality Development	16. Dealing With Difficult Children 17. Miracle of Resiliency 31. Rumors of Inferiority 32. Biology, Destiny, and All That 43. Jealousy and Envy: The Demons Within Us 50. Prime of Our Lives		
Pregnancy	3. Made to Order Babies 4. Sperm Under Siege 5. Clipped Wings 6. What Crack Does to Babies 7. Motherhood On Trial	**Sex Differences/ Roles/Behavior/ Characteristics**	10. Guns and Dolls 32. Biology, Destiny, and All That 38. Those Gangly Years 46. Why Do We Age? 49. Don't Act Your Age!
Prenatal Development	4. Sperm Under Siege 5. Clipped Wings 6. What Crack Does to Babies 7. Motherhood on Trial	**Social Skills/ Socialization**	24. Dr. Spock Had It Right 27. Lasting Effects of Child Maltreatment 31. Rumors of Inferiority 42. Proceeding With Caution 43. Jealousy and Envy: The Demons Within Us 50. Prime of Our Lives
Preschoolers	11. Preschool: Head Start or Hard Push? 12. How Three Key Countries Shape Their Children 13. Day Care Generation 14. Where Pelicans Kiss Seals 28. Children After Divorce	**Stress**	1. Suffer the Little Children: Shameful Bequests to the Next Generation 11. Preschool: Head Start or Hard Push? 28. Children After Divorce 29. Children of Violence 30. Children Under Stress 33. Alienation and the Four Worlds of Childhood 34. Tracked to Fail
Psychoanalytic Theory	33. Alienation and Four Worlds of Childhood		
Puberty	37. Myth About Teen-Agers 38. Those Gangly Years 40. Puberty and Parents: Understanding Your Early Adolescent 43. Jealousy and Envy: The Demons Within Us	**Teratogens**	4. Sperm Under Siege 5. Clipped Wings 6. What Crack Does to Babies
Racism/Prejudice	31. Rumors of Inferiority		

Genetic and Prenatal Influences on Development

Advances in our knowledge of human development are becoming linked to progress in the fields of genetics, biochemistry, and medicine. With this knowledge has come a host of problems society will have to confront, including those generated by new ways of detecting anomalies prior to birth and by improved methods for sustaining life in highly premature infants. Thirty years ago, prematurity generally referred to infants born no more than two months prior to the expected date. Today, infants born at far less than seven months gestational age are frequently brought to term with the assistance of biomedical technology. Unhappily, a substantial proportion of these children suffer from a range of serious medical problems. Many of these babies are victims of substance abuse. The article "Clipped Wings" looks at prenatal exposure to drugs, alcohol, and smoking by the mother. Exposure to any of these can result in low birth weight. As babies, they are more likely to be inattentive, socially withdrawn, and hyperactive. When they reach school age, they are at a higher risk for failure. However, early intervention can greatly improve the intellectual functioning of these children.

The effects of drug and alcohol use during pregnancy are also dependent on what stage of fetal development they were used. An in-depth look at what happens when crack crosses the placental barrier and the resulting damage to the fetus is described in "What Crack Does to Babies." "Motherhood on Trial" goes a step further and addresses the issue of how society should react to women who take illegal drugs during pregnancy, and what will become of the children born to these women.

Until recently, the mother's life-style has been considered responsible for the health of the unborn child. However, as pointed out in "Sperm Under Siege," new findings suggest the father's sperm may also be affected by such factors as smoking, alcohol, drugs, radiation, and chemical exposure, all of which may lead to spontaneous abortion or birth defects.

Advances in genetics has created renewed respect for the role of heredity in human development. "The Gene Dream" describes a colossal new government-backed project which has as its target the mapping of the complete set of instructions for making a human being. This work promises to revolutionize the fields of medicine and biology and perhaps even psychology and sociology. Some scientists foresee the day when individuals can obtain a computer read-out of their complete genetic makeup, along with a diagnosis, prognosis, and suggested treatment.

"Made to Order Babies" describes genetic screening and raises the issue of how this knowledge should be used. The resulting ethical, legal, and social issues will not be easily resolved. While neglect or abuse practiced by parents is in many instances the direct cause of a mounting number of impaired children, critics also take aim at the conditions in society that contribute to the breakdown of parental responsibilities and family structure. "Suffer the Little Children: Shameful Bequests to the Next Generation" presents a forceful argument for a reordering of national priorities regarding the treatment of children in America.

Looking Ahead: Challenge Questions

Consider your current beliefs about abortion, genetic engineering, and socialized medicine. How do you think these views would be challenged if you learned that your baby-to-be was expected to be profoundly retarded?

If manipulation of genetic material can prevent the appearance of physical dysfunctions, might not similar manipulations be used to engineer intellectual abilities, personality traits, or socially desirable behaviors? What factors would constrain a society that attempted to actively and explicitly practice eugenics?

Do you think that a women who uses illegal drugs during her pregnancy should be arrested? Why or why not?

Of the many issues raised in this section—for example, prenatal care, vaccination programs, sex education, prevention of child abuse, and so forth—which would you give the greatest priority to and why?

Unit 1

Article 1

Suffer the Little Children

The world's leaders gather for an extraordinary summit and listen at last to a crying need.

Just how much is a child worth? To a father in northern Thailand, 10-year-old Poo was worth $400 when he sold her to a middleman to work in Madame Suzy's Bangkok brothel. To Madame Suzy, Poo is worth $40 a night while she's still young and fresh. But her price will soon come down.

To a quarry owner outside New Delhi, 12-year-old Ballu is worth 85¢ a day, the amount the child earns breaking rocks in an 11-hour shift. "I wanted to become an engineer," says Ballu. He glances sadly at his callused hands. "But now I have crossed the age for studies and will be a stonecutter all my life."

To local bosses in Mexico City, children are worth about $2.80 a day for scavenging food, glass, cloth and bones from three vast municipal dumps. The walls around the dumps enclose homes, families, even a church and a store. Many of the 5,000 children living there attend school in the dumps; they are not tolerated on the outside because of their smell.

Just how much is the very life of a child worth? A 10¢ packet of salt, sugar and potassium can prevent a child from dying of diarrhea. Yet every day in the developing world more than 40,000 children under the age of five die of diarrhea, measles, malnutrition and other preventable causes. An extra $2.5 billion a year could save the lives of 50 million children over the next decade. That is roughly equal, children's advocates note, to the amount that the world's military establishments, taken together, shell out every day.

Last weekend George Bush joined 34 other Presidents, 27 Prime Ministers, a King, a Grand Duke and a Cardinal, among others, at the United Nations for a meeting unlike any in history: the World Summit for Children. The leaders came to discuss the plight of 150 million children under the age of five suffering from malnutrition, 30 million living in the streets, 7 million driven from their homes by war and famine.

Shamed into action, the leaders endorsed a bold 10-year plan to reduce mortality rates and poverty among children and to improve access to immunizations and education. For once, this was more than a political lullaby of soothing promises; the very existence of the extraordinary summit held out hope to those who have fought to make children's voices heard. To lend support, more than a million people held 2,600 candlelight vigils earlier in the week—in South Korea's Buddhist monasteries, in London's St. Paul's Cathedral, in Ethiopia's refugee camps, around Paris' Eiffel Tower, in 700 villages in Bangladesh.

As the whole word directs its attention, however briefly, to those to whom the earth will soon belong, what kind of leadership can the United States offer? Americans cherish the notion that they cherish their children, but there's woeful evidence to the contrary. Each year thousands of American babies are born premature and underweight, in a country torn by neither war nor famine. The U.S. is one of only four countries—with Iran, Iraq and Bangladesh—that still execute juvenile offenders. And nearly 1 in 4 American children under age six lives in poverty. Congressmen wrestling with budget cuts, policymakers musing about peace dividends, voters weighing their options—all would do well to wonder what sort of legacy they will be leaving to a generation of children whose needs have been so widely ignored. And those needs go far beyond vigils and poignant speeches.

Shameful Bequests to The Next Generation

America's legacy to its young people includes bad schools, poor health care, deadly addictions, crushing debts—and utter indifference

NANCY GIBBS

George Bush knows how to talk about children. With a sure sense of childhood's mythology, of skinned knees and candy apples and first bicycles, he campaigned for office in a swarm of jolly grandchildren and promised justice for all. In this year's State of the Union address, he mentioned families and "kids" more than 30 times—the electronic equivalent of kissing babies on the village green. "To the children out there tonight," he declared as he built to his finale, "with you rests our hope, all that America will mean in the years ahead. Fix your vision on a new century—your century, on dreams you cannot see, on the destiny that is yours and yours alone."

Forget the next century. Just consider for a moment a single day's worth of destiny for American children. Every eight seconds of the school day, a child drops out. Every 26 seconds, a child runs away from home. Every 47 seconds, a child is abused or neglected. Every 67 seconds, a teenager has a baby. Every seven minutes, a child is arrested for a drug offense. Every 36 minutes, a child is killed or injured by a gun. Every day 135,000 children bring their guns to school.

Even children from the most comfort-

From *Time*, October 8, 1990, pp. 41-48. Copyright © 1990 by The Time Inc. Magazine Company. Reprinted by permission.

1. Suffer the Little Children

able surroundings are at risk. A nation filled with loving parents has somehow come to tolerate crumbling schools and a health-care system that caters to the rich and the elderly rather than to the young. A growing number of parents with preschool children are in the workplace, but there is still no adequate system of child care, and parental leaves are hard to come by. Mothers and fathers worry about the toxic residue left from too much television, too many ghastly movies, too many violent video games, too little discipline. They wonder how to raise children who are strong and imaginative and loving. They worry about the possibility that their children will grow wild and distant and angry. Perhaps they fear most that they will get the children they deserve. "Children who go unheeded," warns Harvard psychiatrist Robert Coles, giving voice to a parent's guilty nightmare, "are children who are going to turn on the world that neglected them."

And that anger will come when today's children are old enough to realize how relentlessly their needs were ignored. They will see that their parents and grandparents have left them enormous debts and a fouled environment. They will recognize that their exceptionally prosperous, peaceful, lucky predecessors, living out the end of the millennium, were not willing to make the investments necessary to ensure that the generation to follow could enjoy the same blessings.

The natural case for taking better care of children would be made on moral grounds alone. A society cannot sacrifice its most vulnerable citizens without eroding its sense of community and making a lie of its principles. But having been left behind by a decade of political shortcuts, child advocates have adopted a more practical strategy. "If compassion were not enough to encourage our attention to the plight of our children," declares New York Governor Mario Cuomo, "self-interest should be." Marian Wright Edelman, the crusading founder of the Children's Defense Fund, goes further. "The inattention to children by our society," she warns, "poses a greater threat to our safety, harmony and productivity than any external enemy."

Spending on children, any economist can prove, is a bargain. A nation can spend money either for better schools or for larger jails. It can feed babies or pay forever for the consequences of starving a child's brain when it is trying to grow. One dollar spent on prenatal care for pregnant women can save more than $3 on medical care during an infant's first year, and $10 down the line. A year of preschool costs an average $3,000 per child; a year in prison amounts to $16,500.

But somehow, neither wisdom nor decency, nor even economics, has prevailed with those who make policy in the state houses, the Congress or the White House. "We are hypocrites," charges Senator John D. ("Jay") Rockefeller IV, who is chairman of the National Commission on Children. "We say we love our children, yet they have become the poorest group in America." Nearly a quarter of all children under six live in households that are struggling below the official poverty line—$12,675 a year for a family of four.

In some cases the abandonment of children begins before they are even born. America's infant mortality rate has leveled off at 9.7 deaths per 1,000 births, worse

Between 1978 and 1987, spending on programs for the elderly rose 52%; spending on children dropped 4%

than 17 other developed countries. In the District of Columbia, the rate tops 23 per 1,000, worse than Jamaica or Costa Rica. Fully 250,000 babies are born seriously underweight each year. To keep these infants in intensive care costs about $3,000 a day, and they are two to three times more likely to be blind, deaf or mentally retarded. On the other hand, regular checkups and monitoring of a pregnant woman can cost as little as $500 and greatly increase the chances that she will give birth to a healthy baby.

Every bit as important as prenatal care is nutrition for the child, both before and after birth. "Of all the dumb ways of saving money, not feeding pregnant women and kids is the dumbest," says Dr. Jean Mayer, one of the world's leading experts on nutrition and president of Tufts University. During the first year of life, a child's brain grows to two-thirds its final size. If a baby is denied good, healthy food during this critical period, he will need intensive nutritional and developmental therapies to repair the damage. "Kids' brains can't wait for Dad to get a new job," says Dr. Deborah Frank, director of growth and development at Boston City Hospital, "or for Congress to come back from recess."

Congress understood the obvious benefits of promoting infant nutrition in the 1970s, when it launched the Special Supplemental Food Program for Women, Infants and Children. WIC provides women with vouchers to buy infant formula, cheese, fruit juice, cereals, milk and other wholesome foods, besides offering nutrition classes and medical care. It costs about $30 a month to supply a mother with vouchers—yet government funds are so tight that only 59% of women and infants who qualify for WIC receive the benefits. "A power breakfast for two businessmen is one woman's WIC package for a month," says Dr. Frank. "Why can't public-policy makers see the connection between bad infant nutrition, which is cheap and easy to fix, and developmental problems, which are expensive and often difficult to fix?"

The theme of prevention applies just as forcefully to medicine. This year the U.S. will spend about $660 billion, or 12% of its GNP, on medical services, but only a tiny fraction of that will go toward prevention. For children the most basic requirement is inoculation, the surest way to spare a child—and the health-care system—the ravages of tuberculosis, polio, measles and whooping cough. During the first 20 years after the discovery of the measles vaccine, public-health experts estimate, more than $5 billion was saved in medical costs, not to mention countless lives. And yet these days in California, the nation's richest state, only half of California's two-year-olds are fully immunized. Dallas reported more than 2,400 measles cases from last December through July, eight of them fatal, including one child who lived within six blocks of an immunization clinic.

Even parents who recognize the importance of preventive care are having a harder time affording it for their children. Most Americans over age 65 are covered by Medicare, the federal health-insurance plan under which the elderly—rich or poor—are eligible for benefits. Children's health programs, in contrast, are subject to annual congressional whims and budget cutting. Fewer and fewer employers, even of well-paid professionals, provide health benefits that cover children for routine medical needs. This means that health costs are the responsibility of individual parents, who make do as best they can, often at considerable sacrifice.

Some states and community groups are trying to help. Two years ago, Minnesota pioneered the Children's Health Plan to provide primary preventive care for children. The plan costs the state about $180 per child, but parents pay only $25: in the end everyone saves. Schools in Independence, Mo., established a health-care package to provide drug and alcohol treatment and counseling services for every child in the district. Cost to parents: $10 per child. In Pittsburgh 12,000 children have received free health care through a program crafted by churches, civic

1. GENETIC AND PRENATAL INFLUENCES

groups, Blue Cross and Blue Shield.

But too many kids are denied such care, and that starts a chain reaction. "You can't educate a child unless all systems are go, i.e., brain cells, eyes, ears, etc." says Rae Grad, executive director of the National Commission to Prevent Infant Mortality. A national survey in 1988 found that two-thirds of teachers reported "poor health" among children to be a learning problem. This is why Head Start, the model federal program providing quality preschool for poor children, also includes annual medical and dental screenings. But once again the money is not there: only about 20% of eligible children are fully served by the program.

Head Start and similar preschool strategies improve academic performance in the early grades and pay vast dividends over time. President Bush has promised enough funding to put every needy child in Head Start, which Congress says will require a fivefold increase by 1994 from the present $1.55 billion a year. Both the House and the Senate have approved higher funding levels, and lawmakers will soon meet to reconcile differences between the two bills. But as the deficit mounts, the peace dividend sinks into the Persian Gulf and the savings and loan crisis chews into basic budget items, politicians may have a hard time approving funding increases for a constituency that does not vote. Senator Orrin Hatch of Utah, a proponent of costly child-care legislation, says the outcome of the budget negotiations is "going to be terrible for kids."

Likewise, American society has, in the past generation, abandoned its commitment to providing a world-class system of secondary education. Education Secretary Lauro Cavazos himself calls student performance "dreadfully inadequate." From both the inner cities and the affluent suburbs comes a drumbeat of stories about tin-pot principals who cannot be fired, beleaguered teachers with unmanageable workloads and illiterate graduates with abysmal test scores. If they can possibly afford to, parents choose private or parochial schools, leaving the desperate or destitute in the worst public schools. Teachers, meanwhile, are aware that they are often the most powerful influences in a child's life—and that their job pays less in a year than a linebacker or rock star can earn in a week.

Across the board, people who deal with children are more ill-paid, unregulated and less respected than other professionals. Among physicians, pediatricians' income ranks near the bottom. In Michigan preschool teachers with five years' experience earn $12,000, and prison guards with the same amount of seniority earn almost $30,000. U.S. airline pilots are vigilantly trained, screened and monitored; school-bus drivers are not. "My hairdresser needs 1,500 hours of schooling, takes a written and practical test and is relicensed every year," says Flora Patterson, a foster parent in San Gabriel, Calif. "For foster parents in Los Angeles County there is no mandated training, yet we are dealing with life and death." The typical foster parent there earns about 80¢ an hour.

In France, Belgium, Italy and Denmark, at least 75% of children ages 3 to 5 are in some form of state-funded preschool program

Worst of all is the status of America's surrogate parents: the babysitters and day-care workers who have become essential to the functioning of the modern family. In the absence of anything like a national child-care policy, parents are left to improvise. The rich search for trained, qualified care givers and pay them whatever it takes to keep them. But for the vast majority, child care is a game of Russian roulette: rotating nannies, unlicensed home care, unregulated nurseries that leave parents wondering constantly: Is my child really safe? "Finding child care is such a gigantic crapshoot," says Edward Zigler, director of Yale's Bush Center in Child Development and Social Policy. "If you are lucky, you are home free. But if you are unlucky, well, there are some real horror stories out there of kids being tied into cribs."

The U.S. economy has long been geared to two-income families; many families could not afford a middle-class lifestyle without both parents working. The real median income of parents under age 30 fell more than 24% from 1973 to 1987, according to a study by the Children's Defense Fund and Northeastern University. But social programs rarely reflect those economic realities. Growing financial pressure all too often translates into fewer doctors' visits, more stress and less time spent together as a family. Between 1950 and 1989, the divorce rate doubled: 1.16 million couples split up each year. That makes the need for reliable support services for children all the greater.

In place of responses came rhetoric: a 1986 Administration report on the family titled "Preserving America's Future" called for a return to "traditional values," parental support of children and "lovingly packed lunch boxes." Time and again, Washington has failed to address the needs of working parents—most recently in June, when President Bush vetoed the family-leave bill on the ground that it was too burdensome for business. The bill would have allowed a worker to take up to 12 weeks a year of unpaid leave to care for a newborn, an adopted child or a sick family member.

That is abysmal compared with what other industrialized nations allow. Salaried women in France can take up to 28 weeks of unpaid maternity leave or up to 20 weeks of adoption leave, though they are less likely to need it since day care, health care and early education are widely available in that country. In France, as well as in Belgium, Italy and Denmark, at least 75% of children ages 3 to 5 are in some form of state-funded preschool programs. In Japan both the government and most companies offer monthly subsidies to parents with children. In Germany parents may deduct the cost of child care from their taxes. "Under our tax laws," observes Congresswoman Pat Schroeder of Colorado, "a businesswoman can deduct a new Persian rug for her office but can't deduct most of her costs for child care. The deduction for a Thoroughbred horse is greater than that for children."

If the troubles children face were all born of economic pressure on the family, then wealthy children should emerge unscathed. Yet the problems confronting affluent children are also profound and insidious. Parents who do not spend time with their children often spend money instead. "We supply kids with things in the absence of family," says Barbara MacPhee, a school administrator in New Orleans. "We used to build dreams for them, but now we buy them Nintendo toys and Reebok sneakers." In the absence of parental guidance and affirmation, children are left to soak in whatever example their environment sets. A childhood spent in a shopping mall raises consumerism to a varsity sport; time spent in front of a television requires no more imagination than it takes to change channels.

At Winchester High School in a cozy Boston suburb, clinical social worker Michele Diamond hears it all: the drug use, the alcohol, the eating disorders, the suicide attempts by children who are viewed as privileged. "Kids are left alone a lot to cope," she says, "and they sense less support from their families." Pressured to succeed, to "fit in," to be accepted by top colleges, the students handle their stress however they can. Some just dissolve their problems in a glass. In nearby Belmont, a juvenile officer finds that parents shrug off the danger. When their kids are caught drinking, he notes,

1. Suffer the Little Children

"they say, 'Thank God it isn't cocaine. It's alcohol. We can handle that.'"

All too often it *is* cocaine, the poisonous solace common to the golf club and the ghetto. It is not only the violence of the drug culture that threatens children; it is also the lure of the easy money that turns 11-year-olds into drug runners. "Alienated is too weak a word to describe these kids," says Edward Loughran, a 10-year veteran of the juvenile-justice system in Massachusetts. "They don't value their lives or anyone else's life. Their values system says, 'I am here alone. I don't care what society says.' A lot of these kids are dying young deaths and don't care because they don't feel there is any reason to aspire to anything else."

Violence in the neighborhood is bad enough. Violence in the home is devastating. Reports of child abuse have soared from 600,000 in 1979 to 2.4 million in 1989, a searing testimony to the enduring role of children as the easiest victims. In New York City, half of all abuse reports are repeat cases of children who have had to be rescued before, only to be returned to an abusive home.

When two-year-old "Rebecca" accidentally soiled her underwear, her mother and the mother's boyfriend were not pleased. So they heated up some cooking oil, held Rebecca down and poured it over her. Then they waited a week or so before Rebecca's mother, unable to stand the stench of the child's legs, which were rotting from gangrene, took her to the hospital. After a month's stay that saved her legs, Rebecca was able to move to a foster home. From there she went to live with her paternal grandmother, who had plenty of room: all four of her sons were in state prison.

Around the country there are hundreds of thousands of other children who scream for help from overburdened teachers, understaffed social service agencies, crowded courts and a gridlocked foster-care system. To dismiss child abuse as a personal, private tragedy misses the larger point entirely. If children are not protected from their abusers, then the public will one day have to be protected from the children. To walk through death row in any prison is to learn what child abuse can lead to when it ripens. According to attorneys who have represented them, roughly 4 out of 5 death row inmates were abused as children.

A reordering of priorities toward protecting children would include far higher funding and staffing of Child Protective Services, the organization that investigates charges of abuse and can move to rescue children before the damage is irreparable. But even that would do little good if there is no place to put them. No solution will be possible without an overhaul of the foster-care system, which in many cities is on the verge of collapse. All too often, children are separated from siblings and shuttled from group homes to relatives to foster families, with no sense of the safety, security or stability they need to succeed in school and elsewhere. "If we don't have money for adequate care," says Ruth Massinga, a member of the National Commission on Children, "removing children from their homes is just another devastation."

Failure to make treatment available to drug addicts who seek it will ensure yet another generation of addicted babies and battered kids. In Los Angeles the number of drug-exposed babies entering the foster-care system rose 453% between 1984 and 1987. A survey of states found that drugs are involved in more than 2 out of 3 child abuse and neglect cases. Children born into a family of addicts are left with impossible choices: a life with the abusers they know, or a life at the mercy of a system filled with strangers—lawyers, judges, social workers, foster parents.

It is a common mistake to assume that all abuse is physical. The scars of other forms of abuse—like unrelenting verbal cruelty—can be just as apparent when children grow older, unloved and self-hating. "You can tell kids you love 'em, says April, a runaway in Hollywood. "But that's not the same as showing them. Broken promises is really what tears your heart apart." For April there is not much difference between insult and injury. "Beating kids will hurt kids. Sexual abuse will hurt a kid. But verbal abuse is the worst. I've had all three. If you're not strong enough as a person, and they've been telling you this all your life, that you can never amount to anything, you are going to believe it."

There have always been children who are survivors, who overcome the odds and find some adult—a teacher, a grandparent, a priest—who can provide the anchors the family could not. Touré Diggs, 18, grew up in a rough neighborhood of New Haven, Conn., and is now enrolled at Fairleigh Dickinson University. Since his parents separated three years ago, Touré has tried to help raise his brother Landis, who is 7. In the end Touré knows he is competing with the lure of the street for Landis' soul. "You got to start so young," Touré says. "It's like a game. Whoever gets to the kids first, that's how they are going to turn out."

Schools in particular have come to take that role very seriously, which accounts for the debate over how to teach values and self-discipline to a generation whose boundaries have been loosely drawn. But other institutions are slowly waking up to the implications of writing off an entire generation. The business community, in particular, wonders where it will find a trained, literate, motivated work force in the 21st century. The Business Roundtable, with representatives from the largest 200 companies, has made support for education its highest priority in the '90s. In Dallas, Texas Instruments helps fund the local Head Start program. Eventually, more and more companies may make parental leave a standard benefit, regardless of the messages coming from Washington.

In Des Moines business leaders are sponsoring a program called Smoother Sailing, which sends counselors like "Sunburst Lady" Toni Johansen into the city's elementary schools. National studies have shown that such support helps improve confidence, discipline and attitudes about school. With the extra funding, the city has been able to provide one guidance counselor for every 250 students, in contrast to a national average of one for 850.

But there will be no real progress, no genuine hope for America's children until the sense of urgency forces a reconsideration of values in every home, up to and including the White House. Polls suggest the will is there: 60% of Americans believe the situation for children has worsened over the past five years; 67% say they would be more likely to vote for a candidate who supported increased spending for children's programs even if it meant a tax increase.

When adults lament the absence of "values," it is worth recalling that children are an honest conscience, the perfect mirror of a society's priorities and principles. A society whose values are entirely material is not likely to breed a generation of poets; anti-intellectualism and indifference to education do not inspire rocket scientists. With each passing day these arguments become more apparent, the needs more pressing. Where is the leader who will seize the opportunity to do what is both smart and worthy, and begin retuning policy to focus on children and intercept trouble before it breeds?

—*Reported by Julie Johnson/Des Moines, Melissa Ludtke/Boston and Michael Riley/Washington*

(Article continues)

1. GENETIC AND PRENATAL INFLUENCES

Struggling for Sanity

Mental and emotional distress are taking an alarming toll of the young

ANASTASIA TOUFEXIS

The dozen telephone lines at the cramped office of Talkline/Kids Line in Elk Grove Village, Ill., ring softly every few minutes. Some of the youthful callers seem at first to be vulgar pranksters, out to make mischief with inane jokes and naughty language. But soon the voices on the line—by turns wistful, angry, sad, desperate—start to spill a stream of distress. Some divulge their struggles with alcohol or crack and their worries about school and sex. Others tell of their feelings of boredom and loneliness. Some talk of suicide. What connects them all, says Nancy Helmick, director of the two hot lines, is a sense of "disconnectedness."

Such calls attest to the intense psychological and emotional turmoil many American children are experiencing. It is a problem that was not even recognized until just a decade ago. Says Dr. Lewis Judd, director of the National Institute of Mental Health: "There had been a myth that childhood is a happy time and kids are happy go lucky, but no age range is immune from experiencing mental disorders." A report prepared last year by the Institute of Medicine estimates that as many as 7.5 million children—12% of those below the age of 18—suffer from some form of psychological illness. A federal survey shows that after remaining constant for 10 years, hospitalizations of youngsters with psychiatric disorders jumped from 81,500 to about 112,000 between 1980 and 1986. Suicides among those ages 15 to 19 have almost tripled since 1960, to 1,901 deaths in 1987. Moreover, the age at which children are exhibiting mental problems is dropping: studies suggest that as many as 30% of infants 18 months old and younger are having difficulties ranging from emotional withdrawal to anxiety attacks.

What is causing so much mental anguish? The sad truth is that a growing number of American youngsters have home lives that are hostile to healthy emotional growth. Psyches are extremely fragile and must be nourished from birth. Everyone starts out life with a basic anxiety about survival. An attentive parent contains that stress by making the youngster feel secure and loved.

Neglect and indifference at such a crucial stage can have devastating consequences. Consider the case of Sid. (Names of the children in this story have been changed). When he was three months old, his parents left him with the maid while they took a five-week trip. Upon their return, his mother noticed that Sid was withdrawn, but she did not do anything about it. When Sid was nine months old, his mother left him again for four weeks while she visited a weight-loss clinic. By age three, Sid had still not started talking. He was wrongly labeled feebleminded and borderline autistic before he received appropriate treatment.

As children mature within the shelter of the family, they develop what psychologists call a sense of self. They acquire sensitivities and skills that lead them to believe they can cope independently. "People develop through a chain," observes Dr. Carol West, a child psychotherapist in Beverly Hills. "There has to be stability, a consistent idea of who you are."

The instability that is becoming the hallmark of today's families breeds in children insecurity rather than pride, doubts instead of confidence. Many

> **As many as 7.5 million children—12% of those below the age of 18—suffer from some form of psychological illness**

youngsters feel guilty about broken marriages, torn between parents and households, and worried about family finances. Remarriage can intensify the strains. Children may feel abandoned and excluded as they plunge into rivalries with stepparents and stepsiblings or are forced to adjust to new homes and new schools. Children from troubled homes used to be able to find a psychological anchor in societal institutions. But no longer. The churches, schools and neighborhoods that provided emotional stability by transmitting shared traditions and values have collapsed along with the family.

Such disarray hurts children from all classes; wealth may in fact make it harder for some children to cope. Says Hal Klor, a guidance counselor at Chicago's Lincoln Park High School: "The kids born into a project, they handle it. But the middle-class kids. All of a sudden—a divorce, loss of job, status. Boom. Depression."

Jennifer shuttled by car service across New York City's Central Park between her divorced parents' apartments and traveled by chartered bus to a prep school where kids rated one another according to their family cars. "In the eighth grade I had panic attacks," says Jennifer, now 18. "That's when your stomach goes up and you can't leave the bathroom and you get sweaty and you get headaches and the world closes in on you." Her world eventually narrowed so far that for several weeks she could not set foot outside her home.

The children who suffer the severest problems are those who are physically or sexually abused. Many lose all self-esteem and trust. Michele, 15, who is a manic-depressive and an alcoholic, is the child of an alcoholic father who left when she was two and a mother who took out her rage by beating Michele's younger sister. When Michele was 12, her mother remarried. Michele's new stepbrother promptly began molesting her. "So I molested my younger brother," confesses Michele. "I also hit him a lot. He was four. I was lost; I didn't know how to deal with things."

At the same time, family and society are expecting more from kids than ever before. Parental pressure to make good grades, get into college and qualify for the team can be daunting. Moreover, kids are increasingly functioning as junior adults in many homes, taking on the responsibility of caring for younger siblings or ailing grandparents. And youngsters' own desires—to be accepted and popular with their peers, especially—only add to the strain.

Children express the panic and anxiety they feel in myriad ways: in massive weight gains or losses, in nightmares and disturbed sleep, in fatigue or listlessness, in poor grades or truancy, in continual arguing or fighting, in drinking or drug abuse, in reckless driving or sexual promiscuity, in stealing and mugging. A fairly typical history among disturbed kids, says Dr. L. David Zinn, co-director of Northwestern Memorial Hospital's Ad-

1. Suffer the Little Children

olescent Program, includes difficulty in school at age eight or nine, withdrawal from friends and family and persistent misbehavior at 10 or 11 and skipping school by 15. But the most serious indication of despair—and the most devastating—is suicide attempts. According to a report issued in June by a commission formed by the American Medical Association and the National Association of State Boards of Education, about 10% of teenage boys and 18% of girls try to kill themselves at least once.

Despite the urgency of the problems, only 1 in 5 children who need therapy receives it; poor and minority youngsters get the least care. Treatment is expensive, and even those with money and insurance find it hard to afford. But another reason is that too often the signals of distress are missed or put down to normal mischief.

Treatment relies on therapeutic drugs, reward and punishment, and especially counseling—not just of the youngster but of the entire family. The goal is to instill in the children a feeling of self-worth and to teach them discipline and responsibility. Parents, meanwhile, are taught how to provide emotional support, assert authority and set limits.

One of the most ambitious efforts to reconstruct family life is at Logos School, a private academy outside St. Louis that was founded two decades ago for troubled teens. Strict rules governing both school and extracurricular life are laid out for parents in a 158-page manual. Families are required to have dinner together every night, and parents are expected to keep their children out of establishments or events, say local hangouts or rock concerts, where drugs are known to be sold.

Parents must also impose punishments when curfews and other rules are broken. Says Lynn, whose daughter Sara enrolled at Logos: "My first reaction when I read the parents' manual was that there wasn't a thing there that I didn't firmly believe in, but I'd been too afraid to do it on my own. It sounds like such a cop-out, but we wanted Sara to be happy."

As necessary and beneficial as treatment may be, it makes better sense to prevent emotional turmoil among youngsters by improving the environment they live in. Most important, parents must spend more time with sons and daughters and give them the attention and love they need. To do less will guarantee that ever more children will be struggling for sanity. —*Reported by Kathleen Brady/New York, Elizabeth Taylor/Chicago and James Willwerth/Los Angeles*

Article 2

THE GENE DREAM

Scientists are mapping our complete genetic code, a venture that will revolutionize medicine—and ethics.

Natalie Angier

Natalie Angier, *a New York City–based science writer, is the author of* Natural Obsessions: The Search for the Oncogene.

At first glance, the Petersons* of Utah seem like a dream family, the kind you see only on television. They're devout, traditional and very, very loving. Bob Peterson works at a hospital near home to support the family while he finishes up a master's program in electrical engineering. Diane, who studied home economics at Brigham Young University, is a full-time wife and mother. And her time is certainly full: The Petersons have five sons and two daughters, ranging in age from two to 13. (As Mormons, the parents don't practice birth control.)

The children are towheaded, saucer-eyed and subject to infectious fits of laughter. During the summer months, the backyard pool is as cheerily deafening as the local Y. Says Diane, "Our kids really like just spending time together."

Yet for all the intimacy and joy, the Petersons' story is threaded with tragedy. One of the daughters has cerebral palsy, a nerve- and muscle-cell disorder. The malady isn't fatal, but the girl walks with great difficulty, and she's slightly retarded.

*Not their real name.

Three of the other children suffer from cystic fibrosis, a devastating genetic disease in which the lungs become clogged with mucus, the pancreas fails, malnutrition sets in, and breathing becomes ever more labored. Thus far, their children's symptoms have been relatively mild, but Bob and Diane know the awful truth: Although a person with cystic fibrosis may live to be 20 or even 30, the disease is inevitably fatal.

"Right now, the kids don't act sick," says Bob. "They go on thinking, 'I have a normal life.'" But, he admits softly, "We know it won't last forever. If they do get bad, then we won't have a choice. We'll have to put them in a hospital."

The Petersons realize their children's ailments aren't likely to be cured in the immediate future, but they're battling back the best way possible. Bob, Diane, and their seven children, as well as the three surviving grandparents, have all donated blood samples to biologist Ray White and his team at the University of Utah in Salt Lake City. Scientists are combing through the DNA in the blood, checking for the distinctive chemical patterns present only in cystic fibrosis patients.

Their work is part of a vast biomedical venture recently launched by the government to understand all the genes that either cause us harm or keep us healthy. It's medicine's grandest dream: By comprehending the genome—the complete set of genetic information that makes us who we are—in minute detail, scientists hope to answer the most enigmatic puzzles of human nature. The effort is so immense in its scale and goals that some have called it biology's equivalent of the Apollo moonshot, or the atom bomb's Manhattan project.

In fact, it's the most ambitious scientific project ever undertaken; it will cost a whopping $3 billion and take at least 15 years to complete. By the time researchers are through, they will have deciphered the complete genome. They'll have drawn up a detailed genetic "map," with the size, position and role of all 100,000 human genes clearly marked. And they'll have figured out each gene's particular sequence of chemical components, called nucleotides.

Though there are only four types of nucleotides, represented by the letters A, T, C and G, spelling out all the combinations that make up our total genetic heritage will fill the equivalent of one million pages of text. "What we'll have," says Dr. Leroy Hood, a biologist at the California Institute of Technology in Pasadena, "is a fabulous 500-volume 'encyclopedia' of how to construct a human being." Nobel laureate Walter Gilbert goes so far as to describe the human genome as "the Holy Grail of biology."

From *American Health*, March 1989, pp. 103-106, 108. Copyright © by American Health Partners and the author.

2. Gene Dream

HUMAN GENE MAPS

The latest maps for chromosomes one through six show the location of genes associated with hereditary disorders.

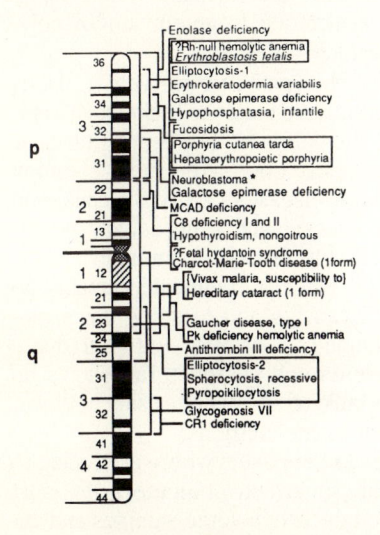

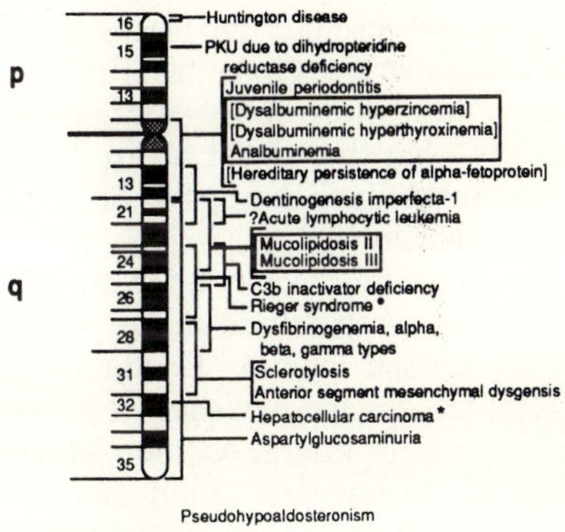

[Chromosome maps 1–6 adapted from gene maps provided by Dr. Victor A. McKusick]

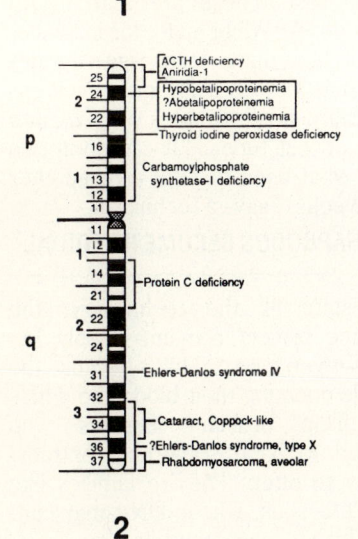

□ Allelic disorders (due to different mutations in the same gene)
[ ] Nondisease
* Cancers
■ Malformation syndrome
{ } Specific infections with a single-gene basis for susceptibility
italics Maternofetal incompatibility

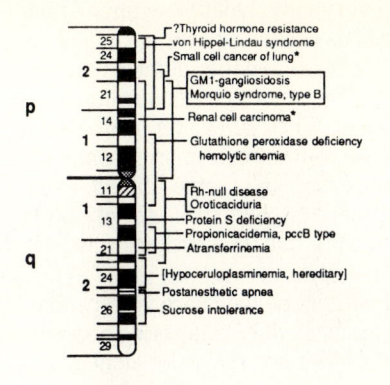

Adapted from gene maps provided by Dr. Victor A. McKusick

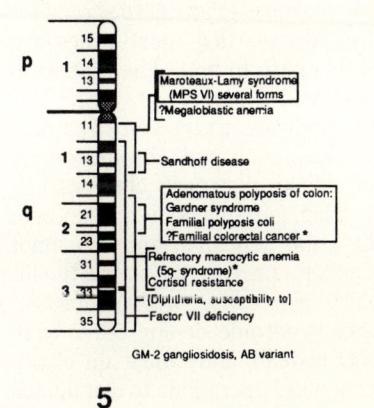

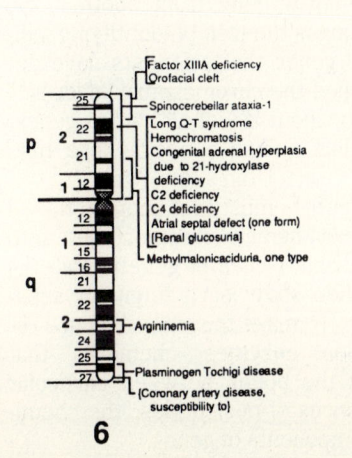

Some scientists, however, think their colleagues are chasing a will-o'-the-wisp. Current genetic engineering techniques, say critics, are too embryonic to attempt anything as massive as sequencing the entire genome. Dr. Robert Weinberg of the Whitehead Institute in Cambridge, MA, calls the whole project "misguided" and doubts that scientists will gain major insights even if they can sequence it.

Still, researchers involved in the Human Genome Initiative insist the knowledge will revolutionize the fields of medicine, biology, health, psychology and sociology, and offer a bounty of applications. Using advanced recombinant DNA techniques, scientists will pluck out the genes that cause the 4,000 known hereditary diseases, including childhood brain cancer, familial colon cancer, manic depression, Huntington's disease—the neurological disorder that killed folk singer Woody Guthrie—and neurofibromatosis, or Elephant Man's disease. Beyond analyzing rare inherited disorders, researchers will glean fresh insights into the more common and complicated human plagues, such as heart disease, hypertension, Alzheimer's, schizophrenia, and lung and breast cancer. Those studies will enable scientists to develop new drugs to combat human disease.

But the Genome Initiative is not restricted to the study of sickness. As biologists decode the complete "text" of our genetic legacy, they'll be asking some profound questions: Are there genes for happiness, anger, the capacity to fall in love? Why are some people able to gorge themselves and still stay slim, while others have trouble losing weight no matter how hard they diet? What genetic advantages turn certain individuals into math prodigies, or Olympic athletes? "The information will be fundamental to us *forever*," says Hood, "because that's what we are."

The most imaginative scientists foresee a day when a physician will be able to send a patient's DNA to a lab for scanning to detect any genetic mutations that might jeopardize the patient's health. Nobel laureate Paul Berg, a biochemistry professor at Stanford, paints a scenario in which we'll each have a genome "credit card" with all our genetic liabilities listed on it. We'll go to a doctor and insert the card into a machine. Instantly reading

1. GENETIC AND PRENATAL INFLUENCES

the medical record, the computer will help the doctor to put together a diagnosis, prognosis and treatment course. Says Caltech's Hood, "It's going to be a brave new world."

Coping with that new world will demand some bravery of our own. Once our genetic heritage has been analyzed in painstaking detail, we'll have to make hard choices about who is entitled to that information, and how the knowledge should be used. This technology is proceeding at an incredible rate, and we have to be sure that it doesn't lead to discrimination in jobs, health insurance or even basic rights, says Dr. Jonathan Beckwith, a geneticist at Harvard Medical School. "We don't want a rerun of eugenics, where certain people were assumed to be genetically inferior, or born criminals."

For better or worse, politicians are convinced that the knowledge is worth seeking. This year, Congress has earmarked almost $50 million for genome studies and, if current trends continue, by 1992 the government should be spending about $200 million annually. Opponents worry the price tag could leave other worthy biomedical projects in the lurch.

Even at that level of funding, the genome project could be beyond the resources of any single country. That's why research teams from Europe, Asia, North America and New Zealand have joined to form the Human Genome Organization. Among other goals, the newly created consortium plans to distribute money for worthwhile projects worldwide. Meanwhile, the Paris-based Center for the Study of Human Polymorphism distributes cell samples to researchers and shares their findings through an international data bank.

In this country, Nobel laureate James Watson, the co-discoverer of the molecular structure of DNA, is in charge of human genome research at the National Institutes of Health. And Dr. Charles Cantor, a highly respected geneticist from New York's Columbia University, has accepted the top spot at the Department of Energy's Human Genome Center.

THE GENETIC HAYSTACK

The Genome Initiative is sure to affect everybody. Doctors estimate that each of us carries an average of four to five severe genetic defects in our DNA. The majority of those mutations are silent: They don't affect you. However, if you were to marry someone who carries the same defect, you could have a child who inherits both bad genes and is stricken with the disease.

Most genetic flaws are so rare that your chances of encountering another silent carrier are slim—let alone marrying and conceiving a child with such a person. But some defects are widespread. For example, five out of 100 people harbor the mutant cystic fibrosis gene; seven out of 100 blacks carry the trait for sickle cell anemia. Bob and Diane Peterson are both cystic fibrosis carriers—but they didn't realize their predicament until they gave birth to afflicted children.

For all the improvements of the last 10 years, prenatal diagnosis techniques remain limited. Doctors can screen fetuses for evidence of about 220 genetic disorders, but most of the tests are so time-consuming and expensive they won't be done unless family history suggests the child may have a disease.

One reason it's difficult to screen for birth defects is that most genes are devilishly hard to find. The 50,000 to 100,000 genes packed into every cell of your body are arrayed on 23 pairs of tiny, sausage-shaped chromosomes, which means that each chromosome holds a higgledy-piggledy collection of up to 4,400 genes. Scientists cannot look under a microscope to see the individual genes for cystic fibrosis, Down's syndrome or any other birth defect; instead, they must do elaborate chemical operations to distinguish one human gene from another. So daunting is the task of identifying individual genes that scientists have determined the chromosomal "address" of only about 2% of all human genes. "It's like finding a needle in a haystack," says Utah's Ray White.

Scientists must first chop up the 23 pairs of human chromosomes into identifiable pieces of genetic material and then study each fragment separately. To make the cuts, they use restriction enzymes—chemicals that break the bonds between particular sequences of nucleotides, the chemical components of genes.

Normally, restriction enzymes snip genetic material at predictable points, as precisely as a good seamstress cuts a swatch of fabric. But scientists have found that the enzymes also cut some fragments at unexpected places, yielding snippets that are longer than normal. It turns out that these variations are inherited, and many have been linked to certain genetic abnormalities. The fragments even serve as reference points for map-making efforts. The DNA segments produced by this technique are nicknamed "riff-lips," for restriction fragment length polymorphisms (RFLPs).

In the past three years, DNA sleuths have used the technique to isolate the genes for Duchenne muscular dystrophy, one of the most common genetic diseases; a grizzly childhood eye cancer; and a hereditary white-blood-cell disease commonly called CGD. But the technique remains labor-intensive and in some ways old-fashioned. Armies of graduate students and postdoctoral fellows do the bulk of the work, using tedious, error-prone methods.

Scientists everywhere are racing to build superfast computers to sort through chromosome samples and analyze RFLP patterns. Until they're devised, researchers are learning to make do. At White's lab, for instance, researchers have jerry-rigged a device that automatically dispenses exceedingly small samples of DNA into rows of test tubes. "It can do in two days what used to take a researcher two weeks," says a technician.

THE HAPGOODS BECOME IMMORTAL

Despite all the technology, the genome project remains deeply human—even folksy. That's because the people donating their blood and genes are from ordinary families who happen to have something extraordinary to offer. They're families like the Petersons, whose DNA may contain clues to cystic fibrosis.

Or they're families like the Hapgoods, whose greatest claim to fame may be their ability to live long and multiply. Brenda and Sam Hapgood,* a Mormon couple in their early 50s, are plump and boisterous, and love to be surrounded by people. That may explain why, although they have five

---
* not their real name.

girls, four boys, three sons-in-law, two daughters-in-law and five grandchildren, they wouldn't mind having a few more kids around. Says Brenda, "I almost wish I hadn't stopped at nine!"

The Hapgoods are one of 40 Utah families helping White construct a so-called linkage map of human DNA. He's trying to find chemical markers in the genome that are "linked" with certain genes. The markers will serve as bright signposts, dividing the snarl of genes into identifiable neighborhoods—just as road signs allow a traveler to pin down his location. Finding those markers is a crucial first step toward identifying the genes themselves, and for providing researchers with a decent chart of the terrain.

That's where the Hapgoods come in. To detect those tiny patches in the DNA that stand out from the background of surrounding genetic material, White must be able to compare the genomes of many related people over several generations. Mormon families are large, and they don't tend to move around much, so it's easy for White to get blood samples from many generations of a given family.

"The researchers told us there are lots of big families around," says Brenda. "What made us special was that all the grandparents were still with us."

In 1984, Brenda, Sam, their parents and nine children all donated blood to White's researchers. Lab technicians then used a special process to keep the blood cells alive and dividing forever—ensuring an infinite supply of Hapgood DNA for study. "Our linkage families are becoming more and more important as we go to the next stage of mapping," says Mark Leppert, one of White's colleagues. "Hundreds of researchers from all over will be using the information from their DNA."

"We're going to go down in medical history!" Brenda says excitedly. "But you know what I'm really worried about?" one son-in-law teases her. "They might decide to clone you!"

Another reason the Hapgoods were chosen for the linkage study is because, in contrast to the Petersons, they didn't seem to have any major hereditary diseases. White wanted his general-purpose map to be a chart of normal human DNA. Ironically, however, two years after the Hapgoods first donated blood, one of the daughters gave birth to a son with a serious genetic defect known as Menkes' disease, a copper deficiency.

The child is two years old but looks like a deformed six-month-old. He has 100 or more seizures a day. Half his brain and most of his immune system have been destroyed. Cradled in his mother Carol's arms, he moans steadily and sadly. "This is as big as he'll get," says Carol. "He'll only live to be four at the very most."

Carol and Brenda hope that the genome project will someday bring relief for Menkes victims. "We originally volunteered for the study to help the scientists out, to help their research," says Brenda. "But now we see that it could be important for people like us."

THE BIG PAYOFF

"You don't need to have the whole project done before you start learning something," says Dr. Daniel Nathans, a Nobel laureate and professor of molecular biology and genetics at Johns Hopkins University in Baltimore. "There are things to be learned every step of the way." The first spin-offs are likely to be new tests for hereditary diseases. Within one to three years, biologists hope to have cheap and accurate probes to detect illnesses known to be caused by defects in a single gene, such as susceptibility to certain kinds of cancers.

Another inherited ailment that could quickly yield to genome research is manic depression, which is also thought to be caused by an error in any one of several genes. The psychiatric disorder afflicts 1% of the population—2.5 million people in the U.S. alone—yet it's often difficult to diagnose. With the gene isolated, experts will be better able to distinguish between the disease and other mood disorders, explains Dr. Helen Donis-Keller, a professor of genetics at Washington University in St. Louis.

Of even greater relevance to the public, the Genome Initiative will give investigators their first handle on widespread disorders such as cancer, high blood pressure and heart disease. Researchers are reasonably certain that multiple DNA mutations share much of the blame for these adult plagues, but as yet they don't know which genes are involved. Only when biologists have an itemized map of the genome will they be able to detect complex DNA patterns that signal trouble in many genes simultaneously.

As the quest proceeds, surprises are sure to follow. "There are probably hundreds or thousands of important hormones yet to be isolated," says Dr. David Kingsbury, a molecular biologist at George Washington University. Among them, he believes, are novel proteins that help nerve cells grow, or *stop* growing. Such hormones could be made into new cancer drugs that target tumors while leaving the rest of the body unscathed.

"I have an intuitive feeling that this is going to open up all sorts of things we couldn't have anticipated," says Donis-Keller. "Even mundane things like obesity and baldness—imagine the implications of having new therapies for them!"

The human genome also holds keys to personality and the emotions. Department of Energy gene chief Charles Cantor says it's estimated that half of our 100,000 genes are believed to be active only in brain cells, indicating that much of our DNA evolved to orchestrate the subtle dance of thought, feeling, memory and desire. "There are genes that are very important in determining our personality, how we think, how we act, what we feel," says Cantor. "I'd like to know how these genes work." Donis-Keller is also curious. "Is panic disorder inherited? Is autism?" she wonders. "These are controversial questions we can start to clarify."

Like the first Apollo rocket, the Human Genome Initiative has cleared the launch pad in a noisy flame of promise. Its crew is international, and so too will be the fruits of exploration. When the human genome is sequenced from tip to tail, the DNA of many people is likely to be represented—perhaps that of the Hapgoods and the Petersons, perhaps that of a Venezuelan peasant family. "It's going to be a genetic composite," predicts Yale professor of genetics Frank Ruddle. "The Indians will work on their genomes, the Russians on theirs, the Europeans on theirs. We'll pool the data and have one great patchwork quilt.

"I get a lot of pleasure out of thinking of this as a world project. No one single person will be immortalized by the research. But it will immortalize us all."

Article 3

Made to Order BABIES

GEOFFREY COWLEY

For centuries, Jewish communities lived Job-like with the knowledge that many of their babies would thrive during infancy, grow demented and blind as toddlers and die by the age of 5. Joseph Ekstein, a Hasidic rabbi in Brooklyn, lost four children to Tay-Sachs disease over three decades, and his experience was not unusual. Some families were just unlucky.

Today, the curse of Tay-Sachs is being lifted—not through better treatments (the hereditary disease is as deadly as ever) but through a new cultural institution called Chevra Dor Yeshorim, the "Association of an Upright Generation." Thanks largely to Rabbi Ekstein's efforts, Orthodox teenagers throughout the world now line up at screening centers to have their blood tested for evidence of the Tay-Sachs gene. Before getting engaged, prospective mates simply call Chevra Dor Yeshorim and read off the code numbers assigned to their test results.

If the records show that neither person carries the gene, or that just one does, the match is judged sound. But if both happen to be carriers (meaning any child they conceive will have a one-in-four chance of suffering the fatal disease), marriage is virtually out of the question. Even if two carriers *wanted* to wed, few rabbis would abet them. "It's a rule of thumb that engagements won't occur until compatibility is established," says Rabbi Jacob Horowitz, codirector of the Brooklyn-based program. "Each day, we could stop many marriages worldwide."

Marriage isn't the only institution being reshaped by modern genetics; a host of new diagnostic tests could soon change every aspect of creating a family. Physicians can now identify some 250 genetic defects, not only in the blood of a potential parent but in the tissue of a developing fetus. The result is that, for the first time in history, people are deciding, rather than wondering, what kind of children they will bear.

Choosing to avoid a horrible disease may be easy, at least in principle, but that's just one of many options 21st-century parents could face. Already, conditions far less grave than Tay-Sachs have been linked to specific genes, and the science is still exploding. Researchers are now at work on a massive $3 billion project to decipher the entire human genetic code. By the turn of the century, knowledge gained through this Human Genome Initiative could enable doctors to screen fetuses—even test-tube embryos—for traits that have nothing to do with disease. "Indeed," says Dr. Paul Berg, director of the Beckman Center for Molecular and Genetic Medicine at Stanford, "we should be able to locate which [gene] combinations affect kinky hair, olive skin and pointy teeth."

How will such knowledge be handled? How *should* it be handled? Are we headed for an age in which having a child is morally analogous to buying a car? There is already evidence that couples are using prenatal tests to identify and abort fetuses on the basis of sex, and there is no reason to assume the trend will stop there. "We should be worried about the future and where this might take us," says George Annas, a professor of health law at Boston University's School of Medicine. "The whole definition of normal could well be changed. The issue becomes not the ability of the child to be happy but rather our ability to be happy with the child."

So far, at least, the emphasis has been on combating serious hereditary disorders. Everyone carries four to six genes that are harmless when inherited from one parent but can be deadly when inherited from both. Luckily, most of these mutations are rare enough that carriers are unlikely to cross paths. But some have become common within particular populations. Five percent of all whites carry the gene for cystic fibrosis, for example, and one in 2,000 is born with the disease. Seven percent of all blacks harbor the mutation for sickle-cell anemia, and one in 500 is afflicted. Asian and Mediterranean people are particularly prone to the deadly blood disease thalassemia, just as Jews are to Tay-Sachs.

Determining whether a person carries one of these mutations can be as easy as running a $5 blood-enzyme test (the Tay-Sachs technique) or as laborious as searching a family's genetic material for disease-related patterns (the method used for Huntington's disease). But carrier tests are often just a first step. If two partners harbor the same bad gene, or the woman's age raises her risk of bearing a child with Down syndrome, the couple can conceive anyway and use prenatal exams to assess the health of the fetus.

Early in the pregnancy, a maternal blood test can signal various problems, but a firm diagnosis requires analyzing fetal blood or cells. Amniocentesis, until recently the quickest cell-sampling technique, takes until the 16th to 20th week to produce results (it involves culturing cells from fluid in the womb). Now, thanks to a new procedure called chorionic villus sampling, tissue snipped from the developing fetal sac can yield the same results as early as the ninth week. Future tests could reveal abnormalities in 4- to 16-cell embryos. "All the pieces are there," says Dr. John Buster, a pioneer in reproductive technology. "We will eventually [discern defects] five days after fertilization, before the embryo even implants in the uterine wall."

Unfortunately, most of the conditions being so deftly diagnosed remain incurable. Though three quarters of all known genetic diseases interfere with a normal life, and half cause early death, fewer than 15 percent can be corrected through treatment. Doctors may someday treat hereditary diseases by inserting palliative genes into defective cells. For now, a positive test result typically leaves parents just two options: abort the fetus or learn to live with the affliction.

When accommodating the disability means watching a toddler die of Tay-Sachs or thalassemia, few couples hesitate to abort, and only the most adamant pro-lifer would blame them. But few of the defects for which fetuses can be screened are so devastating. Consider Huntington's disease, the hereditary brain disorder that killed the folk singer Woody Guthrie. Huntington's relentlessly destroys its victim's mind, and anyone who inherits the gene eventually gets the disease. Yet Huntington's rarely strikes anyone under 40, and it can remain dormant into a person's 70s. What does a parent do with the knowledge that a fetus has the gene? Is some life better than none?

Most carriers think not. "Rather than looking at an individual child and saying, 'This child can have 30 or 40 good years,' most people look at it with a sense of responsibility for future generations," says Dorene Markel, a genetic counselor at the University of Michigan Medical Center who deals exclusively with Huntington's families. "It's an extremely violent, terrible disease, and the seriousness of it overshadows all the good years that have come before." Indeed, Markel says very few couples even go so far as to conceive after learning that each potential child would have a 50 percent chance of inheriting the illness.

With Huntington's, the only question is when the ax will fall. But for other disabilities, there's no telling how hard it will come down. Down syndrome, for example, is easily diagnosed in the

3. Made to Order Babies

Screening New Arrivals

When is a child worth keeping? A recent survey of parents suggests:

■ Only 1 percent would abort on the basis of sex.

■ 6 percent would abort a child likely to get Alzheimer's in old age.

■ 11 percent would abort a child predisposed to obesity.

SOURCE: NEW ENGLAND REGIONAL GENETICS GROUP

womb: if fetal cells contain extra copies of the 21st chromosome, the child will be retarded and bear "mongoloid" physical features. But the diagnosis reveals nothing about how severe the problem will be. Some Down victims grow up happy, well adjusted and able to care for themselves, despite their limitations. Others remain totally dependent. How do parents respond?

A minority are content to prepare for an uncertain future. "I'm glad we had the test done," says Diane Lott of Brethren, Mich., who knowingly gave birth to a Down baby in June. "I can say, 'I have a Down baby,' with a smile on my face." But such parents are the exception. Most abort, according to Barbara Bowles, president of the 700-member National Society of Genetic Counselors, partly because Down syndrome is so noticeable. Parents faced with a disease like cystic fibrosis (which often kills its victims by age 30) are more likely to give birth, she says, "because the child [appears] normal."

As more abnormalities are linked to genes, the dilemmas can only get stickier. Despite all the uncertainties, a positive test for Down or Huntington's leaves no doubt that the condition will set in. But not every disease-related gene guarantees ill health. Those associated with conditions like alcoholism, Alzheimer's disease and manic-depressive illness signal only a susceptibility. Preventing such conditions would thus require aborting kids who might never have suffered. And because one gene can have more than one effect, the effort could have unintended consequences. There is considerable evidence linking manic-depressive illness to artistic genius, notes Dr. Melvin Konner, an anthropologist and nonpracticing physician at Emory University. "Doing away with the gene would destroy the impetus for much human creativity."

The future possibilities are even more troubling when you consider that mere imperfections could be screened for as easily as serious diseases. Stuttering, obesity and reading disorders are all traceable to genetic markers, notes Dr.

17

1. GENETIC AND PRENATAL INFLUENCES

Kathleen Nolan of The Hastings Center, a biomedical think tank in suburban New York. And many aspects of appearance and personality are under fairly simple genetic control. Are we headed for a time when straight teeth, a flat stomach and a sense of humor are standards for admission into some families? It's not inconceivable. "I see people in my clinic occasionally who have a sort of new-car mentality," says Dr. Francis Collins, a University of Michigan geneticist who recently helped identify the gene for cystic fibrosis. "It's got to be perfect, and if it isn't you take it back to the lot and get a new one."

At the moment, gender is the only nonmedical condition for which prenatal tests are widely available. There are no firm figures on how often people abort to get their way, but physicians say many patients use the tests for that purpose. The requests have traditionally come from Asians and East Indians expressing a cultural preference for males. But others are now asking, too. "I've found a high incidence of sex selection coming from doctors' families in the last two years," says Dr. Lawrence D. Platt, a geneticist at the University of Southern California—"much higher than ethnic requests. Once there is public awareness about the technology, other people will use the procedure as well."

Those people will find their physicians increasingly willing to help. A 1973 survey of American geneticists found that only 1 percent considered it morally acceptable to help parents identify and abort fetuses of the undesired sex. Last year University of Virginia ethicist John C. Fletcher and Dr. Mark I. Evans, a geneticist at Wayne State University, conducted a similar poll and found that nearly 20 percent approved. Meanwhile, 62 percent of the geneticists questioned in a 1985 survey said they would screen fetuses for a couple who had four healthy daughters and wanted a son.

Some clinicians abhor this trend. Dr. Robin Dawn Clark, head of clinical genetics at the Loma Linda University Medical Center, says it "affronts the reasons I went into the profession." And one geneticist after another echoes the refrain, "Gender is not a disease." Even so, some feel it is not their place to judge their patients' motives. "Doctors who don't regularly disapprove of abortion, but who decline to perform abortions because of sex selection, are being selective and inconsistent," says Dr. Michael A. Roth, a Detroit obstetrician who has no qualms himself. "The decision to terminate on sex alone is not made flippantly. All of my patients have gone through an agonizing decision process and have come to the delicate decision with their spouses." (Though not yet widely available, new techniques can enable gender-obsessed parents to get what they want without aborting at all. The trick is to segregate sperm according to chromosome type and inseminate the woman only with sperm for the gender of choice.)

Right or wrong, the new gender option has set an important precedent. If parents will screen babies for one nonmedical condition, there is no reason to assume they won't screen them for others. Indeed, preliminary results from a recent survey of 200 New England couples showed that while only 1 percent would abort on the basis of sex, 11 percent would abort to save a child from obesity. As Clark observes, the temptation will be to select for "other features that are honored by society."

The trend toward ever greater control could lead to bizarre, scifi scenarios. But it seems unlikely that prenatal swimsuit competitions will sweep the globe anytime soon: most of the globe has yet to reap the benefits of 19-century medicine. Even in America, many prospective parents are still struggling to obtain basic health insurance. If the masses could suddenly afford cosmetic screening tests, the trauma of abortion would remain a powerful deterrent. And while John Buster's dream of extracting week-old embryos for a quick gene check could ease that trauma, it seems a safe bet many women would still opt to leave their embryos alone.

The more immediate danger is that the power to predict children's medical futures will diminish society's tolerance for serious defects. Parents have already sued physicians for "wrongful life" after giving birth to disabled children, claiming it was the doctor's responsibility to detect the defect in the womb. The fear of such suits could prompt physicians to run every available test, however remote the possibility of spotting a medical problem. Conversely, parents who are content to forgo all the genetic fortune-telling could find themselves stigmatized for their backward ways. When four-cell embryos can be screened for hereditary diseases, failing to ensure a child's future health could become the same sort of offense that declining heroic measures for a sick child is today.

In light of all the dangers, some critics find the very practice of prenatal testing morally questionable. "Even at the beginning of the journey the eugenics question looms large," says Jeremy Rifkin, a Washington activist famous for his opposition to genetic tinkering. "Screening is eugenics." Perhaps, but its primary effect so far has been to bring fewer seriously diseased children into the world. In Britain's Northeast Thames region, the number of Indian and Cypriot children born with thalassemia fell by 78 percent after prenatal tests became available in the 1970s. Likewise, carrier and prenatal screening have virtually eliminated Tay-Sachs from the United States and Canada.

Failing to think, as a society, about the appropriate uses of the new tests would be a grave mistake. They're rife with potential for abuse, and the coming advances in genetic science will make them more so. But they promise some control over diseases that have caused immense suffering and expense. Society need only remember that there are no perfect embryos but many ways to be a successful human being.

With ELIZABETH LEONARD *in New York,*
GREGORY CERIO *in Detroit and*
DANIEL GLICK *in Washington*

1. GENETIC AND PRENATAL INFLUENCES

gory (i.e., from retarded to low average or from low average to average). Generally known as either educational day care or infant day care, these programs provide a developmentally stimulating environment to high-risk babies and/or intensive parent support to prepare the parent to help her child.

In one such program based at the University of California/Los Angeles, weekly meetings were held among staff, parents, and infants over a period of four years. By the project's end, the low-birthweight babies had caught up in mental function to the control group of normal birthweight children (Rauh et al., 1988). The Infant Health and Development Project, which was conducted in eight cities and provided low-birthweight babies with pediatric follow-up and an educational curriculum with family support, on average increased their I.Q. scores by thirteen points and the scores of very-low birthweight children by more than six points. Another project targeted poor single teenage mothers whose infants were at high risk for intellectual impairment (Martin, Ramey and Ramey, 1990). One group of children was enrolled in educational day care from six and one-half weeks of age to four and one-half years for five days a week, fifty weeks a year. By four and one-half years, the children's I.Q. scores were in the normal range and ten points higher than a control group. In addition, by the time their children were four and one-half, mothers in the experimental group were more likely to have graduated from high school and be self-supporting than were mothers in the control group.

These studies indicate that some disadvantages of poverty and low birthweight can be mitigated and intellectual impairment avoided. The key is attention to the cognitive development of young children, in conjunction with social support of their families.

5. Clipped Wings

aged nearly a pound (14.6 ounces or 416 grams) smaller than those born to women who had normal weight gain and did not use cigarettes, marijuana, and cocaine (see Table 1). The effect of these substances on size is more than the sum of the risk factors combined.

Like alcohol use, drug use has different effects at different points in fetal development. Use in very early pregnancy is more likely to cause birth defects affecting organ formation and the central nervous systems. Later use may result in low birthweight due to either preterm birth or intrauterine growth retardation (Kaye et al., 1989; MacGregor et al., 1987; Petitti and Coleman, 1990). While some symptoms may be immediately visible, others may not be apparent until later childhood (Weston et al., 1989; Gray and Yaffe, 1986; Frank et al., 1988).

In infancy, damaged babies can experience problems in such taken-for-granted functions as sleeping and waking, resulting in exhaustion and poor development. In childhood, problems are found in vision, motor control, and in social interaction (Weston et al., 1989). Such problems may be caused not only by fetal drug exposure but also by insufficient prenatal care for the mother or by an unstimulating or difficult home environment for the infant (Lifschitz et al., 1985).

WHAT CAN be done to ameliorate the condition of children born with such damage? Quite a bit, based on the success of supportive prenatal care and the results of model projects that have provided intensive assistance to both baby and mother from the time of birth. These projects have successfully raised the I.Q. of low- and very-low birthweight babies an average of ten points or more—an increase that may lift a child with below-average intelligence into a higher I.Q. cate-

Figure 6
Prevalence of Marijuana Use Among Women Aged 22-44 Years by Race and Education Level

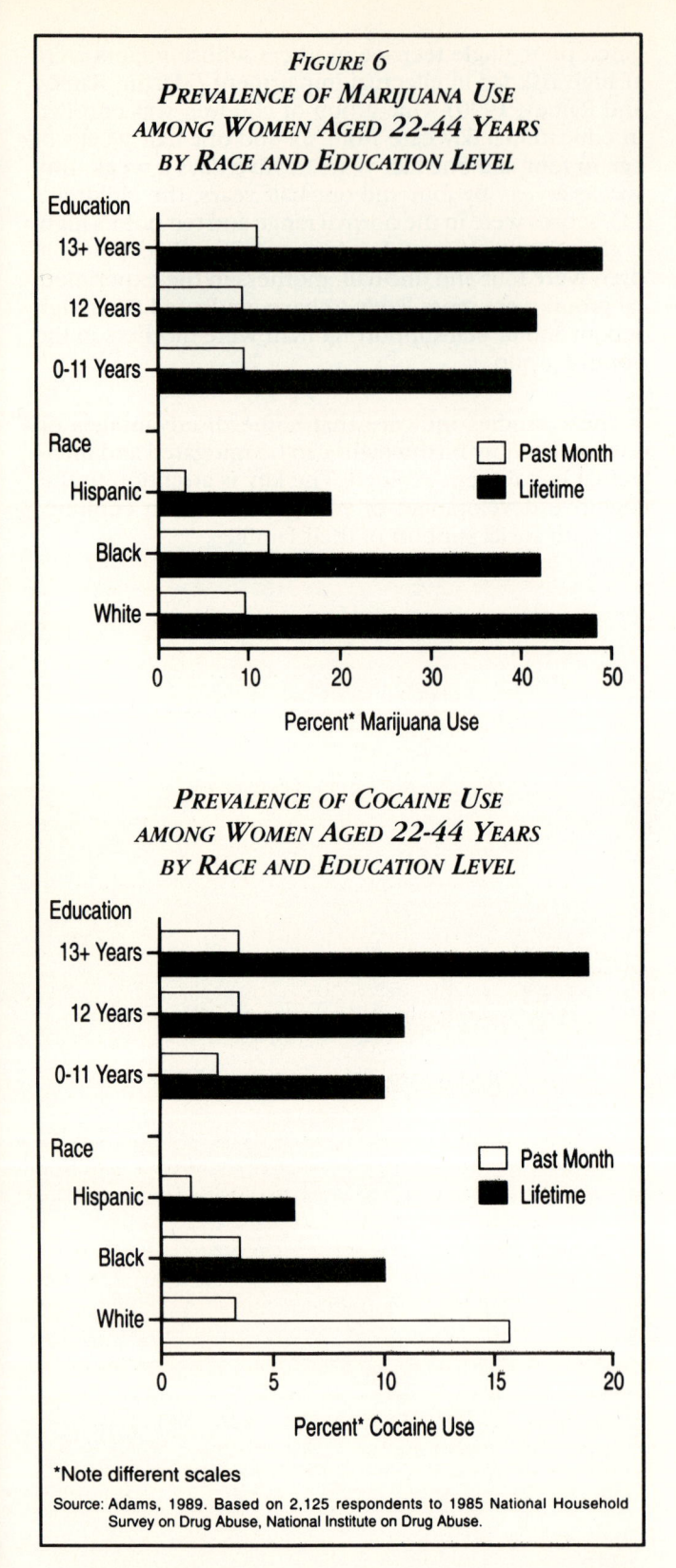

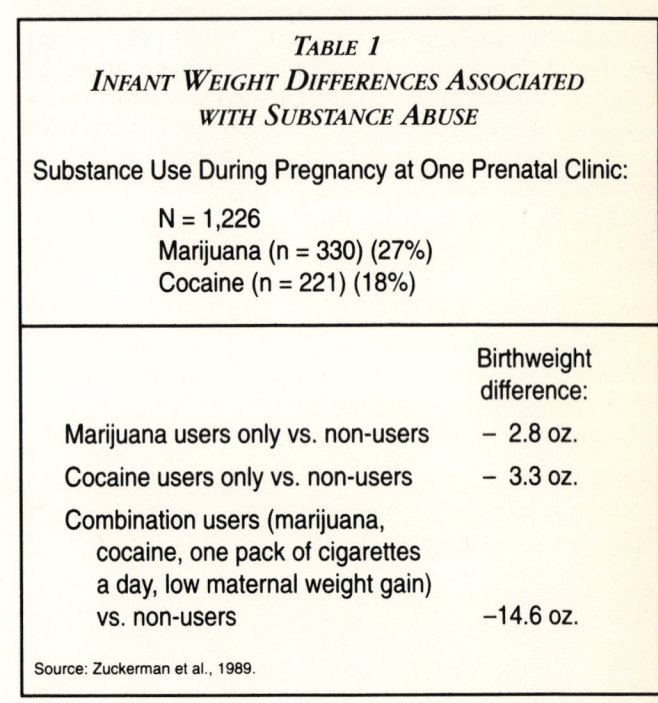

Prevalence of Cocaine Use Among Women Aged 22-44 Years by Race and Education Level

*Note different scales

Source: Adams, 1989. Based on 2,125 respondents to 1985 National Household Survey on Drug Abuse, National Institute on Drug Abuse.

ly to smoke and to gain less weight during pregnancy, two factors associated with low birthweight. The cumulative effect of these risk factors is demonstrated by the finding that infants born to women who gained little weight, who had smoked one pack of cigarettes a day, and who tested positive for marijuana and cocaine aver-

Table 1
Infant Weight Differences Associated with Substance Abuse

Substance Use During Pregnancy at One Prenatal Clinic:

N = 1,226
Marijuana (n = 330) (27%)
Cocaine (n = 221) (18%)

	Birthweight difference:
Marijuana users only vs. non-users	– 2.8 oz.
Cocaine users only vs. non-users	– 3.3 oz.
Combination users (marijuana, cocaine, one pack of cigarettes a day, low maternal weight gain) vs. non-users	–14.6 oz.

Source: Zuckerman et al., 1989.

1. GENETIC AND PRENATAL INFLUENCES

school failure also is connected to a history of fetal alcohol exposure (Abel and Sokol, 1987; Ernhart et al., 1985). Figure 5 shows the drinking habits of women of childbearing age by race and education.

When consumed in pregnancy, alcohol easily crosses the placenta, but exactly how it affects the fetus is not well known. The effects of alcohol vary according to how far along in the pregnancy the drinking occurs. The first trimester of pregnancy is a period of brain growth and organ and limb formation. The embryo is most susceptible to alcohol from week two to week eight of development, a point at which a woman may not even know she is pregnant (Hoyseth and Jones, 1989). Researchers have yet to determine how much alcohol it takes to cause problems in development and how alcohol affects each critical gestational period. It appears that the more alcohol consumed during pregnancy, the worse the effect.

And many of the effects do not appear until ages four to seven, when children enter school.

Nearly one in four (23 percent) white women, eighteen to twenty-nine, reported "binge" drinking (five drinks or more a day at least five times in the past year). This was nearly three times the rate for black women of that age (about 8 percent). Fewer women (around 3 percent for both black and white) reported steady alcohol use (two drinks or more per day in the past two weeks).

4. Fetal Drug Exposure

The abuse of drugs of all kinds—marijuana, cocaine, crack, heroin, or amphetamines—by pregnant women affected about 11 percent of newborns in 1988—about 425,000 babies (Weston et al., 1989).

Cocaine and crack use during pregnancy are consistently associated with lower birthweight, premature birth, and smaller head circumference in comparison with babies whose mothers were free of these drugs (Chasnoff et al., 1989; Cherukuri et al., 1988; Doberczak et al., 1987; Keith et al., 1989; Zuckerman et al., 1989). In a study of 1,226 women attending a prenatal clinic, 27 percent tested positive for marijuana and 18 percent for cocaine. Infants of those who had used marijuana weighed an average of 2.8 ounces (79 grams) less at birth and were half a centimeter shorter in length. Infants of mothers who had used cocaine averaged 3.3 ounces (93 grams) less in weight and .7 of a centimeter less in length and also had a smaller head circumference than babies of nonusers (Zuckerman et al., 1989). The study concluded that "marijuana use and cocaine use during pregnancy are each independently associated with impaired fetal growth" (Zuckerman et al., 1989).

In addition, women who use these substances are like-

FIGURE 4
Relation of Maternal Cigarette Smoking during Pregnancy and Various Measures of School Failure and Learning Deficiency at Age 7

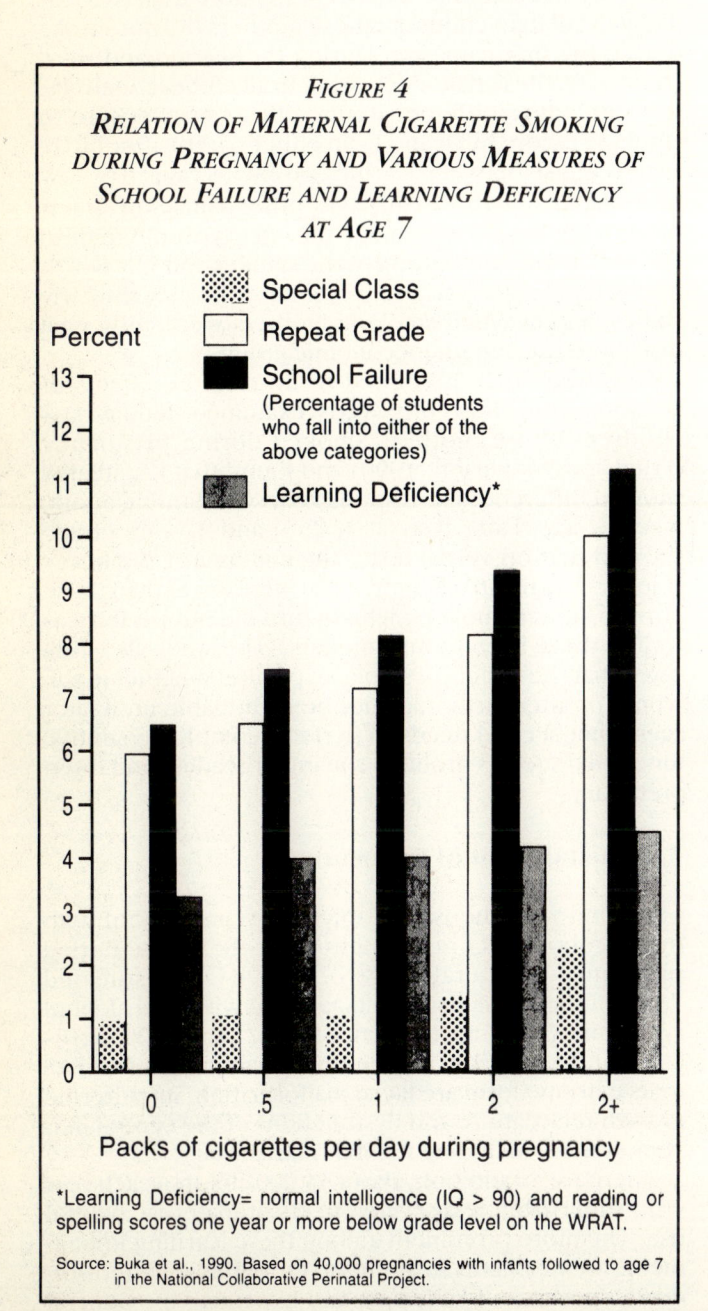

*Learning Deficiency = normal intelligence (IQ > 90) and reading or spelling scores one year or more below grade level on the WRAT.

Source: Buka et al., 1990. Based on 40,000 pregnancies with infants followed to age 7 in the National Collaborative Perinatal Project.

FIGURE 5
Drinking Habits of Women Aged 18-44, by Age, Race, and Education Level, 1985

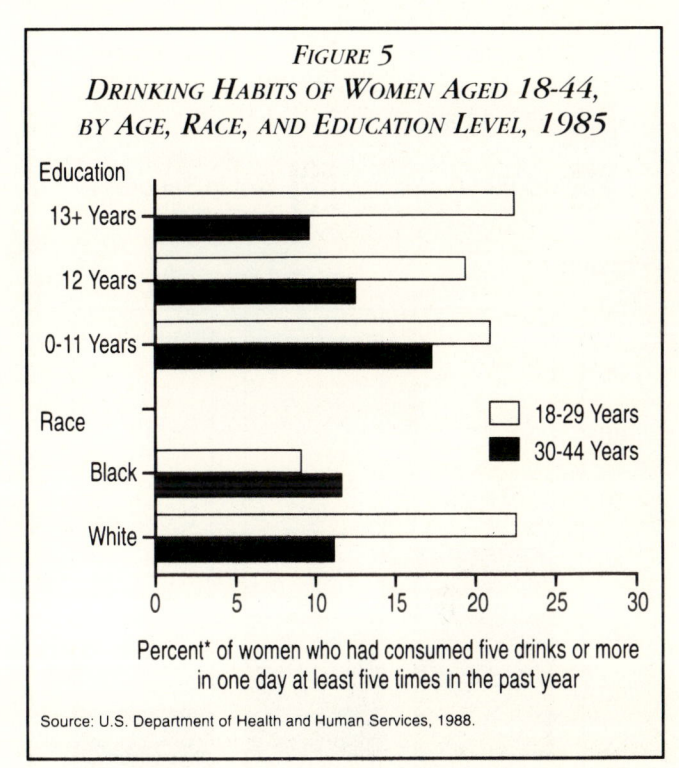

Percent* of women who had consumed five drinks or more in one day at least five times in the past year

Source: U.S. Department of Health and Human Services, 1988.

5. Clipped Wings

to, among other problems, frequent hospitalization and school absence (Streissguth, 1986). A growing number of new studies has shown that children of smokers are smaller in stature and lag behind other children in cognitive development and educational achievement. These children are particularly subject to hyperactivity and inattention (Rush and Callahan, 1989).

Data from the National Collaborative Perinatal Project on births from 1960 to 1966 measured, among other things, the amount pregnant women smoked at each prenatal visit and how their children functioned in school at age seven. Compared to offspring of nonsmokers, children of heavy smokers (more than two packs per day) were nearly twice as likely to experience school failure by age seven (see Figure 4). The impact of heavy smoking is apparently greater the earlier it occurs during pregnancy. Children of women who smoked heavily during the first trimester of pregnancy were more than twice as likely to fail than children whose mothers did not smoke during the first trimester. During the second and third trimesters, these risks decreased. In all of these analyses, it is difficult to differentiate the effects of exposure to smoking before birth and from either parent after birth; to distinguish between learning problems caused by low birthweight and those caused by other damaging effects of smoking; or, to disentangle the effects of smoke from the socioeconomic setting of the smoker. But it is worth noting that Figure 4 is based on children born in the early sixties, an era when smoking mothers were fairly well distributed across socioeconomic groups.

One study that attempted to divorce the effects of smoking from those of poverty examined middle-class children whose mothers smoked during pregnancy (Fried and Watkinson, 1990) and found that the infants showed differences in responsiveness beginning at one week of age. Later tests at 1, 2, 3, and 4 years of age showed that on verbal tests "the children of the heavy smokers had mean test scores that were lower than those born to lighter smokers, who in turn did not perform as well as those born to nonsmokers." The study also indicated that the effects of smoke exposure, whether in the womb or after birth, may not be identifiable until later ages when a child needs to perform complex cognitive functions, such as problem solving or reading and interpretation.

3. Prenatal Alcohol Exposure

Around forty thousand babies per year are born with fetal alcohol effect resulting from alcohol abuse during pregnancy (Fitzgerald, 1988). In 1984, an estimated 7,024 of these infants were diagnosed with fetal alcohol syndrome (FAS), an incidence of 2.2 per 1,000 births (Abel and Sokol, 1987). The three main features of FAS in its extreme form are facial malformation, intrauterine growth retardation, and dysfunctions of the central nervous system, including mental retardation.

There are, in addition, about 33,000 children each year who suffer from less-severe effects of maternal alcohol use. The more prominent among these learning impairments are problems in attention (attention-deficit disorders), speech and language, and hyperactivity. General

---

FIGURE 3
RELATION OF BIRTHWEIGHT TO INTELLIGENCE AND ACHIEVEMENT SCORES AT AGE 7

Legend:
- IQ > 90
- IQ > 90 & "LD"
- IQ 80-90
- IQ 80-90 & LD*
- IQ < 80
- Not in School

[Bar chart showing percentages for Very Low Birthweight, Low Birthweight, and Normal Birthweight categories]

* LD (learning deficiency) refers to academic achievement scores one year or more below grade level according to WRAT reading and spelling tests.

Source: Buka et al., 1990. Based on 40,000 children followed from birth (1960-66) to age 7 in the National Collaborative Perinatal Project.

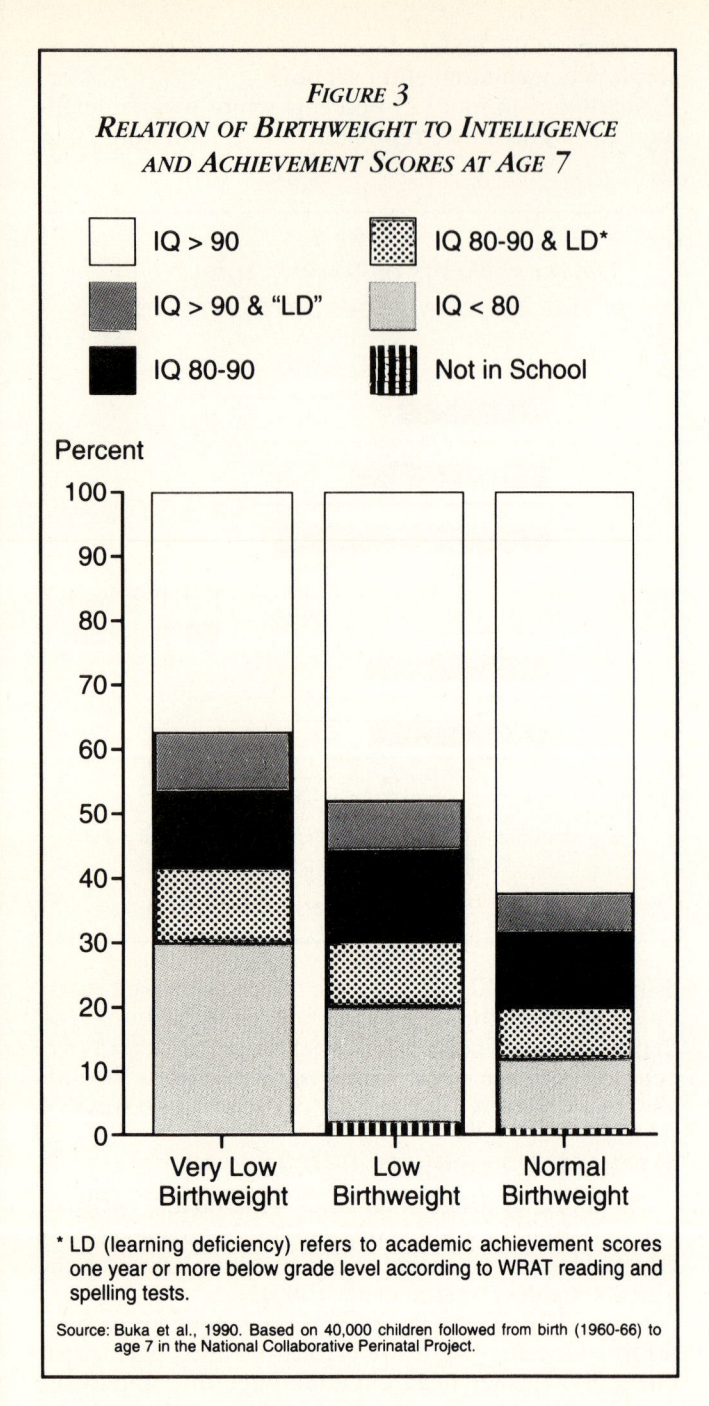

Indeed, follow-up studies of low-birthweight infants at school age have concluded that "the influence of the environment far outweighs most effects of nonoptimal prenatal or perinatal factors on outcome" (Aylward et al., 1989). This finding suggests that early assistance can improve the intellectual functioning of children at risk for learning delay or impairment (Richmond, 1990).

2. Maternal Smoking

Maternal smoking during pregnancy has long been known to be related to low birthweight (Abel, 1980), an increased risk for cancer in the offspring (Stjernfeldt et al., 1986), and early and persistent asthma, which leads

1. GENETIC AND PRENATAL INFLUENCES

brighter side, the evidence that many of these impairments can be overcome by improved environmental conditions suggests that postnatal treatment is possible; promising experiments in treatment are, in fact, under way and are outlined at the end of this article.

1. Low Birthweight

The collection of graphs begins with a set on low birthweight, which is strongly associated with lowered I.Q. and poor school performance. While low birthweight can be brought on by other factors, including maternal malnutrition and teenage pregnancy, significant causes are maternal smoking, drinking, and drug use.

Around 6.9 percent of babies born in the United States weigh less than 5.5 pounds (2,500 grams) at birth and are considered "low-birthweight" babies. In 1987, this accounted for some 269,100 infants. Low birthweight may result when babies are born prematurely (born too early) or from intrauterine growth retardation (born too small) as a result of maternal malnutrition or actions that restrict blood flow to the fetus, such as smoking or drug use.

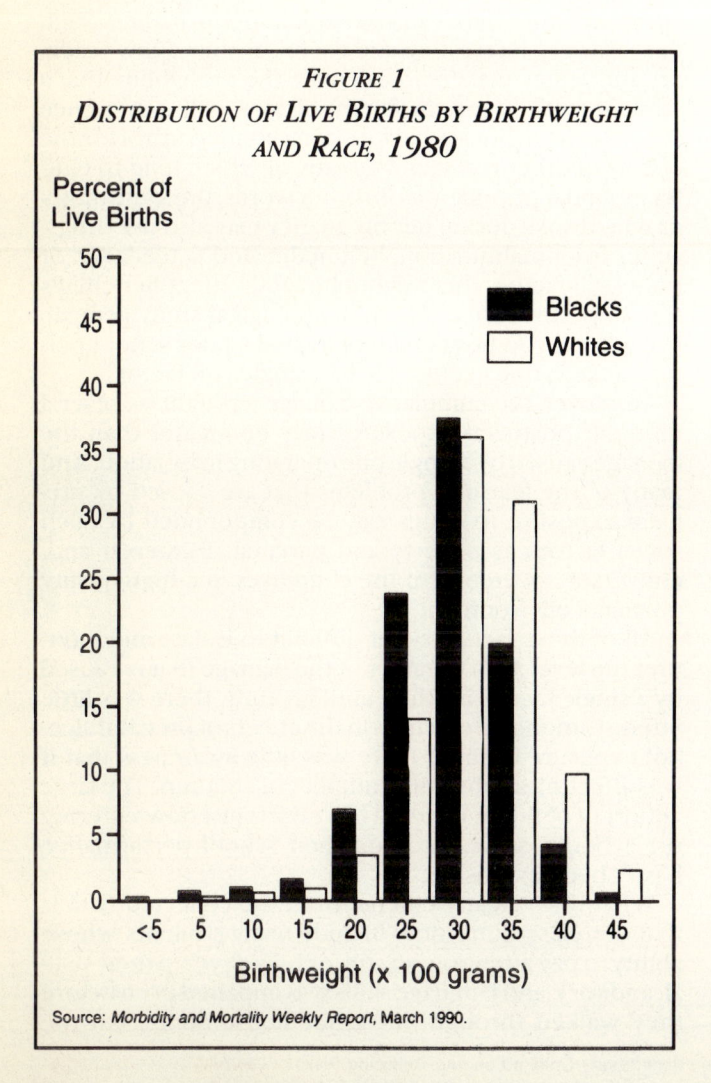

In 1987, about 48,750 babies were born at very low birthweights (under 3.25 lbs. or 1,500 grams). Research estimates that 6 to 8 percent of these babies experience major handicaps such as severe mental retardation or cerebral palsy (Eilers et al., 1986; Hack and Breslau, 1986). Another 25 to 26 percent have borderline I.Q. scores, problems in understanding and expressing language, or other deficits (Hack and Breslau, 1986; Lefebvre et al., 1988; Nickel et al., 1982; Vohr et al., 1988). Although these children may enter the public school system, many of them show intellectual disabilities and require special educational assistance. Reading, spelling, handwriting, arts, crafts, and mathematics are difficult school subjects for them. Many are late in developing

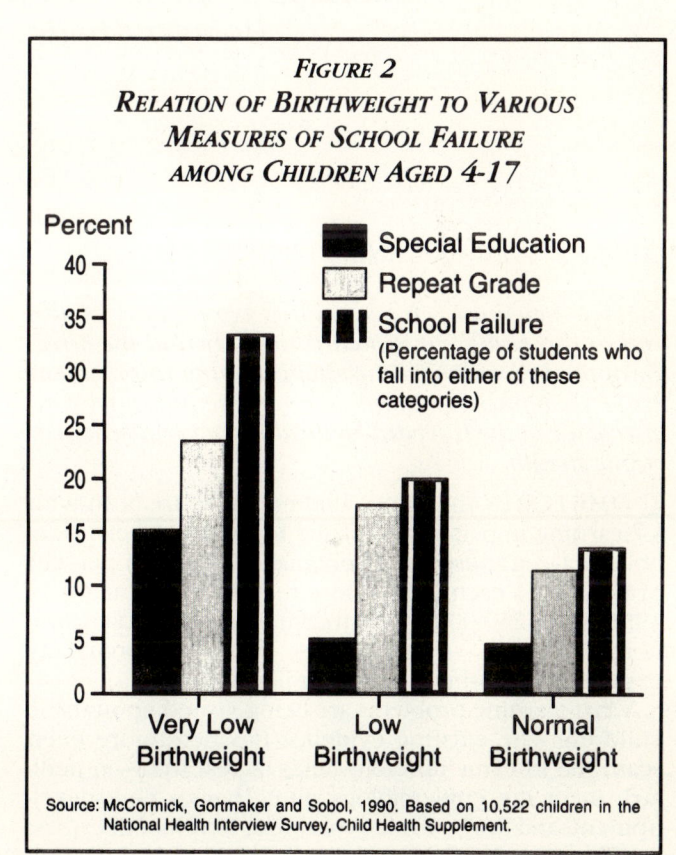

FIGURE 1
DISTRIBUTION OF LIVE BIRTHS BY BIRTHWEIGHT AND RACE, 1980

Source: *Morbidity and Mortality Weekly Report*, March 1990.

FIGURE 2
RELATION OF BIRTHWEIGHT TO VARIOUS MEASURES OF SCHOOL FAILURE AMONG CHILDREN AGED 4-17

Source: McCormick, Gortmaker and Sobol, 1990. Based on 10,522 children in the National Health Interview Survey, Child Health Supplement.

their speech and language. Children born at very low birthweights are more likely than those born at normal weights to be inattentive, hyperactive, depressed, socially withdrawn, or aggressive (Breslau et al., 1988).

New technologies and the spread of neonatal intensive care over the past decade have improved survival rates of babies born at weights ranging from 3.25 pounds to 5.5 pounds. But, as Figures 2 and 3 show, those born at low birthweight still are at increased risk of school failure. The increased risk, however, is very much tied to the child's postnatal environment. When the data on which Figure 2 is based are controlled to account for socioeconomic circumstances, very low-birthweight babies are approximately twice, not three times, as likely to repeat a grade.

CLIPPED WINGS

*The Fullest Look Yet at How
Prenatal Exposure to Drugs, Alcohol, and Nicotine
Hobbles Children's Learning*

LUCILE F. NEWMAN AND STEPHEN L. BUKA

Lucile F. Newman is a professor of community health and anthropology at Brown University and the director of the Preventable Causes of Learning Impairment Project. Stephen L. Buka is an epidemiologist and instructor at the Harvard Medical School and School of Public Health.

SOME FORTY thousand children a year are born with learning impairments related to their mother's alcohol use. Drug abuse during pregnancy affects 11 percent of newborns each year—more than 425,000 infants in 1988. Some 260,000 children each year are born at below normal weights—often because they were prenatally exposed to nicotine, alcohol, or illegal drugs.

What learning problems are being visited upon these children? The existing evidence has heretofore been scattered in many different fields of research—in pediatric medicine, epidemiology, public health, child development, and drug and alcohol abuse. Neither educators, health professionals, nor policy makers could go to one single place to receive a full picture of how widespread or severe were these preventable causes of learning impairment.

In our report for the Education Commission of the States, excerpts of which follow, we combed these various fields to collect and synthesize the major studies that relate prenatal exposure to nicotine, alcohol, and illegal drugs* with various indexes of students' school performance.

The state of current research in this area is not always as full and satisfying as we would wish. Most of what exists is statistical and epidemiological data, which document the frequency of certain high-risk behaviors and correlate those behaviors to student performance. Such data are very interesting and useful, as they allow teachers and policy makers to calculate the probability that a student with a certain family history will experience school failure. But such data often cannot control for the effects of other risk factors, many of which tend to cluster in similar populations. In other words, the same mother who drinks during her pregnancy may also use drugs, suffer from malnutrition, be uneducated, a teenager, or poor—all factors that might ultimately affect her child's school performance. An epidemiological study generally can't tell you how much of a child's poor school performance is due exclusively to a single risk factor.

Moreover, the cumulative damage wrought by several different postnatal exposures may be greater than the damage caused by a single one operating in isolation. And many of the learning problems that are caused by prenatal exposure to drugs can be compounded by such social factors as poverty and parental disinterest and, conversely, overcome if the child lives in a high-quality postnatal environment.

All of these facts make it difficult to isolate and interpret the level and character of the damage that is caused by a single factor. Further, until recently, there was little interest among researchers in the effects of prenatal alcohol exposure because there was little awareness that it was affecting a substantial number of children. The large cohort of children affected by crack is just now entering the schools, so research on their school performance hasn't been extensive.

What does clearly emerge from the collected data is that our classrooms now include many students whose ability to pay attention, sit still, or fully develop their visual, auditory, and language skills was impaired even before they walked through our schoolhouse doors. On the

---

*The full report for the ECS also addressed the effect on children's learning of fetal malnutrition, pre- and postnatal exposure to lead, and child abuse and neglect.

From *American Educator*, Spring 1991, pp. 27-33, 42. Adapted from "Every Child a Learner: Reducing Risks of Learning Impairment During Pregnancy and Infancy," supported by the Exxon Educational Foundation, published by the Education Commission of the States.

1. GENETIC AND PRENATAL INFLUENCES

study of laboratory rats linked heavy alcohol use with infertility because the liquor lowered testosterone levels. Another study, from the University of Washington in Seattle, discovered that newborn babies whose fathers drank at least two glasses of wine or two bottles of beer per day weighed an average of 3 ounces less than babies whose fathers were only occasional sippers—even when all other factors were considered.

Illicit Drugs. Many experts believe that a man's frequent use of substances such as marijuana and cocaine may also result in an unhealthy fetus, but studies that could document such findings have yet to be conducted. However, preliminary research has linked marijuana to infertility. And recent tests at the Yale Infertility Clinic found that long-term cocaine use led to both very low sperm counts and a greater number of sperm with motion problems.

WHAT A DAD CAN DO

The best news about sperm troubles is that many of the risk factors can be easily prevented. Because the body overhauls sperm supplies every 90 days, it only takes a season to get a fresh start on creating a healthy baby. Most experts advise that men wait for three months after quitting smoking, cutting out drug use or abstaining from alcohol before trying to sire a child.

Men who fear they are exposed to work chemicals that may compromise the health of future children can contact NIOSH. (Write to the Division of Standards Development and Technology Transfer, Technical Information Branch, 4676 Columbia Parkway, Mailstop C-19, Cincinnati, OH 45226. Or call [800] 356-4674.) NIOSH keeps files on hazardous chemicals and their effects, and can arrange for a local inspection of the workplace. Because it is primarily a research institution, NIOSH is most useful for investigating chemicals that haven't been studied previously for sperm effects (which is why the Malones approached NIOSH with their concerns about paradichlorobenzene). For better-known pollutants, it's best to ask the federal Occupational Safety and Health Administration (OSHA) to inspect the job site (OSHA has regional offices in most U.S. cities).

There is also advice for men who are concerned over exposure to radiation during medical treatment. Direct radiation to the area around the testes can spur infertility by halting sperm production for more than three years. According to a recent study, it can also triple the number of abnormal sperm the testes produce. Men who know they will be exposed to testicular radiation for medical reasons should consider "banking" sperm before the treatment, for later use in artificial insemination. Most hospitals use lead shields during radiation therapy, but for routine X-rays, even dental X-rays, protection might not be offered automatically. If it's not offered, patients should be sure to request it. "The risks are really, really low, but to be absolutely safe, patients—male or female—should *always* ask for a lead apron to protect their reproductive organs," stresses Martin.

Though the study of sperm health is still in its infancy, it is already clear that a man's reproductive system needs to be treated with respect and caution. Women do not carry the full responsibility for bearing a healthy infant. "The focus should be on both parents—not on 'blaming' either the mother or the father, but on accepting that each plays a role," says Friedler.

Mattison agrees: "Until recently, when a woman had a miscarriage, she would be told it was because she had a 'blighted ovum' [egg]. We never heard anything about a 'blighted sperm.' This new data suggests that both may be responsible. That is not unreasonable," he concludes, "given that it takes both an egg and a sperm to create a baby!"

adding even more controversy to the issue of occupational hazards to sperm. In 1984, employees brought a class-action suit against Milwaukee-based Johnson Controls, the nation's largest manufacturer of car batteries, after the company restricted women "capable of bearing children" from holding jobs in factory areas where lead exceeded a specific level. The suit—which the Supreme Court is scheduled to rule on this spring—focuses on the obstacles the policy creates for women's career advancement. Johnson Controls defends its regulation by pointing to "overwhelming" evidence that a mother's exposure to lead can harm the fetus.

In effect, the company's rule may be a case of reverse discrimination against men. Males continue to work in areas banned to women despite growing evidence that lead may not be safe for sperm either. In several studies over the past 10 years, paternal exposure to lead (and radiation) has been connected to Wilms' tumor, a type of kidney cancer in children. In another recent study, University of Maryland toxicologist Ellen Silbergeld, Ph.D., exposed male rats to lead amounts equivalent to levels below the current occupational safety standards for humans. The rats were then mated with females who had not been exposed at all. Result: The offspring showed clear defects in brain development.

Johnson Controls claims that evidence linking fetal problems to a father's contact with lead is insufficient. But further research into chemicals' effects on sperm may eventually force companies to reduce pollution levels, since *both* sexes can hardly be banned from the factory floor. Says Mattison: "The workplace should be safe for everyone who wants to work there, men and women alike!"

FATHER TIME

Whatever his occupation, a man's age may play an unexpected role in his reproductive health. When researchers at the University of Calgary and the Alberta Children's Hospital in Canada examined sperm samples taken from 30 healthy men aged 20 to 52, they found that the older men had a higher percentage of sperm with structurally abnormal chromosomes. Specifically, only 2 to 3 percent of the sperm from men between ages 20 and 34 were genetically abnormal, while the figure jumped to 7 percent in men 35 to 44 and to almost 14 percent in those 45 and over. "The findings are logical," says Renée Martin, Ph.D., the professor of pediatrics who led the study. "The cells that create sperm are constantly dividing from puberty onwards, and every time they divide they are subject to error."

Such mistakes are more likely to result in miscarriages than in unhealthy babies. "When part of a chromosome is missing or broken, the embryo is more likely to abort as a miscarriage [than to carry to term]," Martin says. Yet her findings may help explain why Savitz's North Carolina study noted a doubled rate of birth defects like cleft palate and hydrocephalus in children whose fathers were over 35 at the time of conception, no matter what the mothers' age.

Currently, there are no tests available to pre-identify sperm likely to cause genetic defects. "Unfortunately there's nothing offered, because [the research] is all so new," says Martin. But tests such as amniocentesis, alpha fetoprotein (AFP) and chorionic villi sampling (CVS) can ferret out some fetal genetic defects that are linked to Mom *or* Dad. Amniocentesis, for example, is routinely recommended for all pregnant women over 35 because with age a woman increases her risk of producing a Down's syndrome baby, characterized by mental retardation and physical abnormalities.

With respect to Down's syndrome, Martin's study provided some good news for older men: It confirmed previous findings that a man's risk of fathering a child afflicted with the syndrome actually drops with age. Some popular textbooks still warn that men over 55 have a high chance of fathering Down's syndrome babies. "That information is outdated," Martin insists. "We now know that for certain."

THE SINS OF THE FATHERS?

For all the hidden dangers facing a man's reproductive system, the most common hazards may be the ones most under his control.

Smoking. Tobacco addicts take note: Smoke gets in your sperm. Cigarettes can reduce fertility by lowering sperm count—the number of individual sperm released in a single ejaculation. "More than half a pack a day can cause sperm density to drop by 20 percent," says Machelle Seibel, M.D., director of the Faulkner Centre for Reproductive Medicine in Boston. One Danish study found that for each pack of cigarettes a father tended to smoke daily (assuming the mother didn't smoke at all), his infant's birthweight fell 4.2 ounces below average. Savitz has found that male smokers double their chances of fathering infants with abnormalities like hydrocephalus, *Bell's palsy* (paralysis of the facial nerve), and mouth cysts. In Savitz's most recent study, children whose fathers smoked around the time of conception were 20 percent more likely to develop brain cancer, lymphoma and leukemia than were children whose fathers did not smoke (the results still held regardless of whether the mother had a tobacco habit).

This is scary news—and not particularly helpful: Savitz's studies didn't record how frequently the fathers lit up, and no research at all suggests why the links appeared. Researchers can't even say for sure that defective sperm was to blame. The babies may instead have been victims of passive smoking—affected by Dad's tobacco while in the womb or shortly after birth.

Drinking. Mothers-to-be are routinely cautioned against sipping any alcohol while pregnant. Now studies suggest that the father's drinking habits just before conception may also pose a danger. So far, research hasn't discovered why alcohol has an adverse effect on sperm, but it does suggest that further investigation is needed. For starters, one

1. GENETIC AND PRENATAL INFLUENCES

to ban women of childbearing age from jobs that entail exposure to hazardous substances. The idea is to protect the women's future children from defects—and the companies themselves from lawsuits. Already, the "fetal protection policy" of one Milwaukee-based company has prompted female employees to file a sex discrimination suit that is now before the U.S. Supreme Court. Conversely, if the new research on sperm is borne out, men whose future plans include fatherhood may go to court to *insist* on protection from hazards. Faced with potential lawsuits from so many individuals, companies may be forced to ensure that workplaces are safe for *all* employees.

Sperm und Drang

At the center of all this controversy are the microscopic products of the male reproductive system. Sperm (officially, spermatozoa) are manufactured by *spermatagonia,* special cells in the testes that are constantly stimulated by the male hormone testosterone. Once formed, a sperm continues to mature as it travels for some 80 days through the *epididymis* (a microscopic network of tubes behind the testicle) to the "waiting area" around the prostate gland, where it is expelled in the next ejaculation.

A normal sperm contains 23 chromosomes—the threadlike strands that house DNA, the molecular foundation of genetic material. While a woman is born with all the eggs she will ever produce, a man creates millions of sperm every day from puberty onwards. This awesome productivity is also what makes sperm so fragile. If a single sperm's DNA is damaged, the result may be a mutation that distorts the genetic information it carries. "Because of the constant turnover of sperm, mutations caused by the environment can arise more frequently in men than in women," says David A. Savitz, Ph.D., an associate professor of epidemiology and chief researcher of the North Carolina review.

If a damaged sperm fertilizes the egg, the consequences can be devastating. "Such sperm can lead to spontaneous abortions, malformations, and functional or behavioral abnormalities," says Marvin Legator, Ph.D., director of environmental toxicology at the department of preventative medicine at the University of Texas in Galveston. And in some cases, sperm may be too badly harmed even to penetrate an egg, leading to mysterious infertility.

Though the findings on sperm's vulnerability are certainly dramatic, researchers emphasize that they are also preliminary. "We have only a very vague notion of how exposure might affect fetal development, and the whole area of research is at a very early stage of investigation," says Savitz. Indeed, questions still far outnumber answers. For starters, there is no hard evidence that a chemical damages an infant by adversely affecting the father's sperm. A man who comes in contact with dangerous substances might harm the baby by exposing his partner indirectly—for example, through contaminated clothing. Another theory holds that the harmful pollutants may be carried in the seminal fluid that buoys sperm. But more researchers are becoming convinced that chemicals can inflict their silent damage directly on the sperm itself.

The Chemical Connection

The most well-known—and most controversial—evidence that chemicals can harm sperm comes from research on U.S. veterans of the Vietnam war who were exposed to the herbicide Agent Orange (dioxin), used by the U.S. military to destroy foliage that hid enemy forces. A number of veterans believe the chemical is responsible for birth defects in their children. The latest study on the issue, published last year by the Harvard School of Public Health, found that Vietnam vets had almost twice the risk of other men of fathering infants with one or more major malformations. But a number of previous studies found conflicting results, and because so little is known about how paternal exposure could translate into birth defects, the veterans have been unsuccessful in their lawsuits against the government.

Scientific uncertainty also dogs investigations into other potentially hazardous chemicals and contaminants. "There seem to be windows of vulnerability for sperm: Certain chemicals may be harmful only at a certain period during sperm production," explains Donald Mattison, M.D., dean of the School of Public Health at the University of Pittsburgh. There isn't enough specific data to make definitive lists of "danger chemicals." Still, a quick scan of the research shows that particular substances often crop up as likely troublemakers. Chief among them: lead, benzene, paint solvents, vinyl chloride, carbon disulphide, the pesticide DBCP, anesthetic gases and radiation. Not surprisingly, occupations that involve contact with these substances also figure heavily in studies of sperm damage. For example, men employed in the paper, wood, chemical, drug and paint industries may have a greater chance of siring stillborn children. And increased leukemia rates have been detected among children whose fathers are medical workers, aircraft or auto mechanics, or who are exposed regularly to paint or radiation. In fact, a study of workers at Britain's Sellafield nuclear power plant in West Cambria found a sixfold leukemia risk among children whose fathers were exposed to the plant's highest radiation levels (about 9 percent of all employees).

Workers in "high-risk" industries should not panic, says Savitz. "The credibility of the studies is limited because we have no firm evidence that certain exposures cause certain birth defects." Yet it makes sense to be watchful for warning signs. For example, if pollution levels are high enough to cause skin irritations, thyroid trouble, or breathing problems, the reproductive system might also be at risk. Another danger signal is a clustered outbreak of male infertility or of a particular disease: It was local concern about high levels of childhood leukemia, for instance, that sparked the investigation at the Sellafield nuclear plant.

The rise in industrial "fetal protection policies" is

SPERM UNDER SIEGE

MORE THAN WE EVER GUESSED, HAVING A HEALTHY BABY MAY DEPEND ON DAD

Anne Merewood

IT DIDN'T MAKE SENSE. Kate Malone's* first pregnancy had gone so smoothly. Yet when she and her husband Paul* tried to have a second child, their efforts were plagued by disaster. For two years, Kate couldn't become pregnant. Then she suffered an ectopic pregnancy, in which the embryo began to grow in one of her fallopian tubes and had to be surgically removed. Her next pregnancy heralded more heartache—it ended in miscarriage at four months and tests revealed that the fetus was genetically abnormal. Within months, she became pregnant and miscarried yet again. By this point, some four years after their troubles began, the couple had adopted a son; baffled and demoralized by the string of apparent bad luck, they gave up trying to have another child. "We had been to the top doctors in the country and no one could find a reason for the infertility or the miscarriages," says Kate.

Soon, however, thanks to a newspaper article she read, Kate uncovered what she now considers the likely cause of the couple's reproductive woes. When it all started, Paul had just been hired by a manufacturing company that used a chemical called paradichlorobenzene, which derives from benzene, a known carcinogen. The article discussed the potential effects of exposure to chemicals, including benzene, on a man's sperm. Kate remembered hearing that two other men in Paul's small office were also suffering from inexplicable infertility. Both of their wives had gone through three miscarriages as well. Kate had always considered their similar misfortunes to be a tragic coincidence. Now she became convinced that the chemical (which has not yet been studied for its effects on reproduction) had blighted the three men's sperm.

Paul had found a new job in a chemical-free workplace, so the couple decided to try once more to have a baby. Kate conceived immediately—and last August gave birth to a healthy boy. The Malones are now arranging for the National Institute for Occupational Safety and Health (NIOSH), the federal agency that assesses work-related health hazards for the public, to inspect Paul's former job site. "Our aim isn't to sue the company, but to help people who are still there," says Kate.

The Malones' suspicions about sperm damage echo the concerns of an increasing number of researchers. These scientists are challenging the double standard that leads women to overhaul their lives before a pregnancy—avoiding stress, cigarettes and champagne—while men are left confident that their lifestyle has little bearing on their fertility or their future child's health. Growing evidence suggests that sperm is both more fragile and potentially more dangerous than previously thought. "There seems to have been both a scientific resistance, and a resistance based on cultural preconceptions, to accepting these new ideas," says Gladys Friedler, Ph.D, an associate professor of psychiatry and pharmacology at Boston University School of Medicine.

But as more and more research is completed, sperm may finally be stripped of its macho image. For example, in one startling review of data on nearly 15,000 newborns, scientists at the University of North Carolina in Chapel Hill concluded that a father's drinking and smoking habits, and even his age, can increase his child's risk of birth defects—ranging from cleft palates to *hydrocephalus,* an abnormal accumulation of spinal fluid in the brain. Other new and equally worrisome studies have linked higher-than-normal rates of stillbirth, premature delivery and low birthweight (which predisposes a baby to medical and developmental problems) to fathers who faced on-the-job exposure to certain chemicals. In fact, one study found that a baby was more likely to be harmed if the father rather than the mother worked in an unsafe environment in the months before conception.

The surprising news of sperm's delicate nature may shift the balance of responsibility for a newborn's wellbeing. The research may also have social and economic implications far beyond the concerns of couples planning a family. In recent years a growing number of companies have sought

*These names have been changed.

WHAT CRACK DOES TO BABIES

BY JANICE HUTCHINSON

Janice Hutchinson is a pediatrician and former senior scientist for the American Medical Association. She is now the medical director of the Child and Youth Services Administration of the District of Columbia Department of Mental Health.

INQUIRING TEACHERS want to know: Who are these kids and how did they get this way? The question refers to the unprecedented numbers of children—estimates range as high as one-half to one million—who are entering the classroom having suffered inutero exposure to cocaine.

Crack, the cooked form of cocaine, became widely available in 1985; the children of the first crack addicts are now in school. Teachers have described them as a new breed, unlike other children with histories of drug exposure. They are often in constant motion, disorganized, and very sensitive to stimuli. Crawling, standing, and walking take longer to develop. They are irritable and hard to please. It is hard for them to make friends. They respond less to the environment. Internal stability is poor. Learning is more difficult. Smiling and eye contact are infrequent. They do not seem to know how to play with toys or with others. And nothing you do for them seems to matter or help.

If teachers are to meet the challenges that these children bring, they may find it helpful to understand the bio-neuro-physiological effects of cocaine on the developing fetus. Scientists are just beginning to understand these effects; research in the area is incomplete and at times conflicting. Thus what we know and what we can speculate about, some of which is summarized below, is just the tip of a rather unknown iceberg. There are surely many more effects—and more complicated avenues of effect—than those so far identified. Nonetheless, there are findings—mainly from research sponsored by the National Institutes of Health and the National Institute on Drug Abuse—that allow us to begin to make some sense of what is happening to the behaviors and learning styles of these children.

IMAGINE THAT a crack molecule has entered the body. It enters into the mucous membranes of the mouth. From there it enters the lungs, where it is absorbed into the bloodstream, and through which it then passes to the heart and, very quickly, to the brain. The immediate effect is an increase in breathing, blood pressure, and heart rate.

Upon arriving in the brain, crack acts at several sites along what is known as the brain's "pleasure pathway"—a collection of sites in the brain that seem in some ways to relate and affect each other. At one point on the pleasure pathway is the limbic system, which is the seat of strong emotional responses, including the very primitive urges to feed, flee, fight, and reproduce. At another point along the pathway is the motor cortex of the brain, which directs the body's movement. Between the limbic system and the motor cortex lies the nucleus accumbens. This is the "attraction center" of the brain; it is what pulls you toward pleasurable activity.

The crack is very active here in the nucleus accumbens; a ripple effect then seems to carry the destruction around to other points along the pleasure pathway. Within the nucleus accumbens, as elsewhere in the brain, are numerous nerve cells; the space between each nerve cell ending is known as the synaptic space. Each of these nerve cells communicates with the others across the synaptic space by sending a variety of neurotransmitters back and forth.

One such neurotransmitter is dopamine. Under normal biological conditions, dopamine, like other neurotransmitters, is continually moving across the synaptic space. In a constantly recurring pattern, the dopamine leaves its home cell, crosses the synaptic space, and reaches receptors on the receiving cell, an action that sends an electrical signal through the receiver cell. The dopamine then disattaches from the receptor cell and returns to its cell of origin where it will be recycled.

But if crack has been ingested, this normal cycle will be disrupted. Crack, upon entering the brain and then the pleasure pathway, seems to settle into the synaptic space between the neurotransmitters. It then acts to pre-

1. GENETIC AND PRENATAL INFLUENCES

vent the dopamine from returning to its home cell. Unable to return home, the dopamine continues to stimulate the receiver cell until the crack has spent itself and dissipated. It is probably this constant stimulation of the receiver cell that causes the euphoric feeling associated with the first few minutes of cocaine ingestion. But the crack high lasts only a few minutes, after which the user will either replenish his intake or experience an often devastating "low." The constant resupply soon leads to a physical addiction, the breaking of which is accompanied by extremely painful withdrawal symptoms.

WHILE THE crack is acting on the mother's brain, what is happening to the fetus? The crack crosses the placental barrier and heads for the inutero brain. The exact effect of the crack on the fetus will depend on the age of the fetus, the dosage of the crack, and probably on other variables that we have not yet identified. But it seems likely that in general a number of things happen. First, the crack probably acts on the fetal nucleus accumbens in the same way that it acts on the user's, leading the fetus to become highly stimulated and, often, addicted. As it stimulates the nucleus accumbens, and surely in other ways as well, the crack damages fetal brain cells and thus causes neurological damage all along the pleasure pathway and in other nearby parts of the brain.

Damage to brain cells in the limbic system, the nucleus accumbens, and elsewhere along the pleasure pathway would likely impair or alter a wide range of the child's normal emotional responses, including, for example, the ability to respond to pleasurable experiences, to form emotional attachments, or to make certain kinds of judgments. Perhaps this explains in part why the crack baby is often unable to proceed through the normal phases of separation-individuation described by child psychiatrist Margaret Mahler; crack babies appear to experience much greater anxiety and difficulty in leaving their mothers when it is time for school.

In addition, the brain's motor cortex may be damaged, which might explain such effects as the slow development of crawling, standing, and walking. The brain location for speech is also nearby, and damage to it may account for the speech impairments suffered by many crack babies. In turn, the speech impairment inhibits the child's ability to communicate, which may, in turn, account for some of the difficulty these children have in forming relationships.

In addition, crack, like nicotine, constricts the adult and fetal arteries, thus slowing the blood—and therefore the oxygen flow—to the fetus and around it. This condition of low oxygenation—known as hypoxia—can also produce brain damage, and it can bring on low birthweight. Low birthweight is, in turn, associated with a wide range of disabling symptoms, including intellectual disabilities.

Reading, mathematics, spelling, handwriting, and the arts are often difficult tasks for low-birthweight babies. Speech and language problems are prominent. Temperamental problems, such as low adaptability, low persistence, and arrhythmicity (for example, the failure to sleep and wake at normal, regular times) may be part of their behavioral style. They typically cry when separated from the mother, have trouble expressing themselves, speak only in short phrases, are very active, and clumsy.

Findings to date suggest that temperament influences both behavior and cognition.

These low birthweight children tend to perform poorly on the Mullen Scale of Early Learning. This test, which consists of four scales, suggests the range of learning abilities that seem to be impaired in the children exposed inutero to crack. The Visual Receptive Organization (VRO) scale assesses visual discrimination, short-term memory, visual organization and sequencing, and visual spatial awareness, including position, size, shape, left/right, and detail. The Visual Expressive Organization (VEO) assesses bilateral and unilateral manipulation, writing, visual discrimination, and visual-motor plan and control. The language receptive organization (LRO) scale assesses auditory comprehension, short- and long-term auditory memory, integration of ideas and visual spatial cues, auditory sequencing, and verbal spatial concepts. The language expressive organization (LEO) scale assesses spontaneous and formal verbal ability, language formulation, auditory comprehension, and short- and long-term memory.

What all of this means ultimately is that these low-birthweight crack children experience the world around them in a very different way from other children. Adults, including teachers, are often unaware that these children see and hear their environment in a completely different manner from adults or even other children. What the teacher often does not realize is that this difficult-to-teach, hard-to manage child is processing information in an unusual way that the child does not determine. Hence, conflict and frustration can arise between teacher and student (and also at home between parent and child).

The combined effects of prenatal drug exposure with a home environment that provides little or no nurturance, understanding, or support for the child create a terrible challenge to teachers. But initial experimental programs suggest that these children can benefit greatly from placement in highly structured, highly tailored educational day care settings beginning in early infancy. In four Washington, D.C.-area therapeutic nurseries that provide such care, two-thirds of the children seem so far to have been successfully mainstreamed into first grade.

Among the characteristics that seem to make such programs successful are early identification of the infants and very low student-teacher ratios. The establishment of an emotionally supportive atmosphere and structure is necessary. Teaching must be intense and focal. Tasks should initially be simple and singular. Too many tasks or activities overstimulate these children, and they cannot respond. Teachers must also provide emotional support and form bonds with the children. Success also depends on aggressively approaching and engaging parents in the psychotherapeutic progress. Consultation with mental health professionals may assist teachers, parents, and students. Intellectually limited students may still require individual tutoring; some students will eventually require special education; and very emotionally disturbed students may require a mental health-based psychotherapy program.

But it does seem clear that with early, appropriate interventions many of these children can improve their behavior and academic performance. Like most childhood problems, the time to act is now; later is too late.

Motherhood on Trial

More and more women are being arrested for taking illegal drugs that may endanger their fetuses' health. But at what point, if any, should a mother-to-be's responsibility to her baby be dictated by the law?

David Ruben

David Ruben is a contributing editor of Parents.

On the morning of October 3, 1989, 18-year-old Monica Young, nearly seven months pregnant with her second child, walked into the Medical University of South Carolina in Charleston complaining of abdominal pains. Her girlfriend, she said, had accidentally kicked her in the stomach. The hospital, a sprawling, state-run complex whose various clinics constitute a medical safety net for many of the area's impoverished residents, admitted Young for observation and tests. Included was a urinalysis, which the hospital had recently begun administering to any woman suspect of taking drugs during her pregnancy. The result: positive for cocaine.

The second of six children, Young was raised and still lives on Charleston's largely poor and black east side. It is a neighborhood of ramshackle houses, dirt yards, and garbage-strewn vacant lots—a place that is easy walking distance yet light-years in spirit from the gracious antebellum homes and elegant shops of the city's postcard-pretty historic district. Young admits that she was using cocaine—but never more than a couple of times a week, she insists. Yes, she had heard that the drug could pose dangers for the baby growing inside her. But she says she was running with an older crowd that seemed to think it was okay—"falling behind the big girls" is how she puts it. "I saw them do it, and I just wanted to experience it. I was just living for myself. I wasn't facing reality."

Two days after her admittance to the hospital, Young was forced to face reality when two Charleston policemen came knocking on her door. "Are you Monica Young?" the officers asked as they entered her seventh-floor room. "Yes," came the surprised response. "We have a warrant for your arrest," they said. The charges: possession of cocaine and distribution of drugs to a person under 18. (In the past, the South Carolina courts have treated a viable fetus as a human being.) And with that, Monica Young, some 28 weeks pregnant, was handcuffed, escorted downstairs, and placed in a patrol car waiting at the hospital door.

Later that day, still in handcuffs, bound now by iron shackles around her ankles as well, Young was led before Summary Court Judge Jack Guedalia. He set her bond at $80,000. "I felt I had to put her under some sort of control so that she would not harm the unborn child," Guedalia would later explain. "I had to think, How can I stop this woman?"

Stop her he did. Unable to make bond, Young was placed in the Charleston County jail. She would remain there for nearly six weeks. "I was treated like I was nothing," she says of her incarceration. "I had stomach pains, but they wouldn't let me see the nurse." She alleges that once when she asked for help to relieve her pain, she was given hot water mixed with cornstarch.

Finally, Patricia Kennedy, a private lawyer who had agreed to handle her case on a *pro bono* basis, convinced Guedalia to reduce the bond to personal recognizance. Young was freed on the strict condition that she live under house arrest, save for daily visits to the county's outpatient drug program for treatment and testing. The message was clear: One way or another, Charleston would stop Monica Young and pregnant women like her from using drugs. Even if it meant locking them up.

The Social Experiment

Monica Young and a growing number of women around the country, nearly all of them poor, many of them black, are guinea pigs in a new and fiercely controversial social experi-

1. GENETIC AND PRENATAL INFLUENCES

TREATMENT PROGRAMS:
The Awful Shortage

The waiting list for drug treatment at the Solid Foundation Mandela House in Oakland, California, bears the names of 80 addicts. Yet this residential rehabilitation center for pregnant women has but six beds. As the only program of its kind in the state, the Mandela House stands as just one example of the woeful disparity between supply and demand: Although definitive drug-use statistics have yet to be gathered, the best estimate is that in the United States some 235,000 pregnant women are in need of treatment for drug or alcohol use, with only 29,000, or 12 percent, actually receiving it. Aside from a handful of federal and state pilot programs and a few small, private facilities such as the Mandela House, assistance for pregnant women who want to kick drugs is hard—if not impossible—to find.

EMPTY COFFERS. During the 1980s, according to Lee Dogoloff, executive director of the American Council for Drug Education in Rockville, Maryland, there was a "large-scale erosion" of funding for treatment programs. And today, only a seventh of President Bush's $9.5 billion war on drugs is spent on treatment; of those funds, just a small fraction is earmarked for long-term, residential programs, which experts believe are far and away the most effective.

Hospitalization, outpatient clinics, and 28-day residential programs have all had a certain amount of success treating chemical dependency, but for pregnant, hard-core addicts, treatment specialists say, four months or more of residential care is what works best: "In many cases, these women grew up in unstable families scarred by drug use and sexual abuse. Their addiction is not the root of the problem; it's a single symptom of a troubled lifestyle, of low self-esteem," says Minnie Thomas, executive director of the Mandela House. "To truly help them," she says, "we've got to address all these issues. And that takes time."

MODEL CARE. Effective programs not only care for the mother's many needs (from preparing her meals to filling out her Medicaid forms), they also put a premium on keeping the family together and provide childcare when possible. Janet Chandler, of the Chicago-based National Association for Perinatal Addiction Research and Education (NAPARE), says that women are often reluctant to enter a program without childcare because "they will be forced to turn their children over to a protection agency. And once the agency has their kids, it's very difficult to regain custody." A good program will also teach women parenting skills and instruct their children about the dangers of drugs and alcohol.

At Baby's Porch, for instance, an innovative, 16-bed residential program in Houston, women receive vocational training in areas like office work and childcare—but their treatment doesn't stop when they walk out the door: A staff member helps them find an apartment and checks up on them periodically; and for two years, the women are expected to attend peer-counseling meetings.

These programs, however, aren't cheap. While some, such as Baby's Porch, charge only $350 per month to house and feed a woman and her two children, other programs can run as much as $10,000 per month. But even at a higher price, says Chandler, the long-term savings should not be underestimated. Using a less expensive method, she says, will mean that, one way or another, society will eventually have to pick up the tab for the offspring of cocaine- and alcohol-abusing mothers.

—*Bill Shapiro*

WHAT PARENTS CAN DO

Increasing the government's funding of residential care facilities is an important step toward guaranteeing that babies start life drug-free. Parents can effect a change by using the tried-and-true tactic of putting pen to paper and letting their elected officials (from city council members to congresspersons to William Bennett, the director of the national drug control policy) know what they think.

Parents can also help lobby local representatives; in Chula Vista, California, for instance, a pregnancy advocacy coalition composed of physicians and other health professionals helped notify county officials about the problems pregnant addicts face, eventually leading the state to loosen its purse strings. Parents interested in starting or joining such a coalition should call their state office of drug and alcohol services for information.

Finally, parents can make tax-deductible contributions to NAPARE, a national, nonprofit organization that trains physicians on issues of substance abuse and pregnancy. Send checks to: NAPARE, 11 East Hubbard Street, Suite 200, Chicago, IL 60611.

---

ment known as "fetal-abuse" prosecution. Spurred by mounting public frustration over the rising toll of "coke babies," more and more prosecutors, in many cases with full cooperation from local doctors and hospitals, are bringing the power of the criminal justice system to bear on what was once considered a purely medical matter: what a pregnant woman chooses to do with her body.

The prosecutors' actions, proponents say, are a bold and necessary response to an increasingly grim social reality: Approximately one in ten newborns in the United States—some 375,000 a year—are exposed to illegal drugs in the womb. In particular, cocaine and its more potent, smokable derivative, crack, have been claiming women of childbearing age at an alarming rate; drug-abuse experts now believe that roughly one-half of the estimated half a million crack users in this country are female. (In contrast, women account for a third or less of all heroin addicts.) As a result, prenatal cocaine abuse, especially in the poverty-stricken inner cities where crack use is concentrated, has skyrocketed. In New York City, for instance, health officials say cocaine abuse among pregnant women has increased 3,000 percent over the past ten years. And although the research on the effects of cocaine in utero is still in the early stages, it is increasingly clear that a woman who uses cocaine—even occasionally—while she is pregnant runs at least some risk of causing serious damage, in some cases death, to her unborn child.

The most serious consequences, researchers believe so far, are impaired fetal growth and premature delivery, both of which hike the odds of infant mortality as well as the slew of complications associated with low birth weight—mental retardation and lung problems, among others. Preliminary studies have also linked cocaine to spontaneous abortion; to small head size and birth defects such as genital urinary deformities, heart defects, and brain damage; and to sudden infant death syndrome. In addition, there is some evidence that maternal cocaine use may put children at greater-than-average risk for a whole host of long-range developmental problems, including learning disabilities, personality disorders, and emotional withdrawal.

So far, few cities have chosen to follow Charleston's lead in actually imprisoning pregnant women who use illegal drugs. But arrests, while still relatively rare, are proliferating. Apart from Monica Young, at least 10 other women have been arrested since October in Charleston's crackdown. At least 8 arrests have occurred in nearby Greenville, South Carolina. And perhaps as many as 20 more women have been seized in Florida, Illinois, Nevada, Massachusetts, Georgia, Colorado, and Michigan, most of them after delivering their babies.

One woman "had a placental abruption and the baby died. She had tested positive for cocaine just before delivery. How would you like to have to go in and talk to her now?"

At stake is not only the fate of substance-abusing women and their babies, but the relationship between every pregnant woman and her unborn child. For while many of us may venture no closer to crack than the six o'clock news, there are plenty of other illegal drugs—marijuana and speed, for instance—that are likely to hit closer to home. Not to mention legal, socially sanctioned drugs like nicotine and caffeine, both of which have been associated with varying degrees of fetal damage. And what about alcohol? More than 35,000 babies are born in the United States each year with alcohol-related defects; up to one-quarter of them may have full-blown fetal alcohol syndrome, a devastating condition that is the leading known cause of mental retardation in this country.

Most of us would agree that a pregnant woman holds a special responsibility to the child growing inside her. But are we ready to invoke the power of the state to enforce it? And if we answer yes for drugs, where do we draw the line?

"This issue has me stumped, and it has a lot of people stumped," says Arlene Bowers Andrews, an assistant professor at the University of South Carolina College of Social Work, who is active in child-welfare issues statewide. "I certainly don't want a police state. But when you look at those babies..."

Case-by-Case Studies

Shirley Brown has taken a good hard look at those babies. As the case manager for high-risk obstetrics at Charleston's Medical University, it is part of her job to deal with the city's drug-baby problem, one infant—and one mother—at a time; Brown was the one who called the police after receiving Monica Young's urinalysis.

"Sometimes I think the people who say what we're doing is wrong should have to go in and talk to a woman about the fact that her baby's dead because she used cocaine. Like I'm about to."

Brown, a compact woman with a steely, no-nonsense manner, is standing in the warm, humid air of the brightly lit neonatal intensive-care unit on the hospital's eighth floor. She is surrounded by impossibly fragile newborns tethered to feeding tubes and blinking electronic probes. Some weigh less than a pound and a half, with feet that would barely cover a single adult finger-joint. "This woman delivered a premature two-pounder while on cocaine two years ago," Brown continues, her soft drawl taking on a stony edge. "Yesterday she came in ready to deliver another. Except she had a placental abruption, and the baby died. She had tested positive for cocaine just before this delivery. How would you like to have to go in and talk to her now?"

It is a challenge, not a question. Brown is one of the chief architects and supporters of the Medical University's new get-tough policy toward pregnant drug abusers. The policy was hatched last summer in response to the escalating numbers of drug-affected babies being born in the hospital—up from 2 to 23 a month in less than a year—and what Brown calls the dismal failure of her efforts to educate pregnant drug users about cocaine's dangers and persuade them to seek treatment voluntarily. Despite the objections of some doctors and staff—who worried about women shunning prenatal care and argued that more should be done to beef up treatment options—the arrests began in October.

Brown and other supporters say the policy is working. Since Monica Young and a handful more were arrested last fall, the number of babies born with drugs in their systems each month has fallen back into the low single digits. "Maybe we're just lucky," says Dr. E. O. Horger, the Medical University's bespectacled director of maternal/fetal medicine and a strong backer of the arrests. "Or, just maybe, we're doing something right."

A History of Arrests

Still, whether or not the Charleston arrests will hold up in court remains to be seen. Throughout the country, prosecutors bringing fetal-abuse cases have met stiff resistance in the courtroom—primarily because judges and juries have reacted cautiously to the politically charged issue of conferring legal rights on the unborn.

In two closely watched cases last year, a grand jury in Rockford, Illinois, refused to indict Melanie A. Green for involuntary manslaughter in the death of her drug-damaged newborn; in Hollywood, Florida, a judge dismissed child-abuse charges that had been brought against Casandra Gethers after she delivered her second cocaine-positive baby. The state is appealing.

In fact, the sole fetal-abuse conviction to date was obtained only via some fancy prosecutorial footwork: Last July in Sanford, Florida, a judge found Jennifer Johnson guilty of delivering a controlled substance to her two children, both of whom were born with cocaine in their systems. The state skirted the fetus/child issue by asserting that the crimes were committed after the babies were born—but before their umbilical cords were severed; thus, the argument ran, Johnson's bloodstream had pumped cocaine into living children through the cords. The judge agreed and sentenced Johnson to a year's house arrest in a drug-rehabilitation center, 200 hours of community service, and 14 years of probation. The case is currently on appeal.

Most fetal-abuse cases, however, have not come to trial at all. That's because prosecutors are, by and large, cutting deals with their defendants, most of whom are arrested only after they deliver babies that test positive for drugs. In return for guilty pleas, these prosecutors are recommending sentences of probation—contingent on these women attending drug-treatment programs and submitting regular urine samples. This, they say, reflects their desire to help women, not jail them.

Raising the Objections

Despite all the nothing-but-good-intentions talk from prosecutors, however, the move to criminalize prenatal behavior is drawing heavy fire on a wide variety of fronts:

■ Fetal-abuse prosecutions have been attacked as discriminatory, because defendants are nearly always poor and disproportionately black—even though there is some evidence that drug use among pregnant women cuts across class and racial lines. For example, a 1989 study of pregnant women in Pinellas County, Florida, conducted by the National Association for Perinatal Addiction Research and Education (NAPARE), found that although the number of black and white women who tested positive for drugs and alcohol was roughly equal, black women were almost ten times more likely to be reported to authorities.

A number of prosecutors concede that they are concerned about the issue; some say they plan to widen their dragnets to include prosperous suburban delivery rooms. Critics, who note that public inner-city hospitals and clinics will always be more likely to report patient drug-use than private physicians, say they are not holding their breath.

"Race and class are front and center in this debate," asserts Wendy Chavkin, a former Rockefeller Foundation fellow at the Columbia School of Public Health in New York City. "These prosecutions reflect age-old attitudes toward poor women and women of color."

■ Feminists see the prosecutions as part of a wider impulse to subordinate women's autonomy to the "rights" of their fetuses. (In fact they object to the very term "fetal-abuse" prosecution and prefer to call this practice "the prosecution of women for their conduct during pregnancy.") "We're building an adversarial relationship between a woman and her fetus," asserts Alison Wetherfield of the National Organization for Women's (NOW) Legal Defense and Education Fund. For instance, judges in several states have ordered delivery of babies via cesarean section over the mothers' objections, based on doctors' assertions that the procedure is in the fetuses' best interest. Widespread opposition has derailed the practice for the time being, but a pivotal test case on the issue has yet to be decided.

And then there's the "slippery slope" argument: If pregnant women are prosecuted for using drugs today, critics ask, which behaviors will be criminalized in the name of fetal health tomorrow? Drinking? Smoking? Eating junk food? Missing doctor's appointments? Not exercising?

Prosecutors generally dismiss such scenarios as farfetched. But earlier this year charges of child abuse were brought against a woman in Wyoming for drinking while pregnant, although the case has since been dropped. And some prosecutors privately say they would consider charging women who smoke while pregnant if it could be proven to have caused fetal damage.

"For a while I thought all this concern over the slippery slope was a bunch of malarkey," says Walter Connolly, Jr., a Detroit lawyer who monitors fetal-abuse cases for the Chicago-based NAPARE. "But now, I'm truly worried."

■ There is also the inescapable link to abortion. It is no accident, many feminists say, that these cases are proliferating at the same time abortion rights are under fire. "I worry that in seeking to elevate the legal status of the fetus, these prosecutors are being used as a backdoor challenge to *Roe v. Wade*," says Kary Moss of the American Civil Liberties Union (ACLU) Women's Rights Project.

But the relationship between the two issues is surprisingly complex. Some who strongly oppose abortion rights are hesitant to back fetal-abuse prosecutions for fear pregnant drug users will choose to abort their fetuses rather than face jail for delivering them. On the other hand, some who have always supported abortion rights also endorse fetal-abuse prosecutions. "It certainly creates a dilemma for those of us who consider ourselves both feminists and children's rights advocates," concedes Patricia Toth, the director of the National Center for the Prosecution of Child Abuse, an Alexandria, Virginia–based organization that advises prosecutors who are involved in fetal-abuse cases. "But if I had to choose whose rights I would come down on, it would always be the children's."

■ Many argue that fetal-abuse prosecutions simply won't solve the problem. "Using a penalty to deter an addict, a person who has a disease, doesn't work," asserts Susan Galbraith, a consultant to the New York City–based National Council on Alcoholism and Drug Dependence. In fact, Galbraith and many others fear the tactic will backfire: As word of the prosecutions spreads, pregnant addicts may shrink from seeking the prenatal care they so desperately need, may even choose to deliver their babies in bedrooms and back alleys, out of fear that medical authorities will report them to the police.

"You're going to read a lot about babies found in the garbage can, under the staircase, in the alleyway," says Leticia Velasquez, a 32-year-old mother of five from the Bronx who kicked a $400-a-day drug habit and now works at a drug-treatment program in New York City. "A lot of these young mothers are not going to go to the hospital. They're not going to tell anybody they're getting high with this baby in their stomach, because they're going to be petrified that they'll get arrested."

■ Finally, opponents of fetal-abuse prosecutions insist it is unfair to pursue a punitive policy when, aside from several federally funded pilot projects, comprehensive inpatient drug-treatment programs for pregnant addicts are virtually nonexistent. To arrest women for harming their fetuses with drugs or alcohol under such conditions, they say, makes about as much sense as punishing a homeless child for staying out too late.

For example: More than half of 78 New York City drug-treatment centers surveyed by Chavkin refuse to help pregnant addicts; nearly 90 percent refuse treatment to pregnant crack addicts on Medicaid. Of those programs that do accept pregnant women, only two provide childcare—a virtual necessity for mothers, experts say. (For more on treatment programs for pregnant women, see "The Awful Shortage.")

Thus, a pregnant woman may find herself trapped in the catch-22 situation that Casandra Gethers of Hollywood, Florida, did: trying to get help, being denied help because she's pregnant, and then being arrested when she delivered a baby with cocaine in her system.

Closing Arguments

In Charleston, South Carolina, Ninth Circuit Solicitor Charles Condon insists that there are enough drug-treatment programs available and that the women who've been arrested simply haven't taken advantage of them. "These women do not care what happens to their child," he says. "They do not respond to warnings; instead they choose to abuse drugs."

Opponents contend that existing programs in Charleston are inadequate; that the women are not purely recreational users, as Condon believes; and that the threat of punishment rarely works with addicts. And as predicted, they say, the hard-line policy has driven some pregnant drug users underground. "We have one woman here who said she didn't show up for her prenatal appointments because she didn't want to be arrested," says Dr. John Emmel, medical director of the Charleston County Substance Abuse Commission, the public drug-treatment program to which the arrested women have been remanded. "The street talk —and you sometimes have to take it with a grain of salt—is that that's what's happening."

Both Shirley Brown and Dr. Horger agree with Charles Condon and report that this is not the case; according to their count, there has been no decline in the normal levels of prenatal appointments and deliveries at the hospital. For his part, Condon, who is rumored to be eyeing a run for state attorney general, says he can think of no bigger success story in his ten years in office.

"We have one woman here," says Dr. John Emmel, "who said she didn't show up for her prenatal appointments because she didn't want to be arrested."

Meanwhile, right before Christmas, Monica Young delivered a daughter at the same hospital in which she had been handcuffed three months earlier. The baby appeared to be healthy, and tests revealed no traces of drugs in her system. As of this writing, however, the charges against Young stand. Distributing drugs to a minor is a felony offense in South Carolina. If convicted, she could face up to 30 years in prison.

Back at the Medical University, Horger says he recently had the opportunity to meet with federal drug czar William Bennett and fill him in on the hospital's fetal-abuse policy. Bennett, Horger reports, expressed keen interest in the program. Whether that could signal new support from Washington for criminal crackdowns against pregnant drug abusers is anybody's guess. It *is* a safe bet, however, that no matter how many Monica Youngs we arrest, until we invest in the social and medical resources to treat female drug users, and muster the political will to confront the poverty and despair that produce so many of them, the terrible flow of poison from mother to fetus will go on.

Development During Infancy and Early Childhood

No period in human development has received more attention during the past quarter century than infancy. A major portion of this research has focused on infants' perceptual skills and cognitive abilities. These studies have made it quite apparent that infants are far from the passive, unknowing beings they were once thought to be. Rather, they are responsive to a wide range of environmental events, including social stimulation and interaction. Indeed, some of the most interesting recent studies of infants have explored the infant's affective and emotional needs. "How Infants See the World" highlights some of these new findings and considers their implications. Daniel Stern, in "Diary of a Baby," takes this theme a step further. Drawing on this new research, he depicts what early social and perceptual experiences may be like, as seen from the perspective of the young infant.

The many skills of the newborn multiply dramatically over the first several years of life, transforming the physically helpless infant into a child who, by age three or even earlier, is capable of thinking, communicating, and skillfully solving problems. Knowledge of the readiness to learn that is now so evident in infants and toddlers has brought with it questions about the best ways to nurture early intellectual and social development. One point of view, supported by many parents eager to give their young children a head start toward academic success, places emphasis on a relatively narrow band of school-related skills. In "Preschool: Head Start or Hard Push?" Philip R. Piccigallo argues that this view is a dangerously narrow one that fails to recognize the importance of everyday experience in creating meaningful learning opportunities for the young child. Many early childhood experts caution that structured learning experiences bring with them the risk of placing too much pressure on the child; they argue that such pressures can, in turn, stunt creativity and turn learning into drudgery rather than a spontaneous process of discovery.

"How Three Key Countries Shape Their Children" is an illuminating look at the ways in which preschools in Japan, China, and the United States are designed to instill somewhat different values. U.S. preschools stress experiences that promote individuality, while those of Japan and China place far greater emphasis on inculcating a sense of community spirit and responsibility.

Cultures also differ widely in their definition of appropriate roles for the two sexes. For more than a century, scientists have been debating the source of gender differences. As gender research has become more sophisticated, it has also become more controversial. However, most scientists agree that the sexes are more alike than different, with greater variation within each sex than between the sexes. Are boys really better at math and girls more verbal? Are girls really kinder and gentler and boys more aggressive? Are gender-related attributes the work of evolutionary processes? "Guns and Dolls" examines recent research on this topic and explores these issues.

Too often we focus on the intellectual development of the child to the near exclusion of the more whimsical and creative aspects of the growing process. In a refreshing change of pace, Ellen Winner describes the charming features of children's artistic development in "Where Pelicans Kiss Seals." It turns out that activities such as drawing can serve as more than a diversion for scientists. Drawings provide another avenue for exploration of many features of intellectual and social development as well.

"The Day Care Generation" considers a critical problem in the United States: the lack of quality supplemental child-care facilities for children. Over fifty percent of working mothers have children under the age of six. This statistic underlines the magnitude of the problem and the urgent need for effective solutions. Parents must choose among various day care options without the aid of any "industry-wide" standards, knowing that day care centers are operated with practically no federal or state regulations or guidelines to ensure quality. The article identifies a number of key factors to consider in evaluating day care arrangements, and reviews some of the studies that have played an important role in the national debate over early child care.

Looking Ahead: Challenge Questions

Discuss the pros and cons of beginning as early as possible to teach young children skills such as reading and math. Explain why you agree or disagree.

What effect does work have on the development of attachment relationships between mother and infant, or on the effectiveness of discipline in school-age children? Does society have a responsibility to provide supplemen-

Unit 2

tary care for children of working mothers? Does industry have this responsibility?

Should we pay more attention to the child's artistic endeavors? Why or why not?

Why has so much effort been devoted to the study of sex differences in the past several decades? How would you summarize the general feelings of this body of research in brief? What is the meaning of Jerome Kagan's statement in "Guns and Dolls" that gender differences are "inevitable but not genetic"?

HOW INFANTS SEE THE WORLD

In recent years, researchers have practically revolutionized the science of infant development, supplementing traditional observational approaches with new techniques that allow them to penetrate the previously inaccessible mind of the baby. The results are intriguing. No longer do scientists see the typical infant as largely passive, appreciating the world from the sidelines. In the emerging view, babies are characterized as socially engaged, curious and surprisingly sophisticated in their emotional responses and intellectual skills.

Ever since the first newborn registered his presence with a howl, the mystery of what goes on in an infant's mind has puzzled and frustrated adults. Do babies see the world as grownups do? Is their experience of emotion—of anger and sadness, surprise and pleasure—the same as ours? Possessed of "no language but a cry," as Alfred, Lord Tennyson once put it, infants are not able to satisfy our curiosity with an accounting of their experiences. Our own perceptions of infancy have been lost to memory.

Given these limitations, scientists investigating how infants think and feel have traditionally relied upon informed guesswork. Some, like the Swiss psychologist Jean Piaget, drew their conclusions from close observation of infant behavior. Others, many of them psychoanalysts working in the Freudian tradition, attempted to reconstruct infancy by talking to adult patients, piecing together the past from the memories, dreams and psychic conflicts of the present.

In recent years, however, new technologies and a growing body of systematic research have transformed the field of infant studies, challenging much of the accepted wisdom handed down by psychoanalysis and early behavioral investigations. In the last decade, says University of Washington psychologist Andrew Meltzoff, "our knowledge of infant development has undergone a scientific revolution."

Turning tables. In this revolution, basic tenets are being overturned. One is the idea that babies are merely passive consumers of experience, that they absorb—but do not seek out—the sights and sounds of the world around them. Indeed, it is clear that scientists vastly underestimated the infant's capacities for active participation. "We now know how much the baby contributes to its own well-being," says Dr. T. Berry Brazelton, clinical professor of pediatrics emeritus at Harvard University Medical School.

Research shows that infants only a few months old are already acting as sophisticated gatherers and interpreters of information. They are capable, for instance, of decoding complex vocal rhythms. They demonstrate distinct preferences for certain sounds, shapes and tastes. They learn, remember and are already beginning to imitate the behavior of others who are important in their lives. In one study, conducted by Meltzoff and his colleague Keith Moore, babies less than 72 hours old copied the actions of adults who stuck out their tongues or opened their mouths. Babies mimic emotional expressions as well: Some researchers have found that 2-day-

8. How Infants See the World

old infants imitate smiles, frowns and looks of surprise modeled by adults.

Scientists know more about what infants can do in part because they have learned to ask questions in ways that babies can answer. One technique, for instance, takes advantage of infants' natural sucking rhythms. An electronically "bugged" nipple is placed in a baby's mouth and hooked up to two tape recorders. Researchers have found that the infant will increase its rate of sucking in response to a recording of its mother's voice but not after hearing a tape of a strange woman saying exactly the same words. Similarly, head turning, heart rate, gaze patterns and leg kicking can all be used as measures of what babies like and dislike, what they see, hear, smell and feel.

Studies like these belie the longstanding view that babies begin life unplugged from the social world and oblivious to the difference between objects and people. Instead, the newest research indicates that infants distinguish the animate from the inanimate at a very early age, and that they may need interaction with other people for the blossoming of some abilities. For example, infants act differently when confronted with a human face than they do when presented with a motionless pattern. They also respond preferentially to "baby talk"—speech that makes use of higher-pitched sounds, melody and softened consonants. Adults, for their part, instinctively talk this way to babies, forming a perfectly designed evolutionary partnership.

Steppingstones. Such findings have renewed investigators' interest in emotions, once thought too imprecise and difficult to study scientifically. As a result, researchers are beginning to construct road maps of emotional development, with milestones paralleling the traditional physical markers of growth, such as crawling and walking. In one version, proposed by Dr. Stanley Greenspan, clinical professor of psychiatry and pediatrics at George Washington University, infants first begin to show particular warmth and pleasure toward parents and other significant adults between 2 and 4 months of age. The ability to communicate a full range of emotions—including curiosity, anger, pleasure and assertiveness—develops gradually between 3 and 8 months.

Early theorists, Sigmund Freud among them, argued that a baby's tie to its mother was predicated upon her ability to satisfy basic biological needs: He was hungry; she fed him. But the bond between mother and infant is more completely explained, many experts believe, by "attachment theory," an approach first proposed by British psychiatrist John Bowlby. Following the lead of animal ethologists such as Konrad Lorenz, Bowlby suggested that the bond between mother and baby is a critical evolutionary survival mechanism. An infant who maintains close, affectionate contact with the mother has a higher chance of healthy survival. Circumstances that disrupt the bond, on the other hand—the loss of a parent, for example—can have a profound effect on later psychological health.

Bowlby and other researchers who have refined and extended his work believe that the early exchanges between infant and caretaker are a testing ground for human relations in later life. Based on the ways in which their parents respond, children develop conceptions of how the adult world operates. This process of adaptation extends well beyond childhood and is never entirely finished, as mothers and fathers are gradually replaced by other important figures such as boyfriends, wives or simply close friends.

Although all infants require mothering, researchers are discovering that babies vary greatly in the kind of mothering they need. Strict standards of "good mothering," once favored by child-rearing specialists, are now being replaced by a more flexible notion: That parenting must be individually tailored and address a child's unique needs. Infants differ greatly from birth, showing wide variation in temperament, activity levels and how strongly they react to sound, light and touch; an overly sensitive baby, one who shrinks back from loud noises or retreats from a hearty embrace, will present a special challenge to parents and require a different parenting strategy than a more easygoing infant.

Different strokes. This new focus on individual differences has expanded the range of what specialists consider "normal" in infancy. "Babies we once might have thought were pathological are now seen as simply different," says Greenspan, a specialist in infant and child development. But experts have also gained important clues to early signs of trouble and are better equipped to intervene before problems become unmanageable. In part, Greenspan says, this means teaching parents to adapt their behavior, thus "turning the infant's vulnerability into an asset." The mother of an extremely sensitive baby, for example, might be counseled to speak in a lower, calmer voice and to avoid sudden movements. Without such intervention, the parents of a "difficult" child may become frustrated and angry, the psychiatrist says, and unintentionally thwart healthy adjustment.

Many mysteries remain, but the gap between the inner life of infants and our understanding of their thoughts and feelings is shrinking. One pioneer in infant development is psychiatrist Daniel Stern, whose work merges the findings of observational studies with the perceptions of psychoanalysis, uniting subjective experience and external behavior. In research carried out at the University of Geneva, where he is professor of psychology, and at Cornell University Medical Center, Stern delves into the most challenging aspects of development: How do infants form a sense of self? When is a baby aware that he is a separate individual, with thoughts and feelings of his own? . . . [such questions are addressed in his book *Diary of a Baby* published by Basic Books, 1990. See next article.]

Erica E. Goode with Sarah Burke

Article 9

*If infants could record
their thoughts and feelings, what
would the entries look like?
Here's one scenario*

Diary OF A Baby

Daniel N. Stern, M.D.

The World of Feelings: *Joey at 6 Weeks*

A Patch of Sunshine: 7:05 a.m. *A space glows over there. A gentle magnet pulls to capture. The space is growing warmer and coming to life. Inside it, forces start to turn around one another in a slow dance.*

Joey has just awakened. He stares at a patch of sunshine on the wall beside his crib.

For Joey at this age, most encounters with the world are dramatic and emotional—a drama whose elements and nature are not obvious to us as adults. Of all the things in the room, it is the patch of sunshine that attracts and holds Joey's attention. Its brightness and intensity are captivating. At 6 weeks of age Joey experiences objects and events mainly in terms of the *feelings* they evoke in him. He can see quite well, though not yet perfectly. He is aware of different colors, shapes and intensities. And he has been born with strong preferences about what he finds pleasing to look at.

Among these preferences, intensity tops the list. It is the most important element in this scene. A baby's nervous system is prepared to evaluate immediately the intensity of a light, a sound—anything accessible to his senses. How intensely he feels about something is probably the first clue he has available to tell him whether to approach it or to stay away. Intensity can lead him to try to protect himself. It can guide his attention and curiosity and determine his internal level of arousal.

How does Joey know that the glowing space is "over there"? How does he know that it is not "over here," close at hand? Even at this age, Joey is able to calculate distances and quadrants of space. Soon he will divide all space into two distinct areas: A near world within the reach of his extended arm, and a far world beyond it. Not for another few months will Joey be able to reach for, and grasp, what he wants with precision. Nonetheless, he is able at 6 weeks to distinguish between reachable and nonreachable space. This ability will help him define which things are actually within reach. It would not be useful if he tried to reach for the moon—or even for things far across the room.

Infants by this age have color vision. The patch of sunshine is, of course, yellowish against the white wall; the latter looks slightly bluish where the sun does not strike it. "Warm," intense colors, like yellow, appear to come forward, and "cooler" colors, like blue, to recede. So to Joey, the sunshine patch appears to advance toward him, while the space surrounding the patch appears to move away. That is part of the dance. Infants love experiences where stimulation and excitation mount—if not too fast or too high. And they tend to get bored by situations where the stimulation is low or stops changing. So after a while, Joey gets bored by the play of appearances he sees in the sun patch. His attention suddenly dies, and he turns his head away from the sunlit wall.

A Hunger Storm: 7:20 a.m. *A storm threatens. Uneasiness spreads from the center and turns into pain. The world is disintegrating.*

It is 4 hours since Joey's last feeding, and he is probably hungry. Suddenly his lower lip protrudes. He starts to fret. Soon the fretting gives way to jerky crying, then moves into a full cry.

Hunger is a powerful experience, a motivation, a drive. It sweeps through an infant's nervous system like a storm, disrupting whatever was going on and temporarily disorganizing behavior and experience. As it grows, it superimposes its own order on the disorganization. First, he breathes faster, stronger, more jaggedly. Soon, his voice—the vocalizations that make the cry sound—comes into play. But while the hunger is building, Joey's breathing and his cry are not yet integrated. Sometimes he breathes without voicing. Sometimes short cries punctuate the end of an expiration but do not yet overlap its full length. Sometimes the crying lasts too long, leaving Joey out of breath.

Finally, the hunger starts to localize within him, somewhere that feels like "the center." (Joey doesn't know yet that it is *his* center; it is simply the center of the entire world.) He eases into a full-throated cry, consisting of fast, deep, gulping

Reprinted from *U.S. News & World Report*, August 20, 1990, pp. 54-59. Adapted from *Diary of A Baby*, by Daniel N. Stern. Copyright © 1990 by Daniel N. Stern. Reprinted by permission of HarperCollins Publishers.

9. Diary of a Baby

inspirations of air and then long expirations accompanied with the loud cry that rides each exhalation to its very end.

As the full cry grows louder, it engulfs and directs all Joey's activity. The powerful expirations of this cry probably give him momentary relief from the pain—just as yelling and jumping up and down "relieve" a stubbed toe. He is now acting in a coordinated way, not just passively experiencing.

Joey's full cry helps deal with the hunger in two ways: It is a beautifully designed signal (police and ambulance sirens are a testament to this) to alert his parents to his distress and to demand a response from them. At the same time, it may help him modulate the intensity of the hunger sensation. Hunger, thus, creates in Joey ways to reach the outside world, and to cope with the one inside.

The Social World: *Joey at 4 Months*

A Face Duet: 9:30 a.m. *I enter the world of her face. It is usually a riot of light and air at play. But this time the world is still and dull. Where is she? Where has she gone?*

Joey is sitting on his mother's lap, facing her. She looks at him intently but with no expression, as if she were absorbed in thought elsewhere. For a long moment, they remain locked in a silent mutual gaze. She finally breaks it by easing into a slight smile. Joey quickly leans forward and they trade smiles.

Then Joey's mother's expression turns to exaggerated surprise, and she leans all the way forward and touches her nose to his, smiling and making bubbling sounds. This facial duet replays itself several times in different forms, and each time, Joey grows more tense and excited. With the second nose touch, his smile freezes. His expression moves back and forth between pleasure and fear. After another suspenseful pause, Joey's mother makes a third approach at an even higher level of hilarity, and lets out a rousing "oooOH!" Joey's face tightens. He closes his eyes, and turns his head to the side. His mother realizes that she has gone too far, and for the moment, she does nothing. Then she whispers to him and breaks into a warm smile. He becomes re-engaged.

Joey has entered a short but extraordinary epoch in his life. Beginning between 8 and 12 weeks, he undergoes a dramatic leap in development. Capacities for interaction blossom: The social smile emerges, he begins to vocalize and he makes long eye-to-eye contact. Almost overnight, he has become truly social. Still, his most intense social interactions are immediate, limited to the face-to-face and the "here and now, between us." In its undiluted form, this intense social world will last until he is about 6 months old. As a way of interacting with others and reading their behavior, it will last all his life.

When Joey is on his mother's lap and looks at her, her face becomes the dominating presence in his world. After about 3 months, when babies know what to expect in a face-to-face encounter with their mother, they get disturbed if she deviates far from the usual. They are particularly perplexed if she suddenly stops interacting, or if they cannot rouse her to expression. In the well-known experiment called a still-face procedure, a mother, in the middle of an interaction, is asked to stop moving, to wipe all expression from her face and just look at the baby. Infants after about 2½ months of age react strongly to a still face. Their smiles die away, and they frown. They make repeated attempts to re-animate the mother by smiling and gesturing and calling. If they don't succeed, they turn away, looking slightly unhappy and confused.

Imagine a mother who is chronically preoccupied (by troubles with her husband, for instance, or her career) and is consequently only partially "there" when she is with her baby face-to-face. Or a mother (or another care giver) who is depressed and rarely available even when present. The child has to learn to have different expectations. He learns to construct a mental picture of a mother who is there as a physical presence but is, as an animated responsive force, present only intermittently or weakly. The child who wanted a high level of arousal and joy would have to avoid direct contact with her, even in her presence. He would learn to look elsewhere for the needed stimulation. Or, he would learn to make extraordinary efforts to charm his mother, to pull her along—to act as an antidepressant to her. But Joey has other expectations of his mother. To be aroused, he does not have to first turn himself on. He can turn to her to find that spark.

It is her eyes that draw him in with their vivid and stimulating qualities: The contrast of light and dark, their curves, angles, brilliance, depth and symmetry. Babies act as if the eyes are indeed windows to the soul. After 7 weeks of age, they treat the eyes as the geographic center of the face and the psychological center of the person. When you play peekaboo with a baby, there is anticipatory pleasure as you lower the blanket to reveal your hair and forehead. But only when the baby sees your eyes does he explode in delight.

Once a pair of smiles has passed between a mother and a baby of this age, a process is set in motion. Such an alternating pattern is common after 3 months. It occurs in vocalizing back and forth as well as in smiling. It is the baby's first and principal lesson in turn taking, the cardinal rule for all later discourse. Thus, this simple playful exchange lays down one of the foundations of social interaction.

Joey can now think in terms of taking an initiative to accomplish a goal. He has a sense of himself as the author of his actions, and that his actions have predictable consequences. He has also come to realize that his mother is a different and separate actor-agent. Usually, it is clear to Joey who is the agent and who is the object of a particular action. But as Joey and his mother trade smiles, he will probably feel that the gestures are jointly initiated. His mother has willed and executed her own smiles, but he has called them forth. Similarly, he has willed and executed his own smiles, but she has evoked them.

There are many such moments of mutual creation. They are the stuff of being with another person that constitute the ties of attachment. So much of attachment consists of the memories and mental models of what happens between you and that other person: How you feel with them. What they can make you experience that others cannot. What you can permit yourself to do or feel or think or wish or dare—but only in their presence. What you can accomplish with their support. What parts, or view, of yourself need their eyes and ears as nourishment.

Parents and care givers need constant feedback to know what to do, and when. Being a "good" parent is in large part knowing how to readjust behavior.

World of Mindscapes: *Joey at 12 Months*

A Shared Feeling: 11:50 a.m. *A wave of delight rises high in me. It swells to a crest. It leans forward, curls, and breaks into musical foam. The foam slips back as the wave passes, and disappears into the quieter water behind.*

Joey and his mother are looking for a stuffed rabbit, his favorite toy, which got hidden under a blanket. Joey finds it. He swings it excitedly into view and looks to his mother with a burst of pleasure. In a smooth crescendo, his face opens up. His eyes grow wider, and his mouth eases into a broad smile, to show her what he found—even more impor-

2. INFANCY AND EARLY CHILDHOOD

tant, to show her how he feels about it. After she sees his face, he lets his face fall back to normal in a smooth diminuendo. She then says "YeaaAAaah!" with a rising, then falling pitch. Joey seems content with her response and goes on playing by himself.

Joey is making two great discoveries, developmental leaps that go hand in hand with his new ability to walk. First, that he has his own private mindscapes: Mental landscapes that are not visible to others unless he makes an attempt to reveal them. Second, that it is possible to share a mindscape of his own with someone else.

This moment between Joey and his mother seems almost too ordinary and fleeting for anything of importance to occur. But it provides a gateway to the world of mutual mindscapes, a sharing of mental worlds. Joey is delighted to find the hidden toy, and this feeling is the real subject of this moment. It occurs in two different "places" simultaneously—one visible, the other invisible. The visible events occur in Joey's face and eyes, which open up and close back down, all in a moment. And Joey is careful to show these events to his mother, as a sign.

The invisible events are those inner sensations of delight that reside somewhere else in the body and mind than on the face. Joey can identify that place no better or worse than adults. It is somewhere "inside." And what happens "inside," during a feeling, is a live event that takes time to unfold. It is not static like a word. It is not abstract like an idea. It is a series of shifting impressions that change all the time, as does music or dance.

Joey wants very much for his mother to perceive his feelings, and she does in the way his face rises and falls with delight. Like any parent, she wants to share his delight at finding the lost toy and showing it to her. How can she communicate that she knows what he has experienced? She could say, "Oh, Joey, I know you felt delighted. I, too, know what that feels like." While Joey might understand some of these words, he doesn't yet understand this expression of the concept. Or she could imitate him. But this wouldn't work either. If Joey's mother raised her empty hand, just as Joey did, and opened and closed her face in an imitation of Joey's delight, it would look ridiculous. More to the point, what would Joey make of it?

What Joey's mother did do was to say, "YeaaAAaah!" and, with her vocal pitch, imitate Joey's inner feeling, the rising and falling wave. The rise in pitch of the first part of her "YeaaAA . . ." lasted exactly as long as the crescendo of Joey's face opening. Likewise, the fall in pitch lasted only as long as his face fall did. She intuitively created the carefully selected and elaborated imitation that has been called "attunement."

Only a human who knows what Joey felt could come up with a "Yeaaah" that is an analogue, not a copy, of his experience. He understands that his message got to her, and answers "Yes!" This kind of matching is done out of awareness, as a special manifestation of empathy. Most of us do it intuitively. And the child of a parent who, for whatever reason, cannot do this, or is inhibited from doing it, will feel psychically more alone with that person and perhaps ultimately in the world.

Joey's mother is starting to let him know which mindscapes of his she can share and which not. Together they have just established that a burst of delight is an inner event they can share. But what about sadness, anger, pride, enthusiasm, fear, doubt, shame, joy, love, desire, pain, boredom? Still to come in Joey's life are experiences of these and other subjective states. Will Joey's mother be fully able to share them, or will she be unable, consciously or unconsciously, to let these feelings become full members of the universe Joey can later expect to share with others?

The World of Words: *Joey at 20 Months*

"Pumpkin" 7:05 a.m. My room is so still. I am all alone here. I want to go where Mommy and Daddy are. There, I wrap myself in the heat that rises and falls. I bathe in the rich tides of our morning world.

Joey gets out of bed. He stands looking about for a moment, as if thinking. He then goes quickly into his parents' bedroom and gets into their bed, slipping under the covers between them. His parents are pretty much awake by now. After a while, his father says to him, "My little pumpkin." Joey answers back from under the covers, "Umpin." His father gently corrects, "Yes, **P**umpkin." Joey tries it again, "**P**umpkin." His father laughs, "That's it, you're my little pumpkin." Joey is quiet for a while. Then he comes out from the covers and announces clearly and firmly, "Me pumpkin!"

As Joey approached 18 months, he began another major leap in maturation that profoundly changed his daily experience: The leap into the world of words, of symbols and self-reflection. Joey is still in the middle of this leap. In some children it starts earlier; in some, later. Like the unfolding of a flower, a uniquely human one, language blossoms overnight when the time is right.

Joey's loneliness upon waking up this morning is different from the acute anxiety of separation he experienced when he was younger. He has some feeling of isolation, of being cut off from human contact. But he knows exactly where to go to find human life in concentrated form, and climbs into his parents' bed. And here, Joey has an important encounter with language.

So far, in respect to language, Joey has heard only its music. He hears the pure sound of the words and feels the emotions the sounds evoke in him, but he hears little of the strict meaning. But on this particular morning, Joey's father's language does not vanish into music and feeling. The word "pumpkin" stands apart, and Joey can start to explore it and play with it. His task is to master the sound and keep it, rather than just letting it wash over him.

To do this, he and his father toss the word back and forth, making it better each time. The game and the rules of alternating turns have been used by Joey and his parents for many months. They used to take turns cooing and smiling to each other when he was only 3 months old. The basic rules of conversation, turn taking, were long established before Joey and his parents even applied them to language. So, once again, Joey and his father follow these tried and true rules and send the word "pumpkin" back and forth between them.

The first time Joey tries his turn, he leaves out the exploded consonants and says, "Umpin." His father then does as most parents intuitively do in this situation: He slowly and clearly enunciates the not yet learned parts of the word, "Pump-kin," and leaves unstressed the already learned parts. Using this teaching technique, Joey quickly gets it right.

What happens now is truly wonderful. When a word is first unlocked, its released meaning is both a gift—in this case a present from Joey's father—and a personal discovery and creation. Once Joey has worked on the word and let it work on him, it is his. He can now use it to refer to a new aspect of himself in the context of his relationship with his father: "Me pumpkin!"

Joey has never heard "Me pumpkin" before. Perhaps no one has ever said it. Thus, Joey is not imitating anybody. Instead, he has created for himself a meaning, by bringing together himself ("me"), a word sound (p.u.m.p.k.i.n) and a special experience, a way of being loved and viewed by his father ("Me pumpkin").

9. Diary of a Baby

The World of Stories: *Joey at 4 Years*

In My Parents' Bedroom: 8:00 a.m. *In the morning, a close friend of the family's arrives to visit Joey and his parents. He asks Joey, "What did you do this morning?" And Joey tells him.*

Joey: *I went into Mommy and Daddy's room to play. They were asleep. Daddy was really pretend asleep. So then we all played "same boat" on the bed.*

The Friend: *How did you play "same boat"?*

Joey: *We live on our boat. And I almost caught a big fish.*

The Friend: *Oh!*

Joey: *Yeah. It pulled and pulled and ran away and came back. I could almost see it. I heard it. And it got away. It's a very special fish, it's called a pumpkin fish. Maybe it's round. And it can skip on top of the water. My friend JoJo can skip, and Marcie, but not Adele. And I'm learning. I never really saw a pumpkin fish, 'cause it gets away just at the end. So no one knows what it looks like. But it's a very special fish. So then, we didn't eat fish for breakfast. But sometimes we did.*

Joey has taken another great leap in development, one that makes him a "different" child, while still remaining himself. He is beginning to be a storyteller, the storyteller of his life. At 4, he can weave together a narrative, recounting the events and experiences that happen to him. This doesn't just mean having words for things—Joey has had that since his second year. Narratives go beyond that. They involve seeing and interpreting human activities in terms of psychological plots, stories that are made up of actors, who have desires and motives, and that take place in a historical context and a physical setting that help interpret the plot. Also, each story has a beginning, middle and end. This transformation in Joey's world view is not unique to him. All children, roughly between 2½ and 4, start to comprehend and make up narratives about their lives.

Joey's story world is a reconstruction of experience. It is a world made to be observed from the outside by someone else. It is a tale told to another. The experience world, on the other hand, is lived from the inside. So Joey's first task of reconstruction is to turn the experience world inside out, so to speak. He does this by turning the perceptions and feelings and internal mental states of the experience world into external actions and activities that others can observe on the open stage of the story world. Children, when beginning to tell autobiographical narratives, use mostly action verbs as Joey does: "I played," "he caught." And there are far fewer references to feeling states.

In normal development, story making has the important role of facilitating the process of self-definition. The child, narrating an autobiographical story, is not only defining his past; he is creating his identity. This process goes on every day, many times: When he recounts what happened at nursery school, or what he had for breakfast, or shopping with Mommy or the fight he had with his sister. Each story making and telling is like a workshop where he can experiment on becoming himself. This is crucial for a child who is rapidly developing and maturing. His identity changes, too. And he must experiment with several versions, ranging from the public to the very private.

Joey's story about the pumpkin fish is a good example of the process of transformation. In the original episode of "Pumpkin," when Joey was 20 months old, he first put into words his father's pet name for him, "Pumpkin." In that moment he pulled together two different things: How his father views and feels about him, and how he sees and feels about himself viewed that way. It is a magical fusion. His father finds Joey lovable and wonderful, and Joey wants to become the way he sees himself in his father's eyes—in his mother's eyes, too. This is one of the more powerful forces in shaping a child's development.

In this curious interchange between parent and child, Joey is loved for who he is and who he isn't yet. And he appreciates himself for who he is, and also for who he may one day become. That is a magical fish! It is both a fish and not a fish. But it is the prize we are after, even though it is uncatchable. Even if it were to be caught for a moment, it would escape because it cannot be held. It is the point in time where present and future meet in the rush forward. It is the becoming yourself.

Joey now has in his hands the power to interpret and reinterpret his life; with control over his own past, he will have much greater control over his present and future. He can fashion his own diary, an oral one. He no longer needs an interpreter: From now on, he will be talking directly to you.

Psychiatrist Daniel N. Stern, M.D., is also the author of *The First Relationship* (1970) and *The Interpersonal World of the Infant* (1985).

Guns and Dolls

Alas, our children don't exemplify equality any more than we did. Is biology to blame? Scientists say maybe—but parents can do better, too.

LAURA SHAPIRO

Meet Rebecca. She's 3 years old, and both her parents have full-time jobs. Every evening Rebecca's father makes dinner for the family—Rebecca's mother rarely cooks. But when it's dinner time in Rebecca's dollhouse, she invariably chooses the Mommy doll and puts her to work in the kitchen.

Now meet George. He's 4, and his parents are still loyal to the values of the '60s. He was never taught the word "gun," much less given a war toy of any sort. On his own, however, he picked up the word "shoot." Thereafter he would grab a stick from the park, brandish it about and call it his "shooter."

Are boys and girls *born* different? Does every infant really come into the world programmed for caretaking or war making? Or does culture get to work on our children earlier and more inexorably than even parents are aware? Today these questions have new urgency for a generation that once made sexual equality its cause and now finds itself shopping for Barbie clothes and G.I. Joe paraphernalia. Parents may wonder if gender roles are immutable after all, give or take a Supreme Court justice. But burgeoning research indicates otherwise. No matter how stubborn the stereotype, individuals can challenge it; and they will if they're encouraged to try. Fathers and mothers should be relieved to hear that they do make a difference.

Biologists, psychologists, anthropologists and sociologists have been seeking the origin of gender differences for more than a century, debating the possibilities with increasing rancor ever since researchers were forced to question their favorite theory back in 1902. At that time many scientists believed that intelligence was a function of brain size and that males uniformly had larger brains than women—a fact that would nicely explain men's pre-eminence in art, science and letters. This treasured hypothesis began to disintegrate when a woman graduate student compared the cranial capacities of a group of male scientists with those of female college students; several women came out ahead of the men,

Girls' cribs have pink tags and boys' cribs have blue tags; mothers and ...

NEWBORNS

... fathers should be on the alert, for the gender-role juggernaut has begun

and one of the smallest skulls belonged to a famous male anthropologist.

Gender research has become a lot more sophisticated in the ensuing decades, and a lot more controversial. The touchiest question concerns sex hormones, especially testosterone, which circulates in both sexes but is more abundant in males and is a likely, though unproven, source of aggression. To postulate a biological determinant for behavior in an ostensibly egalitarian

10. Guns and Dolls

society like ours requires a thick skin. "For a while I didn't dare talk about hormones, because women would get up and leave the room," says Beatrice Whiting, professor emeritus of education and anthropology at Harvard. "Now they seem to have more self-confidence. But they're skeptical. The data's not in yet."

Some feminist social scientists are staying away from gender research entirely— "They're saying the results will be used against women," says Jean Berko Gleason, a professor of psychology at Boston University who works on gender differences in the acquisition of language. Others see no reason to shy away from the subject. "Let's say it were proven that there were biological foundations for the division of labor," says Cynthia Fuchs Epstein, professor of sociology at the City University of New York, who doesn't, in fact, believe in such a likelihood. "It doesn't mean we couldn't do anything about it. People can make from scientific findings whatever they want." But a glance at the way society treats those gender differences already on record is not very encouraging. Boys learn to read more slowly than girls, for instance, and suffer more reading disabilities such as dyslexia, while girls fall behind in math when they get to high school. "Society can amplify differences like these or cover them up," says Gleason. "We rush in reading teachers to do remedial reading, and their classes are almost all boys. We don't talk about it, we just scurry around getting them to catch up to the girls. But where are the remedial math teachers? Girls are *supposed* to be less good at math, so that difference is incorporated into the way we live."

No matter where they stand on the question of biology versus culture, social scientists agree that the sexes are much more alike than they are different, and that variations within each sex are far greater than variations between the sexes. Even differences long taken for granted have begun to disappear. Janet Shibley Hyde, a professor of psychology at the University of Wisconsin, analyzed hundreds of studies on verbal and math ability and found boys and girls alike in verbal ability. In math, boys have a moderate edge; but only among highly precocious math students is the disparity large. Most important, Hyde found that verbal and math studies dating from the '60s and '70s showed greater differences than more recent research. "Parents may be making more efforts to tone down the stereotypes," she says. There's also what academics call "the file-drawer effect." "If you do a study that shows no differences, you assume it won't be published," says Claire Etaugh, professor of psychology at Bradley University in Peoria, Ill. "And until recently, you'd be right. So you just file it away."

The most famous gender differences in academics show up in the annual SAT results, which do continue to favor boys. Traditionally they have excelled on the math portion, and since 1972 they have slightly outperformed girls on the verbal side as well. Possible explanations range from bias to biology, but the socioeconomic profile of those taking the test may also play a role. "The SAT gets a lot of publicity every year, but nobody points out that there are more women taking it than men, and the women come from less advantaged backgrounds," says Hyde. "The men are a more highly selected sample: they're better off in terms of parental income, father's education and attendance at private school."

> **GIRLS**
> Girls are encouraged to think about how their actions affect others...
> **2-3 YEARS**
> ...boys often misbehave, get punished and then misbehave again
> **BOYS**

Another longstanding assumption does hold true: boys tend to be somewhat more active, according to a recent study, and the difference may even start prenatally. But the most vivid distinctions between the sexes don't surface until well into the preschool years. "If I showed you a hundred kids aged 2, and you couldn't tell the sex by the haircuts, you couldn't tell if they were boys or girls," says Harvard professor of psychology Jerome Kagan. Staff members at the Children's Museum in Boston say that the boys and girls racing through the exhibits are similarly active, similarly rambunctious and similarly interested in model cars and model kitchens, until they reach first grade or so. And at New York's Bank Street preschool, most of the 3-year-olds clustered around the cooking table to make banana bread one recent morning were boys. (It was a girl who gathered up three briefcases from the costume box and announced, "Let's go to work.")

By the age of 4 or 5, however, children start to embrace gender stereotypes with a determination that makes liberal-minded parents groan in despair. No matter how careful they may have been to correct the disparities in "Pat the Bunny" ("Paul isn't the *only* one who can play peekaboo, *Judy* can play peekaboo"), their children will delight in the traditional male/female distinctions preserved everywhere else: on television, in books, at day care and preschool, in the park and with friends. "One of the things that is very helpful to children is to learn what their identity is," says Kyle Pruett, a psychiatrist at the Yale Child Study Center. "There are rules about being feminine and there are rules about being masculine. You can argue until the cows come home about whether those are good or bad societal influences, but when you look at the children, they love to know the differences. It solidifies who they are."

Water pistols: So girls play dolls, boys play Ghostbusters. Girls take turns at hopscotch, boys compete at football. Girls help Mommy, boys aim their water pistols at guests and shout, "You're dead!" For boys, notes Pruett, guns are an inevitable part of this developmental process, at least in a television-driven culture like our own. "It can be a cardboard paper towelholder, it doesn't have to be a miniature Uzi, but it serves as the focus for fantasies about the way he is going to make himself powerful in the world," he says. "Little girls have their aggressive side, too, but by the time they're socialized it takes a different form. The kinds of things boys work out with guns, girls work out in terms of relationships— with put-downs and social cruelty." As if to underscore his point, a 4-year-old at a recent Manhattan party turned to her young hostess as a small stranger toddled up to them. "Tell her we don't want to play with her," she commanded. "Tell her we don't like her."

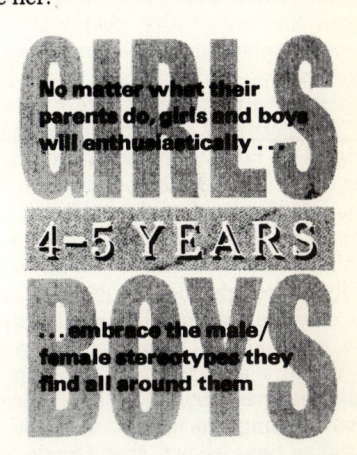

> **GIRLS**
> No matter what their parents do, girls and boys will enthusiastically...
> **4-5 YEARS**
> ...embrace the male/female stereotypes they find all around them
> **BOYS**

Once the girls know they're female and the boys know they're male, the powerful stereotypes that guided them don't just disappear. Whether they're bred into our chromosomes or ingested with our cornflakes, images of the aggressive male and the nurturant female are with us for the rest of our lives. "When we see a man with a child, we say, 'They're playing'," says Epstein. "We never say, 'He's nurturant'."

The case for biologically based gender differences is building up slowly, amid a great deal of academic dispute. The theory is that male and female brains, as well as bodies, develop differently according to the amount of testosterone circulating around

45

2. INFANCY AND EARLY CHILDHOOD

the time of birth. Much of the evidence rests on animal studies showing, for instance, that brain cells from newborn mice change their shape when treated with testosterone. The male sex hormone may also account for the different reactions of male and female rhesus monkeys, raised in isolation, when an infant monkey is placed in the cage. The males are more likely to strike at the infant, the females to nurture it. Scientists disagree—vehemently—on whether animal behavior has human parallels. The most convincing human evidence comes from anthropology, where cross-cultural studies consistently find that while societies differ in their predilection toward violence, the males in any given society will act more aggressively than the females. "But it's very important to emphasize that by aggression we mean only physical violence," says Melvin Konner, a physician and anthropologist at Emory University in Atlanta. "With competitive, verbal or any other form of aggression, the evidence for gender differences doesn't hold." Empirical findings (i.e., look around you) indicate that women in positions of corporate, academic or political power can learn to wield it as aggressively as any man.

Apart from the fact that women everywhere give birth and care for children, there is surprisingly little evidence to support the notion that their biology makes women kinder, gentler people or even equips them specifically for motherhood. Philosophers—and mothers, too—have taken for granted the existence of a maternal "instinct" that research in female hormones has not conclusively proven. At most there may be a temporary hormonal response associated with childbirth that prompts females to nurture their young, but that doesn't explain women's near monopoly on changing diapers. Nor is it likely that a similar hormonal surge is responsible for women's tendency to organize the family's social life or take up the traditionally underpaid "helping" professions—nursing, teaching, social work.

Studies have shown that female newborns cry more readily than males in response to the cry of another infant, and that small girls try more often than boys to comfort or help their mothers when they appear distressed. But in general the results of most research into such traits as empathy and altruism do not consistently favor one sex or the other. There is one major exception: females of all ages seem better able to "read" people, to discern their emotions, without the help of verbal cues. (Typically researchers will display a picture of someone expressing a strong reaction and ask test-takers to identify the emotion.) Perhaps this skill—which in evolutionary terms would have helped females survive and protect their young—is the sole biological foundation for our unshakable faith in female selflessness.

Infant ties: Those who explore the unconscious have had more success than other researchers in trying to account for male aggression and female nurturance, perhaps because their theories cannot be tested in a laboratory but are deemed "true" if they suit our intuitions. According to Nancy J. Chodorow, professor of sociology at Berkeley and the author of the influential book "The Reproduction of Mothering," the fact that both boys and girls are primarily raised by women has crucial effects on gender roles. Girls, who start out as infants identifying with their mothers and continue to do so, grow up defining themselves in relation to other people. Maintaining human connections remains vital to them. Boys eventually turn to their fathers for self-definition, but in order to do so must repress those powerful infant ties to mother and womanhood. Human connections thus become more problematic for them than for women. Chodorow's book, published in 1978, received national attention despite a dense, academic prose style; clearly, her perspective rang true to many.

Harvard's Kagan, who has been studying young children for 35 years, sees a different constellation of influences at work. He speculates that women's propensity for caretaking can be traced back to an early awareness of their role in nature. "Every girl knows, somewhere between the ages of 5 and 10, that she is different from boys and that she will have a child—something that everyone, including children, understands as quintessentially natural," he says. "If, in our society, nature stands for the giving of life, nurturance, help, affection, then the girl will conclude unconsciously that those are the qualities she should strive to attain. And the boy won't. And that's exactly what happens."

Kagan calls such gender differences "inevitable but not genetic," and he emphasizes—as does Chodorow—that they need have no implications for women's status, legally or occupationally. In the real world, of course, they have enormous implications. Even feminists who see gender differences as cultural artifacts agree that, if not inevitable, they're hard to shake. "The most emancipated families, who really feel they want to engage in gender-free behavior toward their kids, will still encourage boys to be boys and girls to be girls," says Epstein of CUNY. "Cultural constraints are acting on you all the time. If I go to buy a toy for a friend's little girl, I think to myself, why don't I buy her a truck? Well, I'm afraid the parents wouldn't like it. A makeup set would really go against my ideology, but maybe I'll buy some blocks. It's very hard. You have to be on the alert every second."

In fact, emancipated parents have to be on

All children have to deal with aggression; girls wield relationships as...

GIRLS 6-7 YEARS BOYS

...weapons, while boys prefer to brandish water pistols.

the alert from the moment their child is born. Beginning with the pink and blue name tags for newborns in the hospital nursery—I'M A GIRL/I'M A BOY—the gender-role juggernaut is overwhelming. Carol Z. Malatesta, associate professor of psychology at Long Island University in New York, notes that baby girls' eyebrows are higher above their eyes and that girls raise their eyebrows more than boys do, giving the girls "a more appealing, socially responsive look." Malatesta and her colleagues, who videotaped and coded the facial expressions on mothers and infants as they played, found that mothers displayed a wider range of emotional responses to girls than to boys. When the baby girls displayed anger, however, they met what seemed to be greater disapproval from their mothers than the boys did. These patterns, Malatesta suggests, may be among the reasons why baby girls grow up to smile more, to seem more sociable than males, and to possess the skill noted earlier in "reading" emotions.

The way parents discipline their toddlers also has an effect on social behavior later on. Judith G. Smetana, associate professor of education, psychology and pediatrics at the University of Rochester, found that mothers were more likely to deal differently with similar kinds of misbehavior depending on the sex of the child. If a little girl bit her friend and snatched a toy, for instance, the mother would explain why biting and snatching were unacceptable. If a boy did the same thing, his mother would be more likely to stop him, punish him and leave it at that. Misbehavior such as hitting in both sexes peaks around the age of 2; after that, little boys go on to misbehave more than girls.

Psychologists have known for years that boys are punished more than girls. Some have conjectured that boys simply drive their parents to distraction more quickly; but as Carolyn Zahn-Waxler, a psychologist at the National Institute of Mental Health, points out, the difference in parental treatment starts even before the difference in behavior shows up. "Girls receive very different messages than boys," she says. "Girls are encouraged to care about the problems of others, beginning very early. By elementary school, they're showing more caregiver behavior, and they have a wider social network."

Children also pick up gender cues in the process of learning to talk. "We compared fathers and mothers reading books to children," says Boston University's Gleason. "Both parents used more inner-state words, words about feelings and emotions, to girls than to boys. And by the age of 2, girls are using more emotion words than boys." According to Gleason, fathers tend to use more directives ("Bring that over here") and more threatening language with their sons than their daughters, while mothers' directives take more polite forms ("Could you bring that to me, please?"). The 4-year-old boys and girls in one study were duly imitating their fathers and mothers in that very conversational pattern. Studies of slightly older children found that boys talking among themselves use more threatening, commanding, dominating language than girls, while girls emphasize agreement and mutuality. Polite or not, however, girls get interrupted by their parents more often than boys, according to language studies—and women get interrupted more often than men.

Despite the ever-increasing complexity and detail of research on gender differences, the not-so-secret agenda governing the discussion hasn't changed in a century: how to understand women. Whether the question is brain size, activity levels or modes of punishing children, the traditional implication is that the standard of life is male, while the entity that needs explaining is female. (Or as an editor put it, suggesting possible titles for this article: "Why Girls Are Different.") Perhaps the time has finally come for a new agenda. Women, after all, are not a big problem. Our society does not suffer from burdensome amounts of empathy and altruism, or a plague of nurturance. The problem is men—or more accurately, maleness.

"There's one set of sex differences that's ineluctable, and that's the death statistics," says Gleason. "Men are killing themselves doing all the things that our society wants them to do. At every age they're dying in accidents, they're being

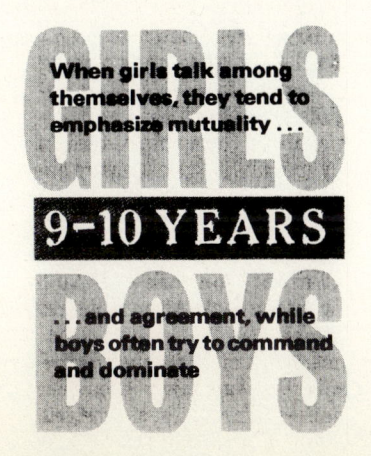

When girls talk among themselves, they tend to emphasize mutuality...

GIRLS
9-10 YEARS
BOYS

...and agreement, while boys often try to command and dominate

10. Guns and Dolls

Where Little Boys Can Play With Nail Polish

For 60 years, America's children have been raised on the handiwork of Fisher-Price, makers of the bright plastic cottages, school buses, stacking rings and little, smiley people that can be found scattered across the nation's living rooms. Children are a familiar sight at corporate headquarters in East Aurora, N.Y., where a nursery known as the Playlab is the company's on-site testing center. From a waiting list of 4,000, local children are invited to spend a few hours a week for six weeks at a time playing with new and prototype toys. Staff members watch from behind a one-way mirror, getting an education in sales potential and gender tastes.

According to Kathleen Alfano, manager of the Child Research Department at Fisher-Price, kids will play with everything from train sets to miniature vacuum cleaners until the age of 3 or 4; after that they go straight for the stereotypes. And the toy business meets them more than halfway. "You see it in stores," says Alfano. "Toys for children 5 and up will be in either the girls' aisles or the boys' aisles. For girls it's jewelry, glitter, dolls, and arts and crafts. For boys it's model kits, construction toys and action figures like G.I. Joe. Sports toys, like basketballs, will be near the boys' end."

The company's own recent venture into gender stereotypes has not been successful. Fisher-Price has long specialized in what Alfano calls "open gender" toys, aimed at boys and girls alike, ages 2 to 7. The colors are vivid and the themes are often from daily life: music, banking, a post office. But three years ago the company set out to increase profits by tackling a risky category known in the industry as "promotional" toys. Developed along strict sex-role lines and heavily promoted on children's television programs, these toys for ages 5 and up are meant to capture kids' fads and fashions as well as their preconceptions about masculinity and femininity. At Fisher-Price they included an elaborate Victorian dollhouse village in shades of rose and lavender, a line of beauty products including real make-up and nail polish, a set of battery-operated racing cars and a game table outfitted for pool, Ping-Pong and glide hockey. "The performance of these products has been very mixed," says Ellen Duggan, a spokesperson for Fisher-Price. "We're now refocusing on toys with the traditional Fisher-Price image." (The company is also independent for the first time in 21 years. Last month longtime owner Quaker Oats divested itself of Fisher-Price.)

Even where no stereotypes are intended, the company has found that some parents will conjure them up. At a recent session for 3-year-olds in the Playlab, the most sought-after toy of the morning was the fire pumper, a push toy that squirts real water. "It's for both boys and girls, but parents are buying it for boys," says Alfano. Similarly, "Fun with Food," a line of kitchen toys including child-size stove, sink, toaster oven and groceries, was a Playlab hit; boys lingered over the stove even longer than girls. "Mothers are buying it for their daughters," says Alfano.

Children tend to cross gender boundaries more freely at the Playlab than they do elsewhere, Alfano has noticed. "When 7-year-olds were testing the nail polish, we left it out after the girls were finished and the boys came and played with it," she says. "They spent the longest time painting their nails and drying them. This is a safe environment. It's not the same as the outside world."

LAURA SHAPIRO *in East Aurora*

2. INFANCY AND EARLY CHILDHOOD

shot, they drive cars badly, they ride the tops of elevators, they're two-fisted hard drinkers. And violence against women is incredibly pervasive. Maybe it's men's raging hormones, but I think it's because they're trying to be a *man*. If I were the mother of a boy, I would be very concerned about societal pressures that idolize behaviors like that."

Studies of other cultures show that male behavior, while characteristically aggressive, need not be characteristically deadly. Harvard's Whiting, who has been analyzing children cross-culturally for half a century, found that in societies where boys as well as girls take care of younger siblings, boys as well as girls show nurturant, sociable behavior. "I'm convinced that infants elicit positive behavior from people," says Whiting. "If you have to take care of somebody who can't talk, you have to learn empathy. Of course there can be all kinds of experiences that make you extinguish that eliciting power, so that you no longer respond positively. But on the basis of our data, boys make very good baby tenders."

In our own society, evidence is emerging that fathers who actively participate in raising their children will be steering both sons and daughters toward healthier gender roles. For the last eight years Yale's Pruett has been conducting a groundbreaking longitudinal study of 16 families, representing a range of socioeconomic circumstances, in which the fathers take primary responsibility for child care while the mothers work full time. The children are now between 8 and 10 years old, and Pruett has watched subtle but important differences develop between them and their peers. "It's not that they have conflicts about their gender identity—the boys are masculine and the girls are feminine, they're all interested in the same things their friends are," he says. "But when they were 4 or 5, for instance, the stage at preschool when the boys leave the doll corner and the girls leave the block corner, these children didn't give up one or the other. The boys spent time playing with the girls in the doll corner, and the girls were building things with blocks, taking pride in their accomplishments."

Little footballs: Traditionally, Pruett notes, fathers have enforced sex stereotypes more strongly than mothers, engaging the boys in active play and complimenting the girls on their pretty dresses. "Not these fathers," says Pruett. "That went by the boards. They weren't interested in bringing home little footballs for their sons or little tutus for the girls. They dealt with the kids according to the individual. I even saw a couple of the mothers begin to take over those issues— one of them brought home a Dallas Cowboys sleeper for her 18-month-old. Her husband said, 'Honey, I thought we weren't going to do this, remember?' She said, 'Do what?' So that may be more a function of being in the second tier of parenting rather than the first."

As a result of this loosening up of stereotypes, the children are more relaxed about gender roles. "I saw the boys really enjoy their nurturing skills," says Pruett. "They knew what to do with a baby, they didn't see that as a girl's job, they saw it as a human job. I saw the girls have very active images of the outside world and what their mothers were doing in the workplace—things that become interesting to most girls when they're 8 or 10, but these girls were interested when they were 4 or 5."

Pruett doesn't argue that fathers are better at mothering than mothers, simply that two involved parents are better than "one and a lump." And it's hardly necessary for fathers to quit their jobs in order to become more involved. A 1965-66 study showed that working mothers spent 50 minutes a day engaged primarily with their children, while the fathers spent 12 minutes. Later studies have found fathers in two-career households spending only about a third as much time with their children as mothers. What's more, Pruett predicts that fathers would benefit as much as children from the increased responsibility. "The more involved father tends to feel differently about his own life," he says. "A lot of men, if they're on the fast track, know a lot about competitive relationships, but they don't know much about intimate relationships. Children are experts in intimacy. After a while the wives in my study would say, 'He's just a nicer guy'."

Pruett's study is too small in scope to support major claims for personality development; he emphasizes that his findings are chiefly theoretical until more research can be undertaken. But right now he's watching a motif that fascinates him. "Every single one of these kids is growing something," he says. "They don't just plant a watermelon seed and let it die. They're really propagating things, they're doing salad-bowl starts in the backyard, they're breeding guinea pigs. That says worlds about what they think matters. Generativity is valued a great deal, when both your mother and your father say it's OK." Scientists may never agree on what divides the sexes; but someday, perhaps, our children will learn to relish what unites them.

Preschool: Head Start or Hard Push?

Philip R. Piccigallo

PHILIP R. PICCIGALLO, a public affairs consultant, formerly was assistant to the president of the American Federation of Teachers.

Not long ago an old friend called to congratulate my wife and me on the birth of our son. During the catch-up conversation, he asked about my then three-year-old daughter's progress. Was she taking gymnastics, ballet, or swimming? Was she enrolled in reading, math, and computer classes? Had I succeeded in placing her name on preliminary lists for testing and admission to selective preschools and private kindergartens? "One can never start too early," he assured. "Oh, and how has she done on early tests?"

Though presumably well intentioned, his matter-of-fact advocacy of a child-development approach that has been seriously questioned by a considerable number of national experts troubled me. Controversial ideas about early childhood education are not new, of course. Since the 1960s debates have raged over whether out-of-home learning programs for young children are harmful or beneficial. My friend's views, however, are reflective of a steadily widening contemporary phenomenon: the near singleminded determination of many baby boomer and other new or prospective parents to "fast track" or "position" their children into high achievement careers through early participation in formal programs of academic instruction. In preschools and kindergartens across America, young children are being "taught" to read, do math, interface with computers, play the violin, compete with peers, and "outsmart" tests.

Yet a substantial body of professional evidence suggests that such an approach may be misguided. According to such research, imposing undue pressure to learn on young children may generate anxieties or even neuroses that impair long-term educational and psychological development. While moderate exposure to rigorous intellectual stimuli may be helpful, the possibility of exposing children to early failure by asking more than they are capable of giving poses substantial risk to future achievement. Finally, a series of national education reports conducted in the 1980s spell out a distinct role for early childhood education—one which fundamentally differs from that sought after by many success-driven parents.

Despite such contrary evidence, many parents are impelled by the notion of propelling Jason/Jennifer on an unbroken path from preschool to Harvard. Consumed by their own competitiveness and fueled by books bearing such titles as *How To Have A Smarter Baby, Teach Your Baby Math,* and *Rais-*

Imposing undue pressure to learn on young children may generate anxieties or even neuroses.

ing Brighter Children, growing numbers of parents scramble madly for the earliest and optimal advantages.

The search now precedes kindergarten. Consider, for example, the extensive application and admission procedures for the most elite, achievement-oriented preschools in the Washington, D.C., and New York City areas. As described by *The Washington Post*, parents must submit an application form, visit the school, and then arrange a child's visit. Parents are asked to provide previous, if any, school recommendation forms. Children are then scheduled for a group visit with other applicants, and are tested by a series of games played while under observation. Many parents interviewed worried that they lacked the "right connections or enough money to make a large bequest and influence the decision." The costliest of such schools start at about $6,500 annually.[1]

In a feature entitled, "Fast Lane Kids," *The New York Times* similarly depicted the efforts of "many hard-driving parents" who anxiously seek admission to elite preschools, and "load up" their children with exhausting academic, music, dance, sports, and crafts programs. Rejection often means broken dreams, jettisoned expectations. Both stories, for example, cited parents "devastated" by their eighteen-month-old's failure to win admission to an elite school.[2]

Significantly, this increasingly growing phenomenon is not restricted to any geographical region or to exclusive

2. INFANCY AND EARLY CHILDHOOD

> "Everyone wants to raise the smartest kid in America rather than the best adjusted, happiest kid."

schools. According to David Elkind, a child psychologist and author of *Miseducation: Preschoolers At Risk,* the proliferation of private and public academic programs for young people "has become a societal norm." Books, lecturers, and the media, he contends, have helped to propagate the idea that only "adultified" or "superkids" can succeed in a highly competitive future America.[3]

This sense that children must academically leapfrog ahead of their peers early to ensure later success has been heightened by demographic changes. Hundreds of school districts across America closed schools in the 1970s and early 1980s as the baby boom ebbed. Now many elementary schools overflow as the children of that postwar generation enroll in increasing numbers.

Kindergartens have been particularly affected. There nationwide enrollment in public schools swelled to 3,310,000 in 1986 from 2,687,000 in 1981, or 23 percent, according to the U.S. Department of Education. Baby boomers are having fewer, not more children than their parents. Nevertheless, their large numbers alone are creating the squeeze. High population growth areas, such as Florida, California, and the Rocky Mountain states, are especially pressed.[4]

Societal stress often filters downward. The incidence of highly pressured parents imposing high-pressure learning situations on their children, observes Elaine Bennett, an instructor of pediatrics at Georgetown University and wife of former U.S. Education Secretary William Bennett, has "really gotten out of hand. . . . It's not good for the children, the parents or the schools." Dina Bray, director of Columbia Grammar School in Manhattan, believes that such parents are "forcing their children into an experience that is the antithesis of childhood. . . . Rather than just letting their kids be kids, it's as if these parents are asking the children, 'How many points did you get today?'"[5]

THE ONGOING DEBATE

But what are the consequences, if any, of imposing high-pressure learning methods and expectations on young children?

There are different schools of thought on the subject. Proponents are best represented, as one would imagine, by the authors of the best-selling "smart baby" books. Siegfried and Therese Englemann, authors of *Give Your Child A Superior Mind,* for instance, propose a step-by-step instructional program for parents to increase their children's intelligence. "Lessons, examples, experiments, and guidelines" for ages birth through five are offered to teach children "the basic rules of language, the alphabet, geometric shapes, telling time, counting, reading, math, algebra, equations, and more."[6]

Glen Doman, founder of the Better Baby Institute, suggests that a child who has not learned to read early may be irreversibly disadvantaged. "A four-year-old learns reading quicker than a five," he says, "a three-year-old quicker than a four, and I dare say a less-than-one-year-old learns quicker than a one-year-old." His book, *How To Teach Your Baby To Read,* has sold two million copies worldwide.[7]

Yet parents should be aware of counterarguments. An impressive body of professionals contend that the pressures of inducing formal, instruction on young children may actually constitute, in Dr. Elkind's phrase, "miseducation." Consequently, they can "invoke internal conflicts and can set the groundwork" for serious psychological problems.[8]

Child psychologists and educators stress children's enjoyment in learning; also their sense of fulfillment in meeting parents' expectations. But gratification derived from growth and learning can turn sour, observes Francis Roberts, superintendent of the Cold Spring Harbor Schools in New York, "when early learning is pressured" and goals exceed yet undeveloped capabilities.[9] A "child pushed too hard" too early, "both before they begin the primary grades and after their formal schooling is already underway," can be "turned off from any zest for learning," caution educational psychologists Julius and Zelda Seigal.[10]

The human infant, it is true, can be "taught" to recite numbers at two, read by three, and cope with early expectations. The important question, according to T. Berry Brazelton, pediatrician and author of *Toddlers and Parents*, is at what price? "Everyone wants to raise the smartest kid in America rather than the best adjusted, happiest kid," he adds.[11]

Virtually all experts, even proponents of preschool academic programs like the Englemanns, agree that "the biggest possible harm that could come from early reading is early failure."[12] If young children "can be saved from the experience of early failure," believe child psychologists David and Barbara Bjorklund, "they quite possibly can be saved from later failure and the accompanying feelings of low self-esteem and poor self-confidence."[13] Nor is reading at an early age necessarily of undiluted benefit. Many children who learned to read early often developed more reading problems than those who started later, concludes Dr. Benjamin Spock and others. The national Commission on Reading recommends, above all, that any systematic reading instructions should be "free from undue pressure" so as not to "turn our kindergartens, and even nursery schools and day-care centers, into boot camps."[14] Perhaps Elkind put the entire matter most assertively: "No authority in the field of child psychology, pediatrics, or child psychiatry advocates the formal instruction, in any domain, of infants and young children."

Similarly, most educators prefer that young children be educated under a "developmental curriculum" rather than a formal one. Based primarily on the research of David Weikart, of the High Scope Foundation in Michigan, developmental curriculum emphasizes experiential, play-oriented, self-driven learning that supports a young child's cognitive processes.[15]

The national education reports, again, embraced the notion of enriched childhood education. But typically their principal accent was on "any well-designed, professionally supervised program to stimulate and socialize" culturally disadvantaged three- and four-year-olds. The U.S. Department of Education advocates preschool educational activities. But, sig-

11. Preschool

> Notions of "genius building" and intimidating, high-powered preschools appear more suited to serving parents' rather than children's needs.

nificantly, in its straightforward booklet, "What Works: Research About Teaching and Learning," it omits reference to formal preschooling altogether. Instead, it urges reliance on informal practices—creating an enriched "curriculum of the home" featuring books, supplies, and a special place for studying, closer parent-teacher-student interaction, increased parental attention to school matters, and disciplined study habits, for example, to help children of all socioeconomic backgrounds learn.[16]

EARLY CHILDHOOD EDUCATION AND REFORM

The current public interest in and support for school improvement stems from a series of studies conducted between 1983 and 1988 by a Presidential Commission on Excellence in Education, the Carnegie Corporation of New York, the Twentieth Century Fund, the Committee for Economic Development (CED), and other public-private organizations. These reports variously reported serious deficiencies in our educational system, and offered proposals for rectifying them.

Many of the reports highly praised and recommended enriched childhood education, as well as early investment in prenatal health care. They drew heavily upon the fruitful experience of the federal government's Head Start program, the successful Perry Preschool program in Ypsilanti, Michigan, and the New York University preschool study in Harlem, New York. Investment in such enterprises, they concluded, will produce multifold dividends for society—via savings in remedial education, health services, and less crime—over future years.[17]

To Barbara Finberg, vice president for programs at the Carnegie Corporation, which helped finance several of the projects, the studies reinforced "the idea that early childhood education, at least for disadvantaged youngsters, really does give them a boost toward better educational achievement. . . . It suggests that as children gain a sense of self-worth and confidence in their ability to learn, that in turn leads teachers to believe they can learn.[18]

The essential point is that the reports uniformly recommended early childhood education programs for a distinct group of youngsters, specifically those deemed to be "at risk." As defined by the CED, such "educationally disadvantaged" children are those who "cannot take advantage of available educational opportunities," or those where the "educational resources available to them are inherently unequal." In short, children who, due to poverty, broken homes, disability, or other environmental deprivation *need* additional and reinforced academic stimulation and instruction just to keep up with sufficiently engaged children.[19] As Mortimer Adler put it in *The Paideia Proposal* in 1982, "preschool" tutelage must be provided for those who do not get such preparation from favorable environments."[20]

This is not to say that sufficiently prepared children do not, or should not, benefit from preschool programs. All evidence suggests otherwise. In fact, in 1985 two-thirds of four-year-olds and 54 percent of three-year-olds in families with incomes of $35,000 or more attended preschool programs. In contrast, fewer than 33 percent of four-year-olds and 17 percent of three-year-olds in families with annual incomes of less than $10,000 were enrolled in such programs in 1985. And Head Start reaches only 18 percent of the 2.5 million children who need its services.[21]

Two salient points thus emerge: 1) not nearly enough of those children who most need early enriched intellectual instruction and stimulation presently receive it; and 2) the overwhelming majority of middle- and upper-income parents who heatedly pursue enhanced academic learning for their children may be administering an unnecessary remedy, one which may carry negative consequences.

THE JAPANESE "EDUCATIONAL MIRACLE"

And what about the "Japanese educational challenge"? Aren't they well along in launching young children on rapid academic and high-tech tracks? Yes. But as the leading American authority on the subject, sociologist Merry White, cautions, while learning from the Japanese we should not view their educational system "as a model to be emulated."[22]

Their system functions within a tightly knit, homogeneous society, one that places great importance on uniquely Japanese values such as conformity and consensus. Japanese teachers, in fact, discourage individual competition. Instead, they emphasize group effort and achievement. Such practice would prove unsuitable to Americans, steeped in individualism and pluralism.

Also, it would be difficult to imagine American mothers adopting the role of their Japanese counterparts in their children's education. Japanese mothers—"the best Jewish mothers in the world," says White—study alongside of and are held directly accountable for students' academic achievements or failures. At a time when more than 50 percent of new mothers remain in the American workforce, such behavior seems impracticable. Furthermore, adds White, Japan's intensely competitive, test-driven educational system has been the source of serious psychological strains on students of all ages. It has been widely said that Japanese youngsters have no childhood to speak of.[23]

"RESPECT THE CHILD"

Educational reform currently heads the national agenda. Educators, business leaders, governmental representatives, and parents now urge more rigorous school assignments, upgraded and enforceable standards, professionalized teaching, and expanded accountability.

Early childhood education too has been targeted for reform. Public efforts include lowering the age at which children start formal schooling. "Now that kindergarten for five-year-olds has become virtually universal in the nation's schools," observes Edward B. Fiske, educational editor of *The New York Times,* "demand is rising to make formal instruction available to all four-year-olds."[24] As noted, prekindergarten programs with heavy emphasis on academic instruction and testing for three- and four-year-olds are increasing.

Many experts are concerned that too much societal emphasis— by schools

2. INFANCY AND EARLY CHILDHOOD

and parents, and now reformers—is being placed on the formal instruction of young children. The Early Childhood Literacy Development Committee of the International Reading Association, comprised of national organizations involved in elementary and early childhood education, expressed misgivings in 1986. Too many pre-first-graders, noted the committee, were then exposed to the excessive pressures of accelerated academic and formal pre-reading programs. Consequently, many participating children tend to avoid experimentation and risk-taking. They are easily frustrated. Many grow to dislike reading.

Such programs, said the committee, offer little attention to individual development and individual learning needs. Also, by overly stressing isolated skill development, such programs focus insufficient attention on the comprehensive reading process, which entails the integration of verbal language, writing, and listening with reading. As a result of such practices, warned the committee, otherwise adequately prepared children could become "at risk."[25]

Recent findings by the National Assessment of Educational Progress, administered by the Educational Testing Service, too are not encouraging. According to NAEP, the youngest children in a class are more likely to repeat a grade than their older classmates, a disadvantage that continues through the eighth grade.[26]

From the moment of birth children possess the ability and desire to learn. Educators and child psychologists agree that parents should nurture this inclination by reading to their children regularly, finding time for nature exploration, museum and library visits, and a generous diet of interesting, two-way conversation. Creative interest in the child's domestic and external activities is essential. Moreover, enriched socialization and play programs outside the home can be highly beneficial, not only to children's awakening to the world around them, but to two-worker households.

The present trend toward inducing stressful, often unachievable, academic performance expectations from young children, however, seems excessive. Notions of "genius building" and intimidating, high-powered preschools, featuring instruction even test-driven curriculums, appear more suited to serving parents' rather than children's needs.

Preschool should be a stressless, stimulating, and socializing experience. Tests or measurements beyond prekindergarten assessments of a child's understanding of basic concepts—big and small, light and heavy, less and more, counting, for example—should be suspect. Peer competition and pressures to excel should be discouraged. Parents and social policymakers intent on priming youngsters for future academic careers must bear in mind, as early childhood educators know, that the "principal source of development in the early years is play."[27]

Children have a lifetime to cope with the struggles of educational and career success. "The secret of Education lies in respecting the pupil," wrote Emerson. "It is not for you to choose what he shall know, what he shall do. . . . [O]nly he holds the key to his own secret. Wait and see the new product of Nature. Respect the child. Be not too much the parent."[28]

NOTES

[1] *The Washington Post* (April 26, 1988), pp. 1, 10.
[2] Anita Shreve, "Fast-Lane Kids," *The New York Times Magazine*, Special Supplement, The Business World (June 12, 1988), p. 55.
[3] David Elkind, *Miseducation: Preschoolers at Risk* (New York: Alfred Knopf, 1987), pp. 3-4.
[4] As reported in *The New York Times* (April 29, 1988), p. B1.
[5] Quoted in *The Washington Post* (April 26, 1988), p. 10, and in Shreve, p. 55.
[6] Siegfried and Therese Engelmann, *Give Your Child a Superior Mind* (New York: Simon & Schuster, 1981), quoted in Robin Marantz Henig, "Should Baby Read?" *The New York Times Magazine* (May 22, 1988), p. 37.
[7] Ibid.
[8] Elkind, p. xiv.
[9] Francis Roberts, "School Days," *Parents* (March 1988), p. 46.
[10] Julius and Zelda Seigal, *Growing Up Smart and Happy* (New York: McGraw-Hill, 1985), p. 71.
[11] Quoted in Ibid., and *New Age Journal* (January 1985), p. 54.
[12] Engelmann, p. 28.
[13] David and Barbara Bjorklund, "Is Your Child Ready for School?" *Parents* (June 1988), p. 112.
[14] C. Anderson et al., *Becoming a Nation of Readers: The Report of the Commission on Reading* (Washington, D.C.: National Institute of Education, 1985), pp. 29-30.
[15] See, Madeline Drexler, "The Kindergarten Game: Is Your Child Ready to Play?" *Bostonia* (Jan./Feb. 1988), p. 32; *Changed Lives: The Effects of the Perry Preschool Program on Youths Through Age 19* (Ypsilanti: High/Scope Educational Research Foundation, 1984).
[16] *What Works: Research about Teaching and Learning*, 2nd edition (Washington, D.C.: U.S. Department of Education, 1987), p. 5.
[17] See, "The Payoffs for Preschooling," *The Chicago Tribune* (Dec. 25, 1984); Ann Crittenden, "A Head Start Pays Off in the End," *The Wall Street Journal* (Nov. 29, 1984); Larry Rohter, "Study Stresses Preschool Benefits," *The New York Times*, (April 9, 1985); and the Committee for Economic Development, *Children in Need: Investment Strategies for the Educationally Disadvantaged* (New York: CED, 1987), chapter 1.
[18] Quoted in *The New York Times* (April 9, 1985), p. C1.
[19] CED, pp. 5-9.
[20] Mortimer J. Adler, *The Paideia Proposal: An Educational Manifesto* (New York: Collier, 1982), p. 38.
[21] Children's Defense Fund, *A Call for Action to Make Our Nation Safe for Children* (Washington, D.C.: CDF, 1988), p. 7.
[22] Merry White, *The Japanese Educational Challenge: A Commitment to Children* (New York: Free Press, 1987), p. 191; "Japanese Education: How Do They Do It?" *In the Public Interest* (Summer 1984).
[23] White, pp. 33, 115, 187-88; "Mothers with Babies and Jobs," *The New York Times* (June 19, 1988).
[24] Edward B. Fiske, "Early School Is Now the Rage," *The New York Times* (April 13, 1986), pp. 24-30.
[25] Early Childhood Literacy Development Committee of the International Reading Association, "Literacy Development and Pre-First Grade," *Young Children* (1986), pp. 10-11.
[26] Cited in Drexler, p. 31.
[27] Joan Moyer, Harriet Egerston, and Joan Isenberg, "The Child-Centered Kindergarten," *Childhood Education* (April 1987), p. 238.
[28] Ralph Waldo Emerson, "Education," in *The Complete Writings of Ralph Waldo Emerson* (New York: Wise and Co., 1929), p. 993.

HOW THREE KEY COUNTRIES SHAPE THEIR CHILDREN

Spending a day with four-year-olds in China, Japan, and the US shows that these cultures have distinctively different aims—and some common goals—as they turn to preschools to preserve values in a time of change.

- **VIDEO SHOT, CHINA:** At 7:45 on a brisk Monday morning in early spring parents pull up to the Dong-feng Kindergarten's front gate on bicycles, each carrying a brightly dressed child riding on the back. In the courtyard, two middle-aged nurses, dressed in white, examine the arriving children, one by one...
- **VIDEO SHOT, JAPAN:** At 9:30 the "clean-up" song is played over loudspeakers audible throughout the Komatsudani preschool. As the children put away toys, balls, and tricycles, the music changes to the equally lively exercise song...
- **VIDEO SHOT, UNITED STATES:** At 7:30 a compact car pulls into the parking lot of St. Timothy's Child Center, and a father and his three-year-old son get out and walk hand in hand across the playground and into one of the classrooms. While Steve Cooper signs in on the attendance sheet, his son, Mark, puts his He-Man lunch box away in his cubby...

Joseph J. Tobin, David Y. H. Wu, and Dana H. Davidson

Joseph J. Tobin, a human development specialist at the University of Hawaii, will soon join the Family Studies faculty at the University of New Hampshire.

Cultural anthropologist David Y. H. Wu is a research associate at the East-West Center in Honolulu.

Dana H. Davidson is an associate professor in Family Resources at the University of Hawaii, Manoa.

IN CHINA, JAPAN, AND THE UNITED STATES preschool is an increasingly common way to provide care, education, and group experience for children between infancy and the start of formal schooling. In all three societies, the rise of the preschool is viewed, for better or worse, as a radical departure from traditional modes of caring for young children—who, in previous eras, were raised in their homes by full-time mothers, taken to the fields by parents who farmed, or cared for by hired country girls, mother's helpers, maiden aunts, grandmothers, or older brothers or sisters.

Yet several years of study lead us to view preschools as agents more of cultural conservation than of change. Using videotapes, interviews, questionnaires, and other means of research, we have found that preschools both reflect and moderate social change in these three cultures.

For example, in China preschools are expected to provide an antidote to the spoiling that Chinese fear is inevitable in an era of governmentally decreed single-child families. Chinese parents, preschool educators, and child-development experts are very worried about the problem they call the "4-2-1 syndrome": four grandparents and two parents lavishing attention on one child.

Japanese parents believe that preschools offer their children their best chance of learning to function in a large group and of becoming, in Japanese terms, truly human. The need for such experience arises from an increasing nuclearization and gentrification of the family brought on by a shrinking birthrate, an ongoing migration of young people from big households in the country to single-family apartments in large cities, and the rise of the mid-

From *World Monitor,* April 1989, pp. 36-45. Adapted from *Preschool in Three Cultures* by Joseph J. Tobin, David Y. H. Wu, and Dana H. Davidson. Copyright © 1989 by Yale University Press. Reprinted by permission.

2. INFANCY AND EARLY CHILDHOOD

dle-class *sarariman* (salaried employee) life style.

In the United States, preschools are being asked to respond to changing patterns of men's and women's work, a high divorce rate, and a growing concern for the needs of single-parent families.

As perceptions of work, marriage, and the family change in all three societies, they look to preschools to provide stability, richness, and guidance to children's lives. We have set out not to rate the preschools in the three cultures but to find out what they are meant to do and to be.

Why, for instance, do nonworking Japanese mothers work so hard to make the perfect school lunch? Why do they take pains to dress up so early in the morning just to meet other mothers?

An American woman living in Kyoto with a son in a local preschool explained the burden of lunch making:

"In America I just make a peanut butter sandwich and put in some chips and a piece of fruit and some carrot sticks. But here it is an entirely different story. One day my son came home crying and told me that the Japanese kids had laughed at his sandwich. The next day I made him a Japanese lunch, a *bentō*. But he came home from school again unhappy, again about his lunch. I said, 'Now what's wrong? I made you a Japanese lunch.' He said, in tears, 'But you didn't cut the apple slices so they look like bunny rabbits like the other mothers do.'"

For these nonemployed Japanese mothers, preparing their children for school is their most important work; their children's education is the center of their interests. And meeting other mothers before and after school is, other than shopping, their only daily personal contact with the outside world. The point is not that Japanese society as a whole is more limiting and restrictive for women than is Chinese or American society but rather that Japanese preschools play a more central role in defining mothers' identities and role demands than do preschools in the United States or China.

Today more than 95% of Japanese children attend preschool before they begin first grade. Japanese mothers today, like those of a century ago, remain the chief caretakers of infants and toddlers and thus the chief source of training in one-to-one relationships. What has changed is that the teaching of social skills and the fostering of an identity as a member of a group have become primarily the responsibility of preschools.

The nonpunitive attitude toward a "bad boy" in the Japanese school we videotaped would not necessarily be the same as that in all Japanese preschools. But it illustrates concern for the group. The teachers are careful not to isolate a disruptive child by singling him out for punishment or censure or excluding him from a group activity. They think their most powerful source of influence over children is their being viewed unambivalently as benevolent figures. They seek to maintain order—without intervening directly in children's disputes and misbehavior—by encouraging in various ways other children to deal with their classmates' troubles and misdeeds.

Isn't the bad boy's disruptiveness hard on the other children? "No," said the school director, laughing, "he makes things interesting." "No," said the assistant principal, "by having to learn how to deal with a child like [the bad boy], they learn to be more complete human beings."

By contrast, most of our Chinese informants told us unapologetically that they see the role of the preschool as teaching children to behave properly and instilling in them an appreciation for the values of self-control, discipline, social harmony, and responsibility. At the school we videotaped, even going to the bathroom was regimented, all 26 children being taken to a single large room at the same time.

"Of course, if a child cannot wait, he is allowed to go to the bathroom when he needs to," said one of our Chinese informants.

"But, as a matter of routine, it's good for children to learn to regulate their bodies and attune their rhythms to those of their classmates."

The word used most frequently in China to refer to teachers' control and regimentation of children is *guan*—literally, "to govern." When Ms. Xiang, as we shall see in a day at her school, tells the children to eat their lunch in silence and finish every bite, that is *guan*. When Ms. Wang criticizes one child for squirming and smiling while praising another for sitting straight with her hands behind her back and serious expression on her face, that, too, is *guan*. Instead of waiting to act until minor indiscretions grow into large ones, a good Chinese preschool teacher intervenes aggressively at the beginning.

At the American school we videotaped, the children talked animatedly as they ate—one sign of the relative freedom at the school, though an erring boy had to take a "time out" alone on a chair until he was ready to correct his behavior. When our questionnaire asked, "What is the most important reason for a society to have preschools?" our American respondents gave as their top answer: "to make young children more independent and self-reliant."

With this much of a hint of the differences and similarities we found, please join us in visiting preschools in Japan, China, and the US by means of transcripts condensed from our videotapes.

A Day at Komatsudani.

Komatsudani Hoikuen, a Buddhist preschool located on the grounds of a 300-year-old temple on a hill on the east side of Kyoto, has 120 students. Twelve of these children are infants, under 18 months, who are cared for in a nursery by four teachers. Another 20 Komatsudani children are toddlers, under three years of age, who are cared for in two groups of 10 by three teachers and an aide. The rest of the children are divided into three-year-old, four-year-old, and five-year-old classes, each with 25 to 30 students and one teacher. Each class has its own homeroom within the rambling old temple.

The school opens each morning at 7 a.m., and soon after, children begin to arrive, brought to school by a parent or grandparent on foot, by bicycle, or, less commonly, by car. By 9 a.m. most of the children have arrived, put their lunch boxes and knapsacks away in the cubby holes in their homerooms, and begun playing with their friends in the classrooms, corridors, or playground. Some of the older children stop by the nursery to play with the babies or to take toddlers for a walk on the playground.

At 9:30 the "clean-up" song is played over loudspeakers audible throughout the entire school area. As the children put away toys, balls, and tricycles, the music changes from the clean-up song to the equally lively exercise song.

Then, with their teachers' encouragement, the children form a large circle on the playground and go through 10 minutes of stretching, jumping, hopping, and running together in a group.

Taiso (morning exercise) complete, the "end-of-exercise-go-to-your-room" song comes over the loudspeakers, and the children, led by their teachers, run in a line into the school building, class by class, each child removing his or her shoes in the entranceway.

Inside, the 28 four-year-olds of *Momogumi* (Peach Class) enter their homeroom, which is identified by pictures of peaches on the door and the word *momogumi* written in *hiragana* (the phonetic alphabet). The Momogumi room has four child-sized tables, each with eight chairs that are covered with gaily embroidered seat covers the children have brought from home.

The *Momogumi-san-tachi* (Peach Class children) come in and stand behind their chairs while their teacher, Fukui-sensei, a 23-year-old university graduate, plays the morning song on a small organ and the two *toban* (daily monitors) lead the class in singing...

After attendance is taken by roll call, a counting song is sung to the tune of "Ten Little Indians" to determine how many children are in school that day...

The housekeeping chores and morning ceremonies completed, the children begin a workbook project which lasts about 30 minutes. Throughout this session there is much laughing, talking, and even a bit of playful fighting among tablemates. Fukui-sensei makes no attempt to stop them but forges ahead with the task at hand...

After they turn in their workbooks, the children begin to play loud chasing games, *janken* (paper-rock-scissors), and to engage in mock karate and sword fights. After 20 minutes or so of this free and raucous play and trips to the bathroom, the children, heeding their teacher, grab their *bentō*

12. How Three Key Countries Shape Their Children

(box lunch) from their cubbies and take their place at the table, arranging their lunch and cups and placemats in front of them. The food from home is supplemented by one warm course provided by the school and by a small bottle of milk. All the children sing in unison, under the direction of the daily *toban* and to the accompaniment of the organ:

> As I sit here with my lunch
> I think of Mom
> I bet it's delicious
> I wonder what she's made?

After the song the children stand, bow their heads, put their hands together, and recite: "Buddha, thank you. Honorable father, honorable mother, we humbly thank you."

Lunch itself is loud and lively, each child eating at his or her own pace, which varies from less than 10 minutes for some to 45 minutes or more for others. Fukui-sensei sits with the children at one of the four tables each day (the children keep careful track of whose turn it is), talking quietly to the children near her and occasionally using her chopsticks to help a child snare a hard-to-pick-up morsel from his *bentō*.

Some girls ask Fukui-sensei for help in properly tying up their lunch things in the large cloth *furoshiki* they have brought from home.

On the narrow covered porch adjoining their classroom four girls stand in a cluster, talking and laughing. Several boys are singing songs from television cartoon shows, engaging in more mock-fighting, and playing a game with flash cards meant to teach the *hiragana* syllabary.

One especially energetic boy, Hiroki, who has been much the noisiest and most unruly child in the class throughout the day (though it must be said that no one has tried very hard to control or quiet him), becomes increasingly raucous in his play. Midori runs inside to tell the teacher of Hiroki's misconduct and is encouraged by Fukui-sensei with a "go get 'em" sort of pat on the back to return to the balcony and deal with the problem herself.

Eventually the fighting ceases, the cards are cleaned up (with Fukui-sensei's help), and the children settle in at their desks, where they sing the after-lunch song ("Thank you. It was delicious...") and then rest with their heads on the table for five minutes or so while Fukui-sensei plays a soothing tune on the organ.

Rest time over, a major origami project begins, the children led by their teacher through a 20-step process resulting in the production by each child of an inflatable ball. ("Can you make a triangle? Good, now take these two ends of the triangle and make a smaller triangle, as I'm doing...") Soon the children, paper balls in hand, run laughing and screaming from the classroom to the playground for an extended period of outdoor play.

Back inside, Fukui-sensei reads a story to the class, using not a book but a *kami shibai*, a series

2. INFANCY AND EARLY CHILDHOOD

of a dozen or so large cards, each with a picture on one side and the narrative to be read by the teacher on the back. A song and a snack round out the schedule. After singing the good-bye song ("Teacher, good-bye, everyone, good-bye..."), the children go outside to the playground once more to play until their parents come for them between 4:30 and 6 p.m.

A Day at Dong-feng.

Dong-feng (East Wind) Kindergarten is a preschool run by a city in southwest China for the children of municipal employees. Occupying the grounds of an old estate, the six red brick one- and two-story buildings provide space for 270 three- to six-year-old children and 60 staff members. Three-quarters of Dong-feng's children are day (ri tuo) *students who attend school from about 8 a.m. to 6 p.m. Monday through Saturday. The other quarter are boarding* (guan tuo—*literally, "whole care") students who go home only on Wednesday evenings and weekends.*

At 7:45 on a brisk Monday morning in early spring parents pull up to the front gate on bicycles, each carrying a brightly dressed child riding on the back. In the front courtyard, two middle-aged nurses *(bao jian yuan)*, dressed in white, examine the arriving children one by one.

When four-year-old Li Aimei finishes her health check, her father, Li Chou, takes her by the hand and leads her down a corridor to the four-year-old boarding students' classroom and dormitory. Martial music, played over loudspeakers, fills the courtyard.

A dozen children are already inside, eating steamed buns. Aimei's grip on her father's hand tightens as one of her teachers, Ms. Xiang, approaches, saying good morning. Aimei whispers earnestly to her father: "Don't forget to come and pick me up on Wednesday evening. Keep that thought in your mind, Dad!" After a minute or two, Mr. Li, with a final good-bye and pat on the back, literally hands Aimei over to her teacher.

Mr. Li, by now outside the room, furtively sticks his head back in. Seeing his daughter involved with a steamed bun, neither crying nor searching for him, Mr. Li smiles and strides across the courtyard.

Two columns of desks, each with an attached bench large enough to seat two students, run down the middle of the classroom. Above a blackboard is a mural of children playing in a field with forest animals.

While 40-year-old Ms. Xiang and her coteacher, 25-year-old Ms. Wang, arrange the children in a circle for morning exercise, their assistant *(bao yu yuan)*, 24-year-old Ms. Chen, takes the breakfast bowls and glasses away to the central kitchen to be washed. With Ms. Wang playing an up-tempo song on a small organ, Ms. Xiang leads the children in calisthenics. All of them participate with enthusiasm and surprising grace.

Ms. Wang announces, "Let's do the 'Little Train Friendship Song.' " The children smile and clap. Ms. Wang begins to dance and sing: "I'm a little train looking for some friends. Who will come and ride on me?" Eventually, all the children are hooked up, snake-dancing around the room, singing along with their teacher.

After singing another song (about ducklings), the children are told to sit down at their desks. The teachers distribute wooden parquetry blocks to each child. Ms. Xiang says:

"We all know how to build with blocks, right? Just pay attention to the picture of the building and build it. When we play games like this, we must use our minds, right? Begin. Do your best. Build according to order."

The children begin to work in silence. Those who are working in a non-orderly way are corrected: A child whose box is placed askew on her desk has it placed squarely in the desk's upper right-hand corner by Ms. Xiang. Ms. Wang says: "Keep still! There is no need to talk while you are working."

It is now 10 o'clock, time for the children to go the bathroom. Following Ms. Wang, the 26 children walk in single file across the courtyard to a small cement building. Inside there is only a long ditch running along three walls. Leaving the toilet, again in single file, the children line up in front of a pump, where two daily monitors are kept busy filling and refilling a bucket with water the children use to wash their hands. Several boys in the back of the line indulge in some mock kung fu while other children talk and laugh. After washing their hands the children line up for a game of tag.

At 10:45 it is bath time for the boarding students. Three or four at a time, the children bathe in large tubs. Most of them are able to dry off and get dressed with minimal help from their teachers.

The children return to the classroom and take their seats. Ms. Wang drills them in addition and subtraction. As she pins hand-painted paper apples onto a large piece of cardboard, the children count out loud in unison. The students participate enthusiastically, each correctly answering at least one problem...

Lunch is delivered from the central kitchen in buckets. Again, the children march outside and wash their hands at the pump. By the time they return, a bowl of soybeans, vegetables, and shredded pork, a steamed bun, and a cup of water that has been boiled have been placed on each desk. Ms. Xiang reminds the children to eat in silence and not to waste any food: "Concentrate on your eating as much as you do on your studying. That's the correct way to eat."

The monitors collect the empty bowls and cups and place them in a bin to be returned to the kitchen. Other children wipe off the desk tops. Ms. Xiang then announces "Naptime." There are 26 beds in the dormitory room, each covered with a brightly colored, embroidered quilt. The children take their cups, scoop water from the bucket on a table, gargle loudly, spit into a spittoon on the floor, and wipe their faces with their washcloths. The children place their shoes neatly under their beds, remove their pants and jackets and place them on a corner of the bed, and crawl under their quilts dressed in T-shirts and underpants.

Naptime lasts from noon to 2:30, although most of the children do not sleep the whole time. While they rest, the teachers catch up on paperwork, eat, and relax in the classroom next door.

After nap, Ms. Chen once again takes the children to the bathroom and then to the pump to wash. Next comes a snack of cookies and reconstituted powdered milk.

Returning to their classroom, the children are taught to recite a patriotic story in unison. After the story, the children move outside for some relay races. Ms. Xiang says:

"Today we are going to play the 'Traffic Rules Game.' When I hold up this green card, you can run as fast as you can. Pretend you are bicycles flying down the street. When I hold up this red card, you must stop. If you don't stop, I will make you go back to the start and your team will fall behind. Cheer for your teammates to help them do their best. Don't let your team down."

At 5 p.m., following another group visit to the toilet, the children sit down at tables for supper. Ms. Xiang says, "Let's do the Puppet Song." Following their teachers' stiff movements, the children imitate marionettes as they recite loudly in unison, "We are wooden puppets. We can neither speak nor move." As they finish the verse they freeze...The children are served their evening meal of meat cooked with vegetables and rice.

It is now 6 p.m.. In the courtyard children can be heard calling out to their parents, who have arrived to pick them up. Inside, the children of the boarding class listen to records. They again rinse out their mouths, wipe off their faces, and struggle out of their clothes, some needing the assistance of a teacher. By 7:45 the children are all in bed under their warm quilts, and by 8 all are quiet and appear to be asleep.

A Day at St. Timothy's.

St. Timothy's Child Center operates a set of programs including full-day and half-day care for children two through five years old, a kindergarten for five- and six-year-olds, and after-school care for elementary school children. The center, a nonprofit institution affiliated with St. Timothy's Episcopal Church of Honolulu, is located on the church grounds in a neighborhood of mixed single-family homes, condominiums, and shopping centers. The preschool program, which serves 95 children, is housed in five large classrooms bordering a central playground.

12. How Three Key Countries Shape Their Children

Linda Rios and Pat McNair, two of St. Timothy's 10 teachers, arrive a few minutes after 7 a.m. to open the school. At 7:30 a compact car pulls into the parking lot and a father and his three-year-old son get out and walk hand in hand to one of the classrooms. While Steve Cooper signs in on the attendance sheet, his son, Mark, puts his He-Man lunch box away in his cubby.

Steve says,"Here you go, Mark; you can help Pat feed Pinky [a rabbit] and clean his cage." Pat says, "Yeah, come on, Marky. We could use your help." Steve gives Mark a pat on the back and says, "Have a good day, Champ," but Mark spins around and grabs his father's arm.

Steve: "I'll stay just a minute, and then I have to go."

Mark: "How long's a minute? Stay millions of minutes."

Pat (picking up Mark in her arms): "Let's walk your father to the gate and say good-bye there."

Steve: "See you later, Buddy. Thanks, Pat."

With a final wave, Steve drives off. As soon as the car is out of sight, Mark stops crying. Mark and Pat walk hand in hand over to Pinky's cage, Pat talking animatedly about how Pinky should be fed.

At 9 a.m. the school day formally begins. Cheryl Takashige calls the children in her class to come inside and sit in a circle on the rug in the middle of the room. Once the 18 four-year-olds are seated, Cheryl (32 years old) and her assistant teacher, Linda Rios (46), say good morning to the children.

Cheryl asks if anyone has anything to "show and tell" to the class today. Three hands go up. Lance, with Cheryl's prompting, relates a weekend family trip to see an active volcano. Next Rose shows the class her newest Care Bear, describing the difficulty she had choosing it over a My Little Pony. Mike proudly exhibits a wooden boat he made with his father.

Cheryl then leads the class in an activity involving a felt board and cutout flannel shapes. Linda prepares a large tray with a book, a paintbrush, a block, a puzzle piece, a toy frying pan, and a small brass ring. While Cheryl puts away the felt and flannel, Linda leads the children in singing a song. After finishing the song, Linda holds out the tray and says: "Look. Here are the [learning] centers for this morning...."

In the housekeeping corner—which includes a small table and chairs, a toy stove, sink, refrigerator, and shelves stocked with empty food boxes, plastic pots and pans, and miscellaneous dress-up clothes—Lisa decides that she will be the mother and Rose the auntie. Derek refuses to play the role Lisa assigns him as the baby, opting instead to be the family dog.

In the story corner, sitting on a big stuffed pillow, Linda reads to Kelly and Suzy, who are leaning against her. Across the room, Pete works at unscrewing parts from an old radio (donated to the school by a parent).

2. INFANCY AND EARLY CHILDHOOD

Cheryl finds a pile of unattended Legos. She looks around the room and, spotting Kerry in a corner with a puzzle, calls in his direction. Kerry doesn't budge. Cheryl walks over and puts her face directly in front of Kerry's:

"Kerry, listen to me. Look at me while I'm talking to you. I want you to go over there and clean up the Legos you dumped out before someone steps on them and gets hurt....You have nothing to say? Then you can sit over there on the time-out chair and think about it until you are ready to clean up."

Kerry walks over to the chair and sits with his head in his hands. After a minute or so, Cheryl calls to him, "Are you ready to clean up now? Good. You can get up now and clean up the Legos."

During the 45-minute learning-center period, the children shift from activity to activity according to their interests. At 10:15 Cheryl flicks the lights to announce clean-up time. Once all the toys have been put away, Linda leads the children to the boys' and girls' bathrooms, after which they run onto the playground for free outdoor play....

Snack time. Children line up for a cup of grape juice and a graham cracker and sit on the grass to eat.

The children return to their classrooms for a second round of learning centers. This time there are four new activities: washing baby dolls in a small basin, playing with Play-Doh (an ersatz clay made from flour, oil, water, and food coloring), stringing beads, and cooking. Cheryl explains about making potato soup:

"What's this? A potato, right? What color is it? Brown....We have to be very careful because we will be cutting and using fire."

Working with dull plastic knives, the children, in shifts of six, laboriously cut up the potatoes and carrots. Cheryl adds the vegetables to a pot and each child takes a turn stirring.

Cheryl flicks the lights to signal clean-up time, and the children put away their toys and then come to sit in their assigned places in a circle. Cheryl sits on the floor beside Kerry and puts her arm around his shoulders. Linda leads them in a song:

One gray elephant went out to play
He went out on a sunny day
He had such enormous fun
He called another elephant: COME!

On the word "Come" a child standing in the center of the circle reaches out for the hand of a child sitting down and pulls him into the middle. The song continues until all the children are standing and singing. Sitting down again, the children are served the soup they helped make. Cheryl reminds them of how the soup was made and explains what makes it nutritious.

When the children finish their soup, they throw away their Styrofoam cups and spoons, grab their lunch boxes (decorated with cartoon and television characters), and go outside to sit around the low table to eat lunch. The children talk animatedly while eating.

At 12:45 the teachers announce naptime. The children go the bathroom to brush their teeth and then return to the classroom, where they unroll the mats they keep in their cubbies and find a spot to stretch out.

Cheryl and Linda walk quietly around the room occasionally whispering to children to keep quiet or rubbing the backs of those having trouble falling asleep. By 1:15 all the children are asleep, and the teachers have about an hour to relax and prepare for afternoon activities.

The afternoon schedule is less structured than the morning. Following a snack of fruit and juice, the children play outside with balls, listen to the record player, look at picture books, and draw and cut and paste. From 4 o'clock on children begin to be picked up by their mothers or fathers.

At 5:55 Nicole's mother, Sandy, finally arrives. Cheryl calls out, "Nicole, your mom's here," and Nicole runs over to give her mother a hug. After sending Nicole inside to get her lunch box, Sandy engages Cheryl in conversation:

Sandy: "Sorry, Cheryl. Am I very late? What time is it, anyway?"

What are the most important things for children to learn in preschool?

300 Japanese, 240 Chinese, and 210 American preschool teachers, administrators, parents, and child-development specialists were asked this question. Here are their first choices.

	CHINA	JAPAN	U.S.
Perseverance	13%	2%	3%
Cooperation and how to be a member of a group	37%	30%	32%
Sympathy/empathy/concern for others	4%	31%	5%
Creativity	17%	9%	6%
Beginning reading and math skills	6%	0%	1%
Self-reliance/self-confidence	6%	11%	34%
Art/music/dance	1%	.3%	1%
Communication skills	4%	1%	8%
Physical skills	1%	.3%	1%
Good health, hygiene, and grooming habits	11%	14%	1%
Gentleness	0%	0%	0%

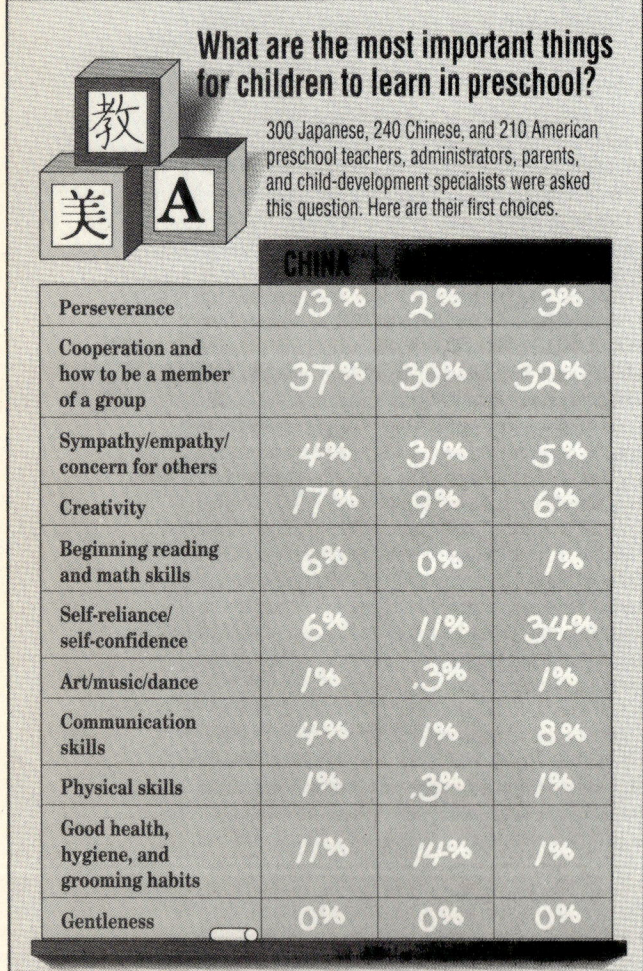

Graphic by Dave Herring

Cheryl: "No, that's okay, you made it."
Sandy: "How'd Nicole do today?"
Cheryl: "Fine, I think...."

AFTERWORD. The emphasis on variety and choice we found at St. Timothy's does seem to be far more characteristic of American than Chinese or Japanese preschools. And Americans tend to view independence as a characteristically Western trait and dependence as characteristically Asian.

Yet our interviews and questionnaires suggest that Chinese view the most important mission of the preschool as making spoiled, overdependent single children less spoiled, more self-reliant, and less dependent on their parents. Japanese children, in classrooms with ratios of 30 students to one teacher, are by both necessity and design more independent of adult supervision than are their peers in American preschool classrooms with much smaller teacher/student ratios. The promotion of self-reliance and independence in young children is therefore American but not uniquely American.

In China and Japan as well as in the United States, helping children develop language skills is believed to be central to the mission of the preschool. But the three systems have very different notions of the power and purpose of words.

In China the emphasis is on enunciation, diction, memorization, and self-confidence in speaking and performing. American and Japanese visitors to Chinese preschools are invariably impressed by the self-possession and command of language of Chinese children who flawlessly deliver long, rehearsed speeches and belt out multi-versed songs.

Language in Japan, both in and out of preschools, is divided into formal and informal systems of discourse. Children in preschools are allowed to speak freely, loudly, even vulgarly to each other during much of the day. But this unrestrained use of language alternates with periods of polite, formal, teacher-directed group recitation of expressions of greeting, thanks, blessing and farewell. Language in Japan—at least the kind of language teachers teach children—is viewed less as a tool for self-expression than as a medium for expressing group solidarity and shared social purpose.

Americans, in contrast, view words as the key to promoting individuality, autonomy, problem solving, friendship, and cognitive development in children. In American preschools children are taught the rules and conventions of self-expression and free speech.

In China citizenship is more widely viewed as something important to teach than it is in Japan and the United States. We see concern for citizenship as part of the more encompassing concern that young children be taught to identify with something larger than themselves and their families. This is perhaps *the* single most important function of preschools in China, Japan, and the United States, since in all three cultures it is the lesson hardest for parents to teach at home.

In Japan the child is taken to preschool to learn to enjoy ties to peers, to learn to transfer some of the warmth of parent-child relations to other relationships, to learn to balance the spontaneity enjoyed at home with formality, emotion with control, and family with society, to learn to become, in other words, truly Japanese.

In China children belong to parents *and* to society. As parents, Chinese naturally have a great desire to cherish and protect their children. But, as citizens, Chinese want to see their children grow up identified with their nation and its struggles, not just with narrower individual and familial concerns. Chinese preschools, and the ongoing debate surrounding them, reflect this search for balance.

American folklore celebrates the loner and the self-made man and looks with scorn on the "ant-colony mentality" seen as characteristic of group-oriented cultures. But some Americans worry that in the celebration of individualism the threads that bind people to one another have been stretched too thin. They are looking to government, church, and community organizations—including preschools—for direction and for a sense of shared purpose and identity.

In all three cultures children enter preschool belonging to their parents and leave with more diffuse, more complex ties to a world still centered on, but now much larger than, their families.

The Day Care Generation

Pat Wingert and Barbara Kantrowitz

Meryl Frank is an expert on child care. For five years she ran a Yale University program that studied parental leave. But after she became a new mother two years ago, Frank discovered that even though she knew about such esoteric topics as staff-child ratios and turnover rates, she was a novice when it came to finding someone to watch her own child. Frank went back to work part time when her son, Isaac, was 5 months old, and in the two years since then she has changed child-care arrangements *nine* times.

Her travails began with a well-regarded day-care center near her suburban New Jersey home. On the surface, it was great. One staff member for every three babies, a sensitive administrator, clean facilities. "But when I went in," Frank recalls, "I saw this line of cribs and all these babies with their arms out crying, wanting to be picked up. I felt like crying myself." She walked out without signing Isaac up and went through a succession of other unsatisfactory situations—a babysitter who couldn't speak English, a woman who cared for 10 children in her home at once—before settling on a neighborhood woman who took Isaac into her home. "She was fabulous," Frank recalls wistfully. Three weeks after that babysitter started, she got sick and had to quit. Frank advertised for help in the newspaper and got 30 inquiries but no qualified babysitter. (When Frank asked one prospective nanny about her philosophy of discipline, the woman replied: "If he touched the stove, I'd punch him.") A few weeks later she finally hired her 10th babysitter. "She's a very nice young woman," Frank says. "Unfortunately, she has to leave in May. And I just found out I'm pregnant again and due in June."

That's what happens when a *pro* tries to get help. For other parents, the situation can be even worse. Child-care tales of woe are a common bond for the current generation of parents. Given the haphazard state of day care in this country, finding the right situation is often just a matter of luck. There's no guarantee that a good thing will last. And always, there's the disturbing question that lurks in the back of every working parent's mind: *what is this doing to my kids?*

The simple and unsettling answer is, nobody really knows for sure. Experts say they're just beginning to understand the ramifications of raising a generation of youngsters outside the home while their parents work. Mothers in this country have always had jobs, but it is only in the past few years that a majority have gone back to the office while their children are still in diapers. In the past, most mothers worked out of necessity. That's still true for the majority today, but they have also been joined by mothers of all economic classes. Some researchers think we won't know all the answers until the 21st century, when the children of today's working mothers are parents themselves. In the meantime, results gathered so far are troubling.

Some of the first studies of day care in the 1970s indicated that there were no ill effects from high-quality child care. There was even evidence that children who were out of the home at an early age were more independent and made friends more easily. Those results received wide attention and reassured many parents. Unfortunately, they don't tell the whole story. "The problem is that much of the day care available

Child care has immediate problems. But what about the long-term effect it will have on kids?

From *Newsweek*, Special Edition, Winter/Spring 1990, pp. 86-87, 89, 92. Copyright 1989 by Newsweek, Inc. All rights reserved. Reprinted by permission.

13. Day Care Generation

Who's Minding the Children?

Even with the sharp rise in working mothers, most children are still cared for at home—their own or someone else's.

Percent of Mothers Working

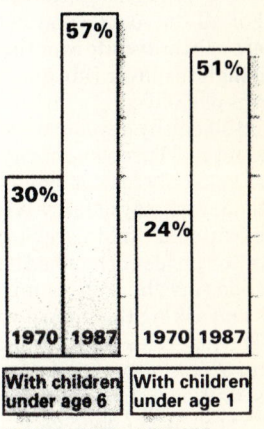

- With children under age 6: 1970: 30%, 1987: 57%
- With children under age 1: 1970: 24%, 1987: 51%

SOURCE: CHILD CARE INC.

Day Care
WHO LOOKS AFTER CHILDREN UNDER AGE 5 WHILE THEIR MOTHERS WORK

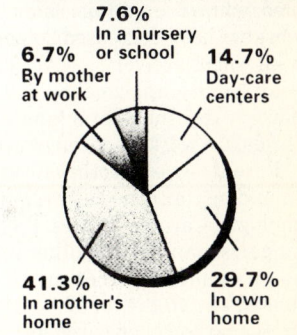

- 7.6% In a nursery or school
- 6.7% By mother at work
- 14.7% Day-care centers
- 41.3% In another's home
- 29.7% In own home

SOURCE: U.S. CENSUS BUREAU

Day care that might be fine for 3- or 4-year-olds may be damaging to infants

---

in this country is not high quality," says Deborah Lowe Vandell, professor of educational psychology at the University of Wisconsin. The first research was often done in university-sponsored centers where the child-care workers were frequently students preparing for careers as teachers. Most children in day care don't get such dedicated attention.

Since the days of these early studies, child care has burgeoned into a $15 billion-a-year industry in this country. Day-care centers get most of the attention because they are the fastest-growing segment, but they account for only a small percentage of child-care arrangements. According to 1986 Census Bureau figures, more than half of the kids under 5 with working mothers were cared for by nonrelatives: 14.7 percent in day-care centers and 23.8 percent in family day care, usually a neighborhood home where one caretaker watches several youngsters. Most of the rest were in nursery school or preschool.

Despite years of lobbying by children's advocates, there are still no federal regulations covering the care of young children. The government offers consumers more guidance choosing breakfast cereal than child care. Each state makes its own rules, and they vary from virtually no governmental supervision to strict enforcement of complicated licensing procedures for day-care centers. Many child-development experts recommend that each caregiver be responsible for no more than three infants under the age of 1. Yet only three states—Kansas, Maryland and Massachusetts—require that ratio. Other states are far more lax. Idaho, for example, allows one caregiver to look after as many as 12 children of any age (including babies). And in 14 states there are absolutely no training requirements before starting a job as a child-care worker.

Day-care centers are the easiest to supervise and inspect because they usually operate openly. Family day care, on the other hand, poses big problems for regulatory agencies. Many times, these are informal arrangements that are hard to track down. Some child-care providers even say that regulation would make matters worse by imposing confusing rules that would keep some potential caregivers out of business and intensify the shortage of good day care.

No wonder working parents sometimes feel like pioneers wandering in the wilderness. The signposts point every which way. One set of researchers argues that babies who spend more than 20 hours a week in child care may grow up maladjusted. Other experts say the high turnover rate among poorly paid and undertrained child-care workers has created an unstable environment for youngsters who need dependability and consistency. And still others are worried about health issues—the wisdom of putting a lot of small children with limited immunities in such close quarters. Here's a synopsis of the current debate in three major areas of concern.

There's no question that the care of the very youngest children is by far the most controversial area of research. The topic so divides the child-development community that a scholarly journal, Early Childhood Research Quarterly, recently devoted two entire issues to the subject. Nobody is saying that mothers ought to stay home until their kids are ready for college. Besides that, it would be economically impossible; two thirds of all working women are the sole support of their families or are married to men who earn less than $15,000 a year. But as the demographics have changed, psychologists are taking a second look at what happens to babies. In 1987, 52 percent of mothers of children under the age of 1 were working, compared with 32 percent 10 years earlier. Many experts believe that day-care arrangements that might be fine for 3- and 4-year-olds may be damaging to infants.

Much of the dispute centers on the work of Pennsylvania State University psychologist Jay Belsky. He says mounting research indicates that babies less than 1 year old who receive nonmaternal care for more than 20 hours a week are at a greater risk of developing insecure relationships with their mothers; they're also at increased risk of emotional and behavioral problems in later childhood. Youngsters who have weak emotional ties to their mothers are more likely to be aggressive and disobedient as they grow older, Belsky says. Of course, kids whose mothers are home all day can have these problems, too. But Belsky says that mothers who aren't with their kids all day long don't get to know their babies as well as mothers who work part time or not at all. Therefore, working mothers may not be as sensitive to a baby's first attempts at communication. In general, he says, mothers are more attentive to these crucial signals than babysitters. Placing a baby in outside care increases the chance that an infant's needs won't be met, Belsky says. He also argues that working parents have so much stress in their lives that they have little energy left over for their children. It's hard to find the strength for "quality time" with the kids after a 10- or 12-hour day at the office. (It is interesting to note that not many people are promoting the concept of quality time these days.)

Work by other researchers has added weight to Belsky's theories. Wisconsin's Vandell studied the day-care histories of 236 Texas third graders and found that youngsters who had more than 30 hours a week of child care during infancy had poorer peer relationships, were harder to discipline and had poorer work habits than children who had been in part-time child care or exclusive maternal care. The children most at risk were from the lowest and highest socioeconomic classes, Vandell says, probably because poor youngsters usually get the worst child care and rich parents tend to have high-stress jobs that require long hours away from home. Vandell emphasizes that her results in the Texas study may be more negative than those for the country as a whole because Texas has minimal child-care regulation. Nonetheless, she thinks there's a "serious problem" in infant care.

Other experts say there isn't enough information yet to form any definitive conclusions about the long-term effects of infant

61

2. INFANCY AND EARLY CHILDHOOD

care. "There is no clear evidence that day care places infants at risk," says Alison Clarke-Stewart, a professor of social ecology at the University of California, Irvine. Clarke-Stewart says that the difference between the emotional attachments of children of working and of nonworking mothers is not as large as Belsky's research indicates. She says parents should be concerned but shouldn't overreact. Instead of pulling kids out of any form of day care, parents might consider choosing part-time work when their children are very young, she says.

For all the controversy over infant care, there's little dispute over the damaging effects of the high turnover rate among caregivers. In all forms of child care, consistency is essential to a child's healthy development. But only the lucky few get it. "Turnover among child-care workers is second only to parking-lot and gas-station attendants," says Marcy Whitebook, director of the National Child Care Staffing Study. "To give you an idea of how bad it is, during our study, we had tiny children coming up to our researchers and asking them, 'Are you my teacher?'"

The just-released study, funded by a consortium of not-for-profit groups, included classroom observations, child assessments and interviews with staff at 227 child-care centers in five cities. The researchers concluded that 41 percent of all child-care workers quit each year, many to seek better-paying jobs. In the past decade, the average day-care-center enrollment has nearly doubled, while the average salaries for child-care workers have decreased 20 percent. Typical annual wages are very low: $9,931 for full-time, year-round employment ($600 less than the 1988 poverty threshold for a family of three). Few child-care workers receive any benefits.

Parents who use other forms of day care should be concerned as well, warns UCLA psychologist Carollee Howes. Paying top dollar for au pairs, nannies and other in-home caregivers doesn't guarantee that they'll stay. Howes conducted two studies of 18- to 24-month-old children who had been cared for in their own homes or in family day-care homes and found that most had already experienced two or three changes in caregivers and some had had as many as six. In her research, Howes found that the more changes children had, the more trouble they had adjusting to first grade.

The solution, most experts agree, is a drastic change in the status, pay and training of child-care workers. Major professional organizations, such as the National Association for the Education of Young Children, have recommended standard accreditation procedures to make child care more of an established profession, for everyone from workers in large for-profit centers to women who only look after youngsters in their neighborhood. But so far, only a small fraction of the country's child-care providers are accredited. Until wide-scale changes take place, Whitebook predicts that "qualified teachers will continue to leave for jobs that offer a living wage." The victims are the millions of children left behind.

When their toddlers come home from day care with a bad case of the sniffles, parents often joke that it's "schoolitis"—the virus that seems to invade classrooms from September unitl June. But there's more and more evidence that child care may be hazardous to a youngster's health.

A recent report from the Centers for Disease Control found that children who are cared for outside their homes are at increased risk for both minor and major ailments because they are exposed to so many other kids at such a young age. Youngsters who spend their days in group settings are more likely to get colds and flu as well as strep throat, infectious hepatitis and spinal meningitis, among other diseases.

Here again, the state and federal governments aren't doing much to help. A survey released this fall by the American Academy of Pediatrics and the American Public Health Association found that even such basic health standards as immunization and hand washing were not required in child-care facilities in half the states. Inspection was another problem. Without adequate staff, states with health regulations often have difficulty enforcing them, especially in family day-care centers.

Some experts think that even with strict regulation, there would still be health problems in child-care centers, especially among infants. "The problem is that caretakers are changing the diapers of several kids, and it's difficult for them to wash their hands frequently enough [after each diaper]," says Earline Kendall, associate dean of graduate studies in education at Belmont College in Nashville, Tenn. Kendall, who has operated four day-care centers herself, says that very young babies have the most limited immunities and are the most vulnerable to the diseases that can be spread through such contact. The best solution, she thinks, would be more generous leave time so that parents can stay home until their kids are a little older.

Despite the compelling evidence about the dark side of day care, many experts say there's a great reluctance to discuss these problems publicly. "People think if you say anything against day care, you're saying young parents shouldn't work, or if they do work, they're bad parents," says Meryl Frank, who is now a consultant on family and work issues. "For a lot of parents, that's just too scary to think about. But we have to be realistic. We have to acknowledge that good day care may be good for kids, but bad day care is bad for kids."

There is a political battle as well. Belsky, who has become a lightning rod for controversy among child-development professionals, says "people don't want working mothers to feel guilty" because "they're afraid the right wing will use this to say that only mothers can care for babies, so women should stay home." But, he says, parents should use these problems as evidence to press for such changes as paid parental leave, more part-time jobs and higher-quality child care. The guilt and anxiety that seem to be part of every working parent's psyche aren't necessarily bad, Belsky says. Parents who worry are also probably alert to potential problems— and likely to look for solutions.

Child-Care Checklist

Questions to ask at day-care centers:

■ **What are the educational and training backgrounds of staff members?**

■ **What is the child-staff ratio for each age?** Most experts say it should be no more than 4:1 for infants, 5:1 for 18 months to 2 years, 8:1 for 2 to 3 years, 10:1 for 3 to 4 years and 15:1 for 5 to 6 years.

■ **What are the disciplinary policies?**

■ **Are parents free to visit at any time?**

■ **Are the center's facilities clean and well maintained?**

■ **Are child-safety precautions observed?** Such as heat covers on radiators, childproof safety seals on all electrical outlets?

■ **Are staff members careful about hygiene?** It's important to wash hands between diaper changes in order to avoid spreading diseases.

■ **Are there facilities and staff for taking care of sick children?**

■ **Is there adequate space, indoors and out, for children to play?**

■ **Most important of all, do the children look happy and cared for? Trust your instincts.**

Where Pelicans Kiss Seals

IN THE SURPRISING WORLD OF CHILDREN'S ART, DELIGHTFUL IMAGES AND ORIGINAL RULES ARE CREATED TO REPRESENT THE WORLD.

Ellen Winner

Ellen Winner is a psychologist at Boston College and at Project Zero of Harvard University.

The story of how children learn to draw seems at first glance to be a simple one: At a very early age they begin by scribbling with any available marker on any available surface. At first the children's drawings are simple, clumsy and unrealistic; gradually they become more technically skilled and realistic.

But the development of drawing is not quite so simple and straightforward. In fact, the story turns out to be quite complex. Watch a 2-year-old scribbling. The child moves the marker vigorously across the page, leaving a tangled web of circular and zig zag lines. It looks as if the marks themselves are an accident—the unintended result of the child's arm movements. But if you replace the child's marker with one that leaves no trace, the child will stop scribbling, as psychologist James Gibson and Patricia Yonas, then a graduate student, showed in 1968. Even though very young children enjoy moving their arms vigorously, they are also interested in making marks on a surface.

If we do not watch a scribble in the making, but only see the final product, it may look like a meaningless tangle of lines. And this is how scribbles have traditionally been viewed—as nonsymbolic designs. But 1- and 2-year-olds are rapidly mastering the concept that words, objects and gestures stand for things. So why shouldn't they also grasp that marks on a page can stand for things? Some of the more recent studies of children as they scribble suggest that these early scrawls are actually experiments in representation—although not purely pictorial representation.

TO A 2-YEAR-OLD, SCRIBBLES AREN'T JUST SCRIBBLES, THEY'RE A PLANE FLYING ACROSS THE SKY.

Psychologist Dennie Wolf, preschool teacher Carolee Fucigna and psychologist Howard Gardner of Project Zero at Harvard University studied how the drawing of nine children developed from age 1 to 7. The researchers took detailed notes on the process of scribbling, and their investigations show us that children have surprising representational abilities long before they spontaneously produce a recognizable form.

At first the representation is almost entirely gestural, not pictorial. Wolf observed a 1½-year-old who took the marker and hopped it around on the page, leaving a mark with each imprint and explaining as she drew, "Rabbit goes hop-hop" (Figure 1). This child was symbolizing the rabbit's motion, not its size, shape or color. The meaning was carried primarily by the marker itself, which stood for the rabbit, and by the process of marking. Someone who saw only the dots left on the page would not see a rabbit. Nonetheless, in the process of marking, the child was representing a rabbit's movement. Moreover, the dots themselves stood for the rabbit's footprints. Here in the child's earliest scribbles we already see glimmerings of the idea that marks on a page can stand for things in the world.

Two-year-olds rarely spontaneously create recognizable forms in their scribbles, but they have the latent ability to do so. When Wolf or Fucigna dictated to 2-year-olds a list of features (head, tummy, arms, legs), these children plotted the features systematically on the page, placing them in correct relative positions (Figure 2). But they lacked the notion that a line stands for the edge of an object and had no way to represent parts of features, since each feature was either a point or a patch. The children clearly understood, however, that marks on a surface can be used to stand for features "out there," off the page, and that they can be used to show the relative spatial locations of features.

Typically at age 3, but sometimes as early as age 2, children's spontaneous scribbles become explicitly pictorial. They often begin by making gestural scribbles but then, noticing that they have drawn a recognizable shape, label and further elaborate it. For example, one 3½-year-old studied by Wolf, Gardner and Fucigna looked at his scribble and called it "a pelican kissing a seal." He then went on to add eyes and freckles so that the drawing would look even more like a pelican and a seal (Figure 3). Another child, on the eve of his second birthday, made some seemingly unreadable marks, looked at his picture and said with confidence, "Chicken pie and noodies" (his

2. INFANCY AND EARLY CHILDHOOD

word for noodles). Clearly he saw the similarity between the lines on the page and noodles on a plate.

Sometimes children between 2 and 3 will use both gestural and pictorial modes at different times. A 2-year-old studied by art educator John Matthews of Goldsmiths College in London drew a cross-like shape, then looked at it and called it "an airplane." One month later, this same child moved his brush all around in a rotating motion while announcing, "This is an airplane." The label was the same, but the processes and products were different. In the first case, the drawing was an airplane because it looked like one. In the second case, it was an airplane because the marker moved like one, leaving a record of the airplane's path.

With pictorially based representations, the child begins to draw enclosed shapes such as circular forms and discovers that a line can be used to represent an object's edge. This major milestone marks the child's invention of a basic rule of graphic symbolization. This invention cannot be attributed to closer observation of nature, since objects don't have lines around them. Nor can it be attributed to the influence of seeing line drawings. As shown by psychologist John Kennedy at the University of Toronto, congenitally blind children and adults, when asked to make drawings (using special equipment), also use lines to stand for an object's boundaries.

Sometime around 3 to 4 years of age, children create their first image of a human—the universal "tadpole"—consisting of a circle and two lines for legs. Figure 4 shows a typical tadpole, with a circle standing for either the head alone, or, more probably, head and trunk fused; it has two legs but no arms. It was drawn by a 3-year-old; by 4, children begin to distinguish the head from the trunk and often add arms to their figures.

The tadpole is indisputably a purely graphic (rather than gestural) representation of a human. But why would children universally invent such an odd image to stand for a person? Many people believe that children draw humans in this queer fashion because this is the best they can do; the tadpole is simply a failed attempt at realistic representation. According to some investigators, including psychologist Jean Piaget, children's drawings are intended to be realistic, but children draw what they know rather than what they see. Hence, the tadpole, with its odd omissions of trunk and arms, must indicate children's lack of knowledge about the parts of the human body and how they are organized.

Psychologist Norman Freeman of the University of Bristol has a different way of accounting for the typical omissions. He notes that children draw a person from top to bottom, in sequence. We know from verbal-memory tasks that people are subject to "primacy" and "recency" effects—that is, after hearing a sequence of words, they recall best the words they heard first and last and tend to forget those that came in between. The child, Freeman argues, is showing such effects in drawing, recalling the head (drawn first) and legs (drawn last) and forgetting the parts in the middle. As Freeman sees it, tadpoles result from deficient recall, not deficient concepts.

But other research suggests that we should look on the tadpole more positively. Psychologist Claire Golomb of the University of Massachusetts in Boston suggests that children know more about the human body than

Figure 1: While hopping a marker around the page, a 1½-year-old said, "Rabbit goes hop, hop, hop" and made these marks.

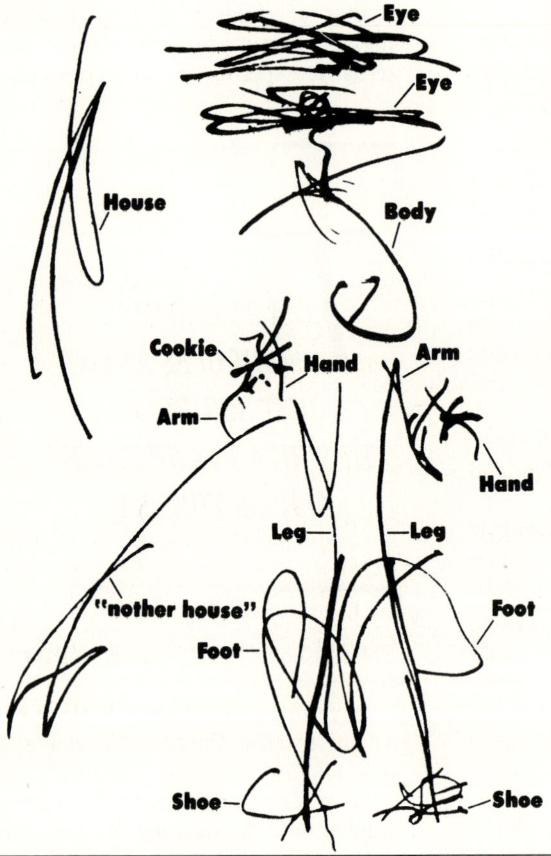

Figure 2: When someone dictates a list of body parts, even 2-year-olds can show them in their correct positions.

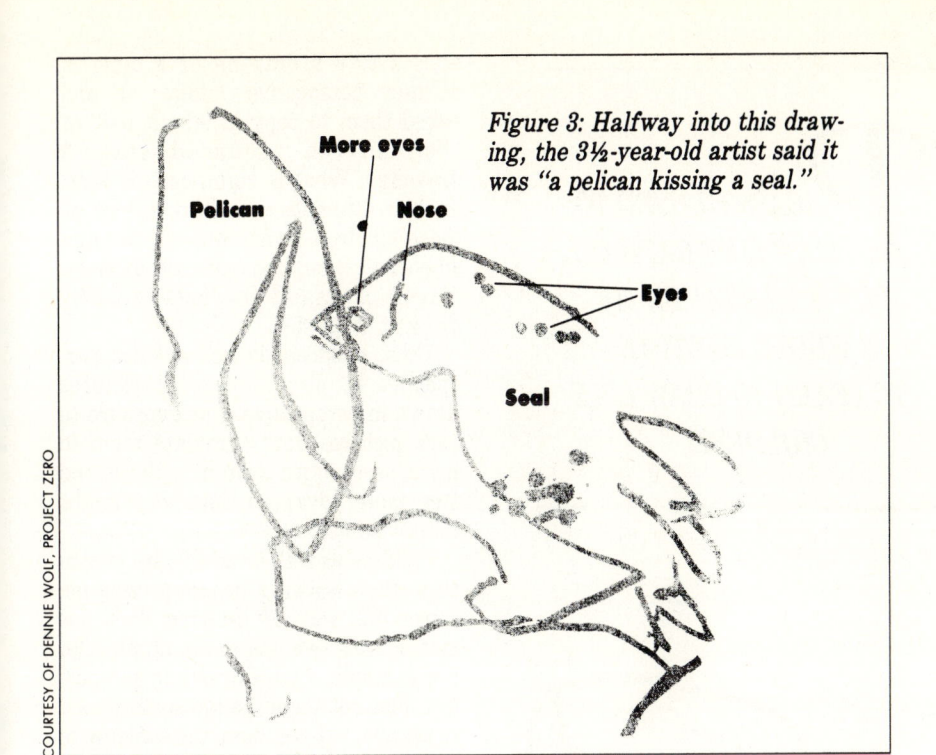

Figure 3: Halfway into this drawing, the 3½-year-old artist said it was "a pelican kissing a seal."

seem to draw simple geometric shapes in part because realism does not spark their interest. Once they do catch the desire for realism, Golomb says, that desire begins to overcome their natural tendency to simplicity, leading them toward more complex, graphically differentiated drawing.

By late childhood or early adolescence, children in our culture begin to master linear perspective, the Western system for creating the illusion of three-dimensional depth on a two-dimensional surface, invented and perfected during the Renaissance.

Many people believe that the ability to use perspective is taught, either explicitly in art class or tacitly through exposure to pictures showing such perspective. But something far more creative on the part of children may be happening, says psychologist John Willats, formerly of the North East London Polytechnic.

Willats seated groups of children of different ages in front of a table with objects on it and asked them to draw what they saw (Figure 6a). The children, 108 in all, who were from 5 to 17 years old, used six different systems of perspective; these, Willats found, formed a developmental sequence:

The 5- and 6-year-olds were entirely unable to represent depth. They simply drew a rectangular box for the tabletop and let the objects float above it (stage 1: Figure 6b). Seven- and 8-year-olds drew the tabletop as a straight line or thin surface and placed the objects on that line (stage 2: Figure 6c). Again, their pictures contained no recognizable strategy for representing depth.

At about age 9, children made their first readable attempts to depict the third dimension. They drew the tabletop as a rectangle and placed the objects inside or on top of the rectangle (stage 3: Figure 6d). These children had invented a system for representing depth: To depict near objects, draw them on the bottom of the page; to depict far objects, draw them on the top of the page. In other words, transform near or far in the world into down or up on paper. No one teaches a child to draw this way. Moreover, no child actually sees a tabletop as a rectangle (unless looking at it from a bird's-eye view). Hence, this strategy is a genuine invention.

Younger adolescents drew the table-

about how to draw it, and their body-part omissions are not due to forgetting. She found that when 3-year-olds were asked to name body parts, they almost always mentioned arms, although they typically omitted them from their tadpoles. She also discovered that such children were less likely to create tadpoles when they made Play-Doh people or when they were given a two-dimensional assemble-the-parts task. Even on a drawing task, when Golomb gave children a drawing of a head with features to complete, 3- and 4-year-olds (who ordinarily drew tadpoles) typically differentiated a head and a trunk. Finally, if children were asked to draw someone playing ball—a task implicitly requiring that they draw arms—they were likely to include them (Figure 5).

One 3-year-old drew a tadpole but described it in full detail as she drew it, naming parts that were not there, such as feet, cheeks and chin. Clearly, this child was not trying to show all of these parts, because she made no special marks for them. Instead, her simple figure stood for an entire human in all its complexity.

Although adults make drawings that are far more complex, they, too, do not (and cannot) draw all that they see. Young children are more selective than adults, no doubt because drawing is difficult for them, but also because they have not yet been fired up by the peculiarly Western pictorial ideal of realism.

Psychologist Rudolf Arnheim, formerly of the University of Michigan, argues that children try to create the simplest form that can still be "read" as a human. Because they have a limited repertoire of forms, they reduce them to simple geometric shapes. In the case of the tadpole, they usually reduce the human body to a circle and two straight lines.

Adult artists often deliberately select a limited repertoire of forms and, like children, reduce natural forms to a few simple geometric shapes. They recognize that realism is but one ideal among many and may choose not to be realistic. Young children, however,

WHY DO 3-YEAR-OLDS FIRST DRAW A PERSON AS AN ODD ARMLESS 'TADPOLE'— A CIRCLE WITH TWO LINES?

2. INFANCY AND EARLY CHILDHOOD

top as a parallelogram rather than as a rectangle (stage 4: Figure 6e). As in the previous stage, they incorrectly drew the lines parallel, not converging, but they now correctly used oblique lines to represent edges receding in space. Again, such a system for representing depth is neither taught nor based on visual experience, since using parallel lines is not optically correct.

In the last two stages, older adolescents drew in perspective, making the lines of the tabletop converge. Some made the lines converge only slightly (naïve perspective—stage 5: Figure 6f); others achieved geometrically correct perspective (stage 6: Figure 6g).

This sequence cannot be explained simply by a growing desire and ability to draw objects as they really appear, because as children develop, their drawings actually get less realistic before they get more realistic. A tabletop, viewed from eye level, might be seen as an edge (stage 2), or from a bird's-eye view it might be seen as a rectangle (stage 3). But no one ever sees a tabletop as a parallelogram (stage 4) or with incorrectly converging edges (stage 5).

Willats believes this sequence does not result from copying pictures in perspective. After all, he argues, in our culture drawings with perspective rarely depict a rectangular surface as either a rectangle (stage 3) or a parallelogram (stage 4).

But I believe these two stages may indeed be attempts to copy the perspective seen in pictures. A tabletop drawn in perspective shows its surface (and stage 3 is an advance over stage 2 because it shows the surface), and its lines are at nonright angles (and stage 4 is an advance over stage 3 because the angles are oblique). But if children are trying to copy perspective drawings, they are doing it at their own developmental level. For example, in stage-4 drawings, children reduce what should be a trapezoid to a simpler, more regular parallelogram.

One way to test the effect of exposure to pictorial perspective is to ask children to copy such pictures. Freeman did just this, finding that although children could not copy the model's perspective system accurately, they could adopt a system more advanced than the one they used spontaneously. Freeman showed children

PICASSO: I USED TO DRAW LIKE RAPHAEL, BUT IT HAS TAKEN ME A WHOLE LIFETIME TO LEARN TO DRAW LIKE CHILDREN.

Figure 4: The 3-year-old's first drawing of a person: a "tadpole" consisting of a circle with two lines for legs.

ages 5 to 8 a drawing of a table in oblique perspective (stage 4) and asked them to copy it. About half of the children produced stage-3 drawings. What is significant is that, in their attempts at imitation, half of these children—who were at the age when they would be expected to make stage-1 or stage-2 drawings—actually drew at stage 3.

Thus, children do not acquire perspective by directly copying pictures drawn in perspective. But exposure to such pictures does stimulate them to try a perspective system at least one step more advanced than they might otherwise use.

Children as well as adults are in conflict when drawing in perspective because this way of drawing does not match what we know about objects. For example, a table drawn in perspective does not show the top surface as a rectangle, yet we know the tabletop is rectangular. Although we would like to show things as we see them, we also want to show them as we know they are. Perhaps this is why, as shown by psychologist Margaret Hagen and Harry Elliot, then a graduate student at Boston University, adults prefer drawings with stage-4 perspective to those with stage-6 perspective; stage-6 drawings make objects look too distorted.

Knowing that a good pictorial likeness is not necessarily an exact copy of a scene as it actually appears, artists often deliberately break the rules of perspective. For example, to correct for the size distortion called for by the rules of perspective, they may draw a distant mountain larger than it would appear in a photograph. Perhaps for similar reasons, children may not at first draw objects with optical realism; they are interested in showing things in the most informative way rather than showing exactly how things look.

Do children improve further as they get older? If realism is the standard, the answer is clearly yes. For example, their figures become more complex, and they can represent depth through linear perspective. But I believe, on esthetic grounds, that children's drawings actually get worse with age.

Because preschool children are unconcerned with realism, their drawings are free, fanciful and inventive. Suns may be green, cars may

Figure 5: A young child, asked to draw people playing ball, includes arms.

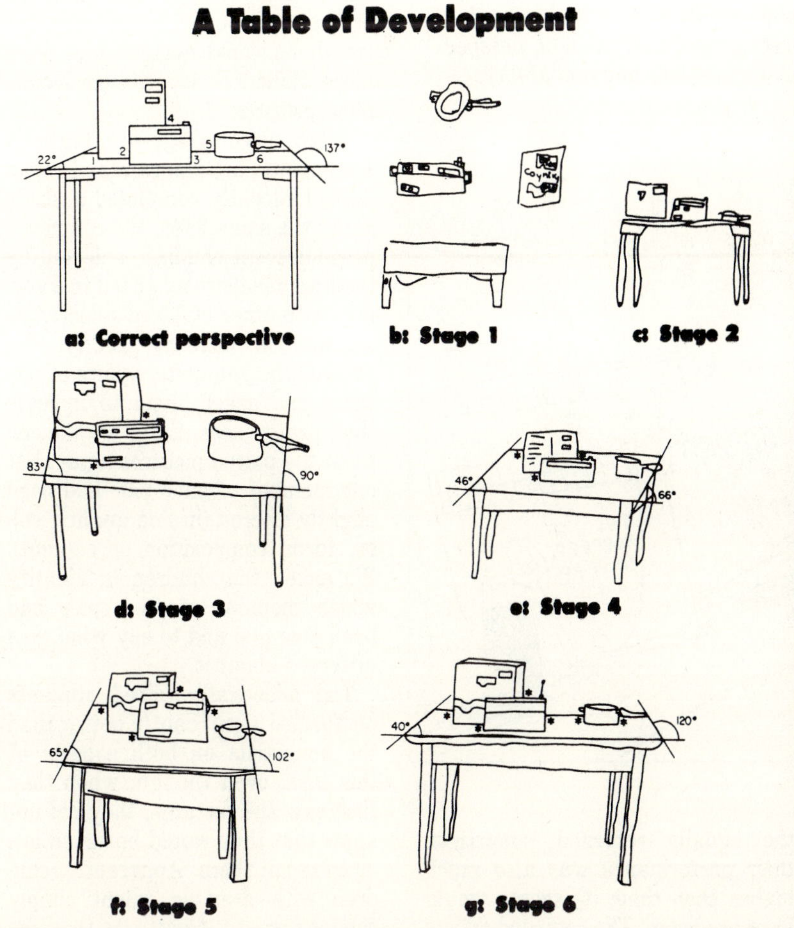

A Table of Development

a: Correct perspective
b: Stage 1
c: Stage 2
d: Stage 3
e: Stage 4
f: Stage 5
g: Stage 6

Figure 6: As children develop, they become better at representing depth. The first table (a), drawn in correct perspective, shows what the children were trying to copy.

14. Where Pelicans Kiss Seals

ARE SMARTER CHILDREN BETTER ARTISTS?

Some children show more drawing ability than others and produce elaborate, realistic drawings at an early age. When others are still drawing tadpoles, these children are drawing human figures with differentiated body parts in correct proportion and are even putting depth cues into their drawings. People often assume that these advanced artists are brighter than those who lag behind and produce primitive, undifferentiated, unrealistic drawings.

Indeed, psychologists have developed intelligence tests based, in part, on the assumption that drawing level reflects cognitive level, IQ or both. For instance, as part of the Stanford-Binet Intelligence Scale, children are asked to copy shapes, and the Goodenough-Harris Draw-a-Man test uses drawing as a measure of IQ, with more parts and details yielding higher scores.

But studies of both normal and abnormal people show that drawing ability is independent of ability in other areas. The most dramatic evidence comes from studies of idiot savants who, despite severe retardation, autism or both, draw at an astonishingly sophisticated level.

The best-known case, studied by psychologist Lorna Selfe, formerly of the University of Nottingham, is that of Nadia, an autistic child who, as early as age 3½, drew in an optically realistic style reminiscent of Leonardo da Vinci. In addition to her studies of Nadia, Selfe compared retarded autistic children who were gifted in drawing with normal children of the same mental age. She found that the retarded children were better able to depict proportion, depth and the overlap of objects in space.

In a similar vein, psychologists Neil O'Connor and Beate Hermelin of the Medical Research Council Developmental Psychology Project in London studied five young-adult idiot savants who had special drawing ability but very low verbal and performance IQ scores and com-

2. INFANCY AND EARLY CHILDHOOD

ARE SMARTER CHILDREN BETTER ARTISTS? (continued)

Top: Nadia, 5½, extremely autistic and artistic, made the drawing (left) that resembles a Leonardo da Vinci sketch (right). Below: An early elementary school child drew in perspective, revealing unusual ability.

pared them with other equally mentally retarded people who had no special drawing ability. They gave these two groups a battery of tests, including the Draw-a-Man test. On this test the retarded artists performed at a much higher level than the equally retarded nonartists; their performance was also much higher than their IQ scores would have predicted. The retarded artists particularly excelled in their ability to depict body proportion, rather than in depicting specific features, providing details or in their levels of motor control and coordination.

If gifted artists—both retarded and nonretarded—aren't necessarily more intelligent, what skills enable them to draw better than other people? To address this question, O'Connor and Hermelin used a battery of other tests, all of which assessed visual memory. When retarded artists were compared with retarded nonartists, the artists outperformed the nonartists on all tests.

Visual-memory skill, independent of IQ, also seems to help normal children to excel at drawing. Recently, Hermelin and O'Connor compared artistically gifted normal children with nongifted children matched in IQ. They found that the artists had superior memory for two-dimensional designs and were more skilled at identifying incomplete pictures.

With psychologist Elizabeth Rosenblatt at Harvard's Project Zero, I recently completed a study along the same lines. We compared preadolescent children, selected by their art teachers as gifted in drawing, with other children selected as average in drawing ability. We showed the youngsters pairs of pictures and asked them to indicate their preferences. Later we showed them the paired pictures again, but one member of each pair had been slightly altered (in line quality, color, form, composition or content). We asked the children to identify which member of each pair had been changed and to say what was different about it.

The artistically gifted students performed significantly better than the nonartists on both aspects of this task, even though, when they first saw the pictures, they did not know that they would be asked later to recall them. Apparently, children with drawing talent simply cannot forget the patterns they see around them, just as musicians often report being unable to get melodies out of their minds.

14. Where Pelicans Kiss Seals

Figures 7 and 8: This 6-year-old's drawing (top) is free, fanciful and inventive. An 11-year-old's drawing (bottom) is neater and more realistic but also less inventive.

float in the sky and complex, irregular forms in nature are reduced to a few regular geometric shapes. They produce simple, strong pictures that evoke the abstractions found in folk, "primitive" and contemporary art (Figure 7).

The older child's drawing may be more realistic, neat and precise, but, in my opinion, it is also less imaginative and less striking (Figure 8). Suns are now appropriately yellow and placed carefully in the corner of the picture, and cars now rest firmly on the ground.

This development is inextricably tied up with acquiring the technical skills essential for adult artistic activity. Nonetheless, once such skills are mastered, artists often turn back to young children's drawings for inspiration and may work hard to do consciously and deliberately what they once did effortlessly and because they had no choice. "I used to draw like Raphael," Picasso is quoted as saying, "but it has taken me a whole lifetime to learn to draw like children."

Development During Childhood

- Social and Emotional Development (Articles 15-17)
- Cognitive and Language Development (Articles 18-20)
- Developmental Problems (Articles 21-23)

Most of the changes that occur during the transition to childhood—as well as those that occur during childhood itself—involve social, cognitive, and language development. The articles in this section touch upon contemporary themes pertaining to each of these developmental domains.

Two selections in the subsection *Social and Emotional Development* offer recommendations to parents regarding personality development. Bruce A. Baldwin tells us in "Building Confidence" that many parents use academic achievement as a measure of their child's motivation. However, the way parents respond to their children can facilitate or destroy the development of positive motivation. Baldwin suggests that parents can help create a healthy atmosphere for achievement motivation by implementing his "Confidence Builder's Checklist."

In a classic study, Thomas, Chess, and Birch identified three basic types of temperament: easy, difficult, and slow-to-warm-up children. Most children display difficult behavior on occasion, as is pointed out in "Dealing With Difficult Young Children." A child's temperament has a great influence on how individuals react to him or her and this, in turn, can affect the child's personality development. Anne K. Soderman recommends strategies for parents and teachers to help deal with difficult behaviors.

All life transitions create some degree of stress. Although some children react to stress by striking out against its perceived source, and others respond to it by withdrawing and attempting to isolate themselves, some children seem to take stress-producing situations in stride. These "invulnerable" or "resilient" children, as pointed out in "The Miracle of Resiliency," have developmental histories that may include extreme poverty or chronic family stress. Most people would consider both factors to be potentially damaging to the development of personal and social competence skills. Yet these resilient children do not become victims; instead, they develop effective interpersonal skills. Without question, studies of these children will contribute important knowledge to our understanding of personality development and child-rearing practices.

The articles in the *Cognitive and Language Development* subsection focus on developmental changes in cognitive processing, challenges to IQ definitions of intelligence, the way children learn, and issues related to reading and language acquisition. During the past two decades, the study of cognitive development was dominated by Piaget's theory. Although this theory provides a rich description of what a child can and cannot do during a particular stage of development, it is less adequate for explaining how a child acquires various cognitive skills. Thus, many developmentalists have turned to information-processing models in an effort to integrate cognitive psychology with cognitive developmental theory. Information-processing research has directed attention to individual differences in skill acquisition, challenging traditional concepts of intelligence; this is addressed in "Three Heads Are Better Than One."

Young children learn best when they can touch, explore, and move about. Knowing what is best and actually implementing these findings into the classroom are two different matters. In the "back to basics" movement of the 1980s, homework, drills, and discipline were emphasized. However, children at the age of 6 are not physically ready to sit for sustained periods; they think concretely rather than abstractly. The article "How Kids Learn" points out that young children should be taught by different methods than those used for older children, and illustrates teaching methods especially suitable for younger pupils.

Most children begin talking at sometime between 10 and 14 months of age. The article "Now We're Talking!" encourages parents to make language fun for their children. This can be accomplished by talking, reading, and playing with your child, as well as also listening to your child. Parents are cautioned not to cross the barrier from being a parent to being a teacher.

The *Developmental Problems* subsection addresses issues relating to the identification and treatment of development disorders. Hundreds of thousands of American children (mostly boys) are given prescription drugs to control a behavior problem that has been called hyperactivity, minimal brain damage, ADHD, and a variety of other labels over the years. In "Suffer the Restless Children," the author raises some very disturbing questions about the reality of the disorder; these questions extend beyond diagnosis of ADHD to the "theory and practice of mental health in the U.S."

Children with varying degrees of disabilities—some

Unit 3

quite severe—are learning to communicate via computers. For some children, as pointed out in "Tykes and Bytes," this is not only the first medium they have had for communicating their needs and feelings, but also the first opportunity they have had to play.

Children who have difficulty learning to read for no apparent reason are termed dyslexic. Though the exact cause is not known, some possible causes are discussed in "Facts About Dyslexia"; treatment and prognosis are also considered.

Looking Ahead: Challenge Questions

Children who are described as resilient are an enigma. How would you explain such strength of personality in the face of so many potentially disruptive influences in their lives? What coping mechanisms seem to provide resilient children with such strength of character?

The ability to solve problems ranks high among lay definitions of intelligence. Yet problem-solving ability is not exactly the same as the intelligence measured by IQ tests. Which do you believe is the better measure of intelligence? The fact that all human beings are not equally intelligent, any more than they are not equally tall, suggests biological variation in the distribution of intelligence. What characteristics do you think of when you refer to someone as being intelligent? How does dyslexia mask intelligence?

It is said that socialization is a two-way street. How does this apply to child temperament?

Should hyperactivity be labeled a disorder? Should it be treated with drugs? Why or why not?

Building Confidence

Discover Your Child's Natural Motivation

Dr. Bruce A. Baldwin

Dr. Bruce A. Baldwin is a practicing psychologist who heads Direction Dynamics, a consulting service specializing in promoting professional development and quality of life in achieving men and women. He responds each year to many requests for seminars on topics of interest to professional organizations and businesses. This article has been adapted from a chapter of Baldwin's book on parenting, Beyond the Cornucopia Kids: How to Raise Healthy Achieving Children. *For busy achievers, he has also written* It's All in Your Head: Lifestyle Management Strategies for Busy People. *Both are available in bookstores or from Direction Dynamics in Wilmington, North Carolina.*

Tim is in the fifth grade. So far, he's been a rather poor student. Testing has revealed no learning disabilities, but Tim isn't achieving to his potential. He just doesn't seem to care about his homework. He'd rather spend his time making forts with his friends or just tinkering in the garage.

Annette lives just a few doors away. She's a very bright student and loves school. Always willing to tackle new projects, she's a rising star. When not at school, she reads and does extra projects. Her teachers are as proud of her accomplishments as her parents are. As dedicated as she is, she is expected to go far by all who know her.

Now for the compelling question: "Which of these two students is motivated and which is not?" If you answered that Annette has lots of motivation and Tim has virtually none, you would be very wrong. Actually, both of these young people are highly motivated. The difference lies in where their energies are focused. Tim likes to tinker and work outdoors, while Annette likes to study and learn within a classroom environment. The contrast between these two young people also debunks the myth of the "unmotivated" person. As an adult or as a child, there is no such thing.

While everyone is motivated, their directions for personal motivation vary greatly. What parents are most concerned about in an "unmotivated" child is that the child is not "achieving to potential" in school. They realize how important learning is to future success. They fear that their child, whose energies are focused primarily on nonacademic endeavors, will never "make it." Sometimes such parental fears are well founded. Sometimes they are not. Close examination of the emotional reasons for achieving is required to tell the difference.

It may be that both Tim and Annette have healthy achievement motivation. But for the sake of argument, let's look at some other possibilities. Tim, for example, may not be academically oriented because of past experiences that have been turned into emotionally painful failures as the result of his parents' responses. Thus, his nonacademic inclinations may be motivated by the fear of more failures. On the other hand, Annette may be driven to high accomplishments because she has learned that this is the only way she can get any kind of positive response from her parents. For her, success has become heavily linked to self-esteem, so she can't stop achieving.

In these typical situations, appearances are deceptive. Depending on prior experiences, the academic motivations of Tim and Annette could be healthy or unhealthy. The point is that parents' understanding of their children is absolutely essential to developing healthy achievement motivation. In order to gain a clear perspective on the maturation of their children, parents must overcome at least three emotional barriers that tend to distort their perceptions. These barriers are:

Parental frustration with the child's irresponsibility. A familiar lament of parents heard everywhere: "We try so hard, but no matter what we do to teach our kids sound values and responsible behavior, it just doesn't seem to take." The fact is that children will learn values if they are reinforced consistently and modeled well by their parents. Far too often, parents let their frustration and anger determine their responses to their children, and those responses are often emotionally destructive. Parents must be calm, cool and consistently positive in responding to their children's endeavors.

Fears about the child's ultimate ability to "make it." Parental frustration, without an understanding of the childhood learning process, can give way to deep fears about a child's future. "I don't want my son to fail," or "I want my daughter to have all that I have and more." These fears often lead parents to begin pressuring children at an early age to achieve. Such responses, however, may cause some children to rebel and resist learning. A vicious cycle often ensues in which the parents push harder and harder, provoking a child to resist all the more stubbornly.

Parental fantasies about what a child should become as an adult. Virtually every parent has a fantasy about what a child can be and should be. But if parents' fantasies are allowed to become expectations, the child can be lost in the parents' grand vision. A child's directions become what the parent wants them to be, and in the process, a child's interests, aptitudes and aspirations may be completely ignored. A far healthier strategy is for parents to encourage their child to develop career directions and personal potentials without imposing their own.

The dynamics of healthy achievement motivation are not difficult to understand. The hard part for overburdened parents is to change their negative responses to healthy ones.

The first step is to begin to see each of your children as unique. The second is to

15. Building Confidence

move beyond your own frustrations, fears and fantasies that can distort your responses. These shifts will help you to gain a better perspective on the developmental process that will eventually lead to emotional maturity, personal responsibility and ultimately, to success. Then, to keep your momentum for change going, here are some additional ideas about how to respond in ways that will encourage the development of healthy achievement motivation in your children.

The Confidence Builder's Checklist
Volumes have been written on the specifics of creating healthy achievement motivation in children. A consistent core of suggestions, however, appears time and again. It is a mistake to think that it costs money to build achievement motivation in children. It doesn't. What it does cost is some thinking through, some understanding and some change in the way you communicate with your children. Essentially, the values communicated by parents' responses to their children either encourage the development of positive motivation or destroy it.

Let's face it—creating confidence is never easy; its source lies in positive achievement experiences. While all children will not become high achievers as adults, with help all children *can* develop their personal potential in this area. Here is the Confidence Builder's Checklist to help you get started.

Understand that learning takes place in many different contexts besides schoolwork. It occurs in free and unstructured play, through trial and error in accomplishing personally meaningful tasks, through relating to peers, and in projects that are self-initiated and carried to completion. Although there are no grades for these activities, you should realize their importance to the total development of your children and make time for them.

Make a clear distinction between "helping out" and "taking over" when your children have a project. You should make helpful suggestions and get involved with what your children are doing, but not too much. Be wary of getting conned into taking over and doing everything for your children. Instead, try to let your child set the tone and make his own mistakes without any "I told you so" recriminations.

Treat failure as an inevitable and necessary part of success. This positive value should always be evident in your parental responses. Avoid name-calling for goofing up or making a mistake. Instead, try to help your child analyze what went wrong and thereby develop strategies to overcome the problem. Never forget to keep an upbeat "work smarter, not harder" attitude.

There should be no question that you value quality over quantity. Children often become overinvolved in outside activities these days. Keep a watchful eye on your children, and place reasonable limits on their extracurricular involvements. Mandate some time for your children to relax—it will help them gain deeper benefits from the activities that remain.

Reinforce personal commitment and perseverance in spite of obstacles. In other words, one of your achievement values should be that you finish what you start, or at least take it to a reasonable end point. Don't permit your child to quit an activity or a project just because the going gets rough. This teaches the child to confront obstacles, not avoid them.

Help your child set attainable goals. Often, children who are left to their own devices set unrealistic goals for themselves. Some of their objectives may be impossible, while other mistakes may result from overestimating their capabilities. Your input can help to place personal effort and project practicality in perspective.

Recognize that cooperation is often more important than competition. You already know that very competitive parents sometimes teach their own "win at all costs" orientation to their children. Parents should know better. The adversarial relationships that are born of competitiveness often interfere with developing positive attitudes toward teamwork, cooperative efforts and getting along well with others. Emphasize cooperation and working well with people instead.

Reinforce adequate advance preparation for meeting any personal challenge. "Anything worth doing is worth doing well" is the value behind preparation. Get your children to think through what they will need to carry out a realistic plan; then help them get ready. Discourage children from thoughtlessly rushing into a challenging situation which may result in needless mistakes or even danger.

Unconditional positive acceptance should always be present when you are involved with your children. Your children should know from experience that you love them no matter what, that your acceptance of them is *not* contingent on their being the best or always successful. If given that awareness, children are emotionally freed to do their best because they're not afraid to fail.

Help your child to keep achievements in healthy perspective. In other words, when your child comes out number one or successfully meets a challenge, it should be understood that this does not mean he or she is better than anyone else. Success means that, in this instance, your child was simply more dedicated or more skilled, or tried harder. Avoid creating "good, better, best" hierarchies.

The positive feedback you give should be behaviorally specific. Far too much of the time, criticism is quite detailed, but positive feedback is given in very general terms. To be useful, positive feedback must be behaviorally specific. And the result will be that your children will consistently receive from you usable information about what they did right.

"Personal best" is the frame of reference in which your child evaluates accomplishments. Avoid the trap of the perfectionist who always finds something wrong. Instead, encourage your child to reach for his or her personal best and to recognize differences in personalities and capabilities. No matter what mistakes are made, always try to find something right about what a child has accomplished.

In whatever your child does, reinforce personal responsibility for actions taken. Don't permit your child to blame circumstances or other people for his or her failures. In all ways, mandate good sportsmanship—and that means accepting failures gracefully. Only naive parents encourage scapegoating instead of a careful examination of personal performance to spot needed areas of improvement.

Talk to your children about their work, but do it in casual and optimistic ways. You can not only make achieving a creative challenge, but you also spark enthusiasm by the way you talk about it. Tell interesting stories about yourself and help your children to conceptualize ideas and strategies through your dialogue. As a result, they will be encouraged to ask questions and communicate with you.

Become a "resource development specialist" for your children. If your children don't know something, help them to find the answer without doing it for them. You may suggest specific books, people to call and talk to, or even particular places where they can watch how something is done. Encourage this kind of resourcefulness and outreach by your children, and take pride in seeing how quickly they learn to do it themselves.

Attention to detail should be encouraged as part of all achievement endeavors. Equipment should be inspected, homework checked and the finer points of an activity examined. Children are notorious for global perception, but they can gain an edge by attending to details. Most won't learn this skill unless it is taught to them. It is one of several important base line skills that ensure quality work. It also helps to prevent mistakes of omission.

In broader perspective, healthy achievement motivation is a personal asset that enhances every part of life. Wise parents

3. CHILDHOOD: Social and Emotional Development

recognize the importance of achievement motivation in more than just career terms. It is a value system that becomes a primary means for personal expression and gaining emotional fulfillment. As it develops in children, healthy motivation brings with it four confidence qualities that make day-to-day living easier for a lifetime:

Change is not threatening. If a healthy achiever has an inner sense of control, there is no perception of self as a victim of circumstance, of other people, or of the vicissitudes of life. This person is capable not only of "going with the flow," but also of determining the direction of the flow most of the time.

Challenge is not feared. The healthy achiever does not fear carefully considered risks because failure is not personally threatening. In fact, failure is seen as an opportunity to learn. As a result, healthy achievers see opportunities for personal growth in both their failures and their successes. It's a win-win situation.

There is no need for defensiveness. The healthy achiever is aware of and accepts his own personal strengths and weaknesses, and personal growth is directed toward further developing strengths and removing weaknesses or vulnerabilities. One result of this orientation is that healthy achievers, secure within, have no need to be defensive. Positive feedback and helpful advice are gracefully accepted.

Self-esteem is internally based. The healthy achiever is not emotionally driven to win always, to be number one, or to have the most of everything in order to feel good about himself. Because self-esteem and positive acceptance are based within, that person feels less personal vulnerability when he or she encounters a failure. This quality permits achievement to be an expression of self rather than a way to allay deep insecurity.

The "I can make things happen!" attitude is a powerful quality that enables a person to make choices no matter how adverse the circumstances. As such, healthy achievement motivation is an index of deep personal strength, adaptive coping and emotional hardiness.

Someone has said that luck is where opportunity and preparation meet. Because the healthy achiever is well prepared, opportunities are found even in adversity. As a wit once commented: "When opportunity knocks, most people complain about the noise." That's characteristic of those who are insecure about meeting challenges. But it doesn't *have* to happen that way. It all depends on your perspective. The healthy achiever opens the door.

Dealing with Difficult Young Children

Strategies for Teachers and Parents

Anne K. Soderman

Anne K. Soderman, Ph.D., is Assistant Professor in the Department of Family and Child Ecology, Michigan State University, East Lansing, Michigan.

Young children's personality development remains one of the most intriguing and widely debated issues for both parents and teachers. This article will review what we know about individual temperament and recommend ways to understand and build upon children's strengths.

Individual Temperament

Generations of parents and child development professionals have recognized that children exhibit individual personality traits. As early as 1924, Gesell wrote that

> The personality of the child grows like an organic structure.... Original nature ... provides certain tendencies or materials, but the final patterns of personality are the result of education and experience. (p. 3)

Much of the research on temperament has been conducted only in the past two decades. Thomas, Chess, and Birch (1968) collected parent interviews and observed 136 children in natural settings from infancy to preadolescence. They identified nine characteristics of temperament that they clustered into three basic types.

- *Easy* children were moderately low in intensity, adaptable, approachable, predictable with body functions, and positive in mood (40 percent of the sample).
- *Difficult* children were often negative in mood, adapted slowly to change, had unpredictable biological functions, and frequently exhibited intense reactions (10 percent of the sample).
- *Slow-to-warm-up* children took longer to adapt than the easy children and demonstrated low activity and intensity levels (15 percent of the sample). Children who did not appear to fit in any of these categories comprised the remaining 35 percent of the group.

A flurry of responses followed, many of which either challenged or extended this pioneering work. For example, Bates (1980) criticized the research methodology of asking parents to assess their own children. He held that difficult temperament in children is an adult social perception with only a "modest empirical foundation."

Early infant behaviors (such as susceptibility to emotional stimulation; strength and speed of response; and quality, fluctuation, and intensity of mood) were

> Children's temperament affects how others react to them and influences their personality development.

examined closely for indications that they might predict later personality traits (Goldsmith and Gottesman 1981; Als 1981; Dunn and Kendrick 1980). Plomin (1982) praised the Thomas, Chess, and Birch study because it balanced the tendency throughout the 1950s and 1960s to blame children's behavioral problems on their mothers.

In a rebuttal supporting the validity of their earlier findings, Thomas, Chess, and Korn (1982) argued that their within-the-child definition of difficult temperament was not only valid but also practical in looking at later behavioral difficulty in children. They noted that

> the concept of difficult temperament as a within-the-infant characteristic in no way implies a fixed immutable trait... but can be modified or altered by postnatal genetic or maturational influences, situational context, or the effect of the child-environmental interactional process. (p. 15)

Meanwhile Brazelton (1978) observed that an infant's personality may be a determining factor in whether adults react to the child positively or negatively. In an effort to identify infants likely to encounter difficulties in relationships because of their temperament, Brazelton and his colleagues created the Neonatal Behavioral Assessment Scale (NBAS) (1973). This scale elicits and measures whether infants can calm themselves, whether they can shut out disturbing stimuli, their need for stimulation, their level of sociability, and the organization and predictability of their behavior. Although the instrument does not necessarily predict later development, it does offer reliable information about newborn behavior that may cause potential bonding problems (Gander and Gardiner 1981).

Another tool for assessing temperaments in infants and preschool children is a questionnaire designed by Carey and McDevitt (1978). It is based on the nine characteristics identified by Thomas and Chess (1977). According to Powell (1981), principal uses of the questionnaire are to serve as a basis for a general discussion of a child's temperament so that the child's needs can be met better, and to help with clinical problems "of which infant temperament can be a part" (p. 111).

In summary, we know that
- very early in life infants have identifiable personalities
- some traits of children's personalities produce behaviors that are perceived by others as ranging from easy to difficult
- young children's behaviors are con-

3. CHILDHOOD: Social and Emotional Development

tinually being modified or intensified depending upon adults' responses

Influences on Behavior

The child's own traits and behaviors, environmental influences, and adults' perceptions and reactions all affect long term personality development. Thomas, Chess, and Korn (1982) remind us that

> To debate whether a child's characteristics or parental perceptions or other environmental influences are more important is antithetical to the view which sees them *all* as all-important in a constantly evolving sequence of interaction and mutual influence. (p. 15)

When children are difficult, less confident adults will doubt themselves, feel guilty, and be anxious about the child's future and their relationship. Unless the difficult behaviors (or perceptions of difficult behaviors) can be satisfactorily modified, a sense of helplessness often begins to influence all interactions with the child. Dreams of being a competent parent or teacher may yield to the harsh reality that the child is unhappy, out of control, and not developing to full potential. Initial pride over the child's assertiveness may turn to disapproval and even rejection. Burned out adults are often the result. Such adult responses, in turn, have a marked effect on whether additional stress will be put upon the child.

Does labeling children *difficult* have a harmful effect in itself? Such identification may damage children if parents and teachers do not approach the child's behavior in a positive way that will build on the child's strengths (Rothbart 1981).

While this concern about the effect of labeling is a valid issue, there is no question that some children *are* more pleasant and sociable than others. Elkind (1981) has pointed out that children are members of a hurried society. Children who cannot adapt to tight schedules, rigorous learning programs, and a fast pace, or who become easily frustrated when stressed, will most likely have difficulty functioning well.

Knowing what to look for in children's behavior, and how to interpret it, can help us find ways to help children cope.

Temperamental Characteristics

Regardless of the cause of the behavioral difficulty, the practical implications for difficult children appear very early. Children are active agents in their own socialization, and their temperament affects how others react to them and influences their further personality development. Several temperamental characteristics, and adult perceptions or misperceptions of them, all influence children's personality. Important questions to keep in mind when considering each of these traits are:

- Is the behavior developmentally appropriate for the child's age? Does the behavior exceed what might reasonably be expected when the child's experiences are taken into account?
- Does the child consistently exhibit one or more of these difficult behaviors?
- How much does the adult's behavior style contrast with that of the child?

Unless we know what to expect of children at each age, it is easy to label a child difficult when in reality the child is exhibiting normal behaviors which may vary from day to day and are related to the child's experiences. Likewise, if the adult's personality or energy level is very different from the child's, it is easy to misperceive the meaning of the child's behaviors. The behaviors described here represent extremes along a continuum. Readers are cautioned to carefully consider each of the questions above when reviewing these possible indicators of behavioral difficulty.

Activity Level

This trait refers to the proportion of a child's motor activity and inactivity. Are quiet and active behaviors balanced? A difficult child may be viewed as excessively active, always on the go, unable to sit still, or always touching something. At the other extreme, a child may be seen as too passive, with no spirit or motivation, who just sits. Young children are generally full of healthy energy!

Rythmicity

The child's internal biological timeclock is the key to rythmicity. Are sleep, hunger, eating, and elimination patterns predictable or unpredictable? Does a toddler or preschool child have difficulty settling down easily for scheduled rest or sleep, only to fall asleep at other times? Children who have a healthy sense of independence can often frustrate adults who are eager to maintain a tight schedule!

Approach/Withdrawal

A child's *initial* responses to new objects, people, or foods may indicate that the child is approachable or withdrawn. Children who are too approachable may seem to have no fear or may be willing to accept any suggestion regardless of its appropriateness. Repeated crying, clinging, or avoiding almost all new experiences may indicate that the child is too withdrawn.

This trait affects children's decision making patterns; their ability to prevent overstimulation; and their need to escape uncomfortable, difficult, or unpleasant situations. It can be particularly problematic for adults and children who function differently from each other, or who have a contrasting base of experiences.

Adaptability

Adaptability is an indication that the child responds positively to change. Some children will find it difficult to accept a change in routine, a new sibling, or other important events, even when given time to adjust. Others may hold grudges, or be considered stubborn or resistant. Refusal to care for themselves, or to advance in skill level of an activity, may indicate that the child has a great deal of difficulty adapting. Early school adjustment may be particularly problematic.

Adults should give children every reasonable chance to assess the situation, figure out what should be done, and then experience the consequences of their decisions.

Intensity of Reaction

Energy level and response to stimuli are related to impulse control. Situations which call for an emotional response are especially indicative of the child's ability to cope. Some children may act without thinking, react quickly with anger, grieve too long after a loss, or become too excited when happy. These children are most often described as aggressive. At the other extreme are children who have a bland exterior—they hide their emotions or exhibit little intensity in emotional situations.

Responsiveness Threshold

The level of stimulation necessary to evoke a response may also indicate difficult behaviors. Some children are extremely sensitive to or overstimulated by noise, touch, smell, temperature, light, color, or similar factors, while others will respond slowly or not at all to the same stimuli. Experience can play a large part in a child's responsiveness threshold. A child living on a busy city street may not hear a blaring siren, while such a sound in more isolated areas may rightly cause a frightened reaction.

Mood

This trait deals with a balance between behaviors that are joyful, friendly, and satisfied with those that are not positive. Some children will express a range of moods, while others will react similarly to nearly every situation. They may complain, whine, cry frequently, or fuss. Bad days are frequent, and adults may go out of their way to avoid conflict. Parents and teachers may feel guilty because the child rarely looks happy or joyful.

Attention Span and Resistance

How much time does the child spend pursuing an activity, especially one with obstacles? Can the child return to the activity if interrupted? Some children seem scattered, cannot attend to any task, and move rapidly without any true involvement. Others may be obsessed with one specific activity, such as completing puzzles or playing with a single toy. Age of the child and appropriateness of the activities are important factors here.

Distractibility

Closely related to attention span and resistance is the child's vulnerability to interference when pursuing any activity. Some children may become so completely absorbed that they forget to use the bathroom, or may resist distraction even from a prohibited activity. Others will be unable to shut out typical distractions such as noise. In infancy and toddlerhood distractibility may be an asset when children need to be distracted! In older children, however, distractibility may interfere with the child's ability to attend to responsibilities.

Strategies for Parents and Teachers

You probably recognize some of these difficult behaviors in nearly every child you know! Many of these behaviors can be difficult for an adult to deal with at the moment, but may not indicate that the child has any long term personality problem. A pattern of these behaviors may indicate more serious difficulties, however, and consultation with a specialist is recommended for those children. The recommendations here will focus on ways adults can help children cope with occasional difficulties.

The way in which parents and teachers deal with young children's difficult behavior can have a lasting effect on the children's emerging personality. May (1981) discusses the fragile balance between power and powerlessness in altering interpersonal relationships. Adults can intensify children's difficulties and increase stress if they respond inappropriately. Examples of inappropriate responses would be for the adult to
• ignore difficult behaviors
• coerce children to comply with adult expectations
• shame or compare children to others
• label children with derogatory words
• inflict verbal or physical punishment

Positive change requires that the adult keep children's self-esteem intact while helping them become knowledgeably involved in the process of modifying their own behavior. Adults can help children cope with their own behaviors by using a variety of strategies.

Respect the Child

Individual differences exist as a result of both heredity and the environment. Children's difficult behaviors do not necessarily indicate that the child is intentionally being difficult, is stubborn, or that the adult is inadequate. Instead, the adult can respect the child by closely assessing the situation and working out a plan to help the child cope.

Evaluate Behavior Objectively

Our own temperamental style and personality play a critical role when we evaluate any child's positive or negative qualities. How do our behavior and expectations affect the child? As adults, we may need to change our behaviors before we can effectively work with a difficult child.

Structure the Environment

A close look at the environment may reveal the causes of some difficult behaviors. Are children asked to sit quietly for longer periods than they can legitimately manage? Are noise levels too high or is lighting too harsh? Changing the pace of the day to balance vigorous and calm times, providing a place for quiet activities, or rearranging traffic patterns, for example, may establish a climate for more appropriate behaviors.

Set Effective Limits

Unless children know the limits, and those limits are reasonable, difficult behaviors are likely to follow. Review the rules you have set to make sure they are appropriate, and then discuss them with the children. Natural and logical consequences should result when children do not observe the limits, such as loss of the privilege to use an item.

Often a subtle cue from the adult is all that is needed to remind children before their behavior becomes unacceptable.

Use Positive Interactions

Once we understand how the child interacts with others, we can help children modify their behaviors. With very young children who may be egocentric, for example, adults can state how the child's behavior affects the feelings of others. Rather than expecting an immediate change, you may want to identify steps for making progress.

Be Patient

Helping children to modify their behavior, or modifying your own, is a long range goal. During the process, there must be opportunity to correct behaviors in a nonthreatening atmosphere. Instead of redirecting children too quickly or making decisions for them, adults should give children every reasonable chance to assess the situation, figure out what should be done, and then experience the consequences of their decisions.

Work with Colleagues and Parents

Most teachers and parents are doing the best job they know how to do. If, however, we see negative interaction styles between other adults and a difficult child, we need to approach those adults as partners in trying to help the child cope more effectively. We can note areas of interaction that are difficult for us and stress the normal variations that may be found in young children. It is important for us to understand that a child's behavior, while it may be irritating, can be quite innocent. Our reactions to that behavior, therefore, can sow feelings of inferiority and insecurity or, on the other hand, competence and strong self-esteem.

Once we go beyond feeling guilty or trying to blame someone about a child's behavior, we can work more cooperatively with others.

Conclusions

Many of the recommendations here are common sense techniques that apply to all human relationships—it is the consistent and intentional use of these strategies that makes a difference in children's personality development. Our attitudes toward children, adults, and ourselves play a key role in ensuring successful collaborative efforts to help children become more effective in their relationships.

Bibliography

Als, H. "Assessing Infant Individuality." In *Infants at Risk: Assessment and Intervention*, eds. C. C. Brown and T. B. Brazelton. Boston: Johnson & Johnson, 1981.

Bates, J. E. "The Concept of Difficult Temperament," *Merrill-Palmer Quarterly* 26, no. 4 (October 1980): 299–319.

Brazelton, T. B. "Introduction," In *Organization and Stability of Newborn Behavior: A Commentary on the Brazelton Neonatal Behavior Assessment Scale*, ed. A. Sameroff. *Monographs of the Society*

3. CHILDHOOD: Social and Emotional Development

for Research in Child Development 43, no. 5–6 (1978): 1–13.

Buss, A. H. and Plomin, R. A. *A Temperamental Theory of Personality*. New York: Wiley, 1975.

Carey, W. B. and McDevitt, S. C. "Revision of the Infants' Temperament Questionnaire," *Pediatrics* 61, no. 5 (May 1978): 735–739.

Dunn, J. and Kendrick, C. "Studying Temperament and Parent-Child Interaction: Comparison of Interview and Direct Observation." *Developmental Medicine and Child Neurology* 22, no. 4 (August 1980): 484–496.

Escalona, S. K. *The Roots of Individuality: Normal Patterns of Development in Infancy*, Chicago: Aldine, 1958.

Elkind, D. *The Hurried Child*. Reading, Mass.: Addison-Wesley, 1981.

Freedman, D. G. "Infancy, Biology, and Culture," In *Developmental Psychobiology: The Significance of Infancy*, ed. L. P. Lipsitt. Hillsdale, N.J.: Erlbaum, 1976.

Gander, M. J. and Gardiner, H. W. *Child and Adolescent Development*. Boston: Little, Brown, 1981.

Gardner, H. G. *Developmental Psychology*. Boston: Little, Brown, 1981.

Gesell, A. "The Development of Personality: Molding Your Child's Character." *The Delineator* (April 1924).

Goldsmith, H. H. and Gottesman, I. I. "Origins of Variation in Behavioral Style: A Longitudinal Study of Temperament in Twins." *Child Development* 52, no. 1 (March 1981): 91–103.

May, R. *Power and Innocence: A Search for the Sources of Violence*. New York: Dell, 1981.

McDevitt, S. C. and Carey, W. B. "The Measurements of Temperaments in 3–7 Year-old Children." *Journal of Child Psychology and Psychiatry* 19, no. 3 (July 1978): 245–253.

Plomin, R. "The Difficult Concept of Temperament: A Response to Thomas, Chess, and Korn." *Merrill-Palmer Quarterly* 28, no. 1 (January 1982): 25–33.

Powell, M. *Assessment and Management of Developmental Changes and Problems in Children*. St. Louis: Mosby, 1981.

Rothbart, M. K. "Measurement of Temperament in Infancy." *Child Development* 52, no. 2 (June 1981): 569–678.

Sheldon, W. H. *The Varieties of Human Physique*. New York: Harper & Row, 1940.

Sugarman, G. I. and Stone, M. N. *Your Hyperactive Child*. Chicago: Henry Regnery Co., 1974.

Thomas, A. and Chess, S. *Temperament and Development*. New York: New York University Press, 1968.

Thomas, A.; Chess, S.; and Birch, H. G. *Temperament and Behavior Disorders in Children*. New York: New York University Press, 1968.

Thomas, A.; Chess, S.; and Korn S. J. "The Reality of Difficult Temperament." *Merrill-Palmer Quarterly* 28, no. 1 (January 1982): 1–20.

Torgersen, A. M. and Kringlen, E. "Genetic Aspects of Temperamental Differences in Infants." *Journal of American Academy of Child Psychiatry* 17, no. 3 (Summer 1978): 433–444.

the MIRACLE OF RESILIENCY

DAVID GELMAN

A prominent child psychiatrist, E. James Anthony, once proposed this analogy: there are three dolls, one made of glass, the second of plastic, the third of steel. Struck with a hammer, the glass doll shatters; the plastic doll is scarred. But the steel doll proves invulnerable, reacting only with a metallic ping.

In life, no one is unbreakable. But child-health specialists know there are sharp differences in the way children bear up under stress. In the aftermath of divorce or physical abuse, for instance, some are apt to become nervous and withdrawn; some may be illness-prone and slow to develop. But there are also so-called resilient children who shrug off the hammer blows and go on to highly productive lives. The same small miracle of resiliency has been found under even the most harrowing conditions—in Cambodian refugee camps, in crack-ridden Chicago housing projects. Doctors repeatedly encounter the phenomenon: the one child in a large, benighted brood of five or six who seems able to take adversity in stride. "There are kids in families from very adverse situations who really do beautifully, and seem to rise to the top of their potential, even with everything else working against them," says Dr. W. Thomas Boyce, director of the division of behavioral and developmental pediatrics at the University of California, San Francisco. "Nothing touches them; they thrive no matter what."

Something, clearly, has gone right with these children, but what? Researchers habitually have come at the issue the other way around. The preponderance of the literature has to do with why children fail, fall ill, turn delinquent. Only recently, doctors realized they were neglecting the equally important question of why some children *don't* get sick. Instead of working backward from failure, they decided, there might be as much or more to be learned from studying the secrets of success. In the course of looking at such "risk factors" as poverty, physical impairment or abusive parents, they gradually became aware that there were also "protective factors" that served as buffers against the risks. If those could be identified, the reasoning went, they might help develop interventions that could change the destiny of more vulnerable children.

At the same time, the recognition that many children have these built-in defenses has plunged resiliency research into political controversy. "There is a danger among certain groups who advocate nonfederal involvement in assistance to children," says Duke University professor Neil Boothby, a child psychologist who has studied children in war zones. "They use it to blame people who don't move out of poverty. Internationally, the whole notion of resiliency has been used as an excuse not to do anything."

The quest to identify protective factors has produced an eager burst of studies in the past 10 or 15 years, with new publications tumbling off the presses every month. Although the studies so far offer no startling insights, they are providing fresh perspectives on how nature and nurture intertwine in childhood development. One of the prime protective factors, for example, is a matter of genetic luck of the draw: a child born with an easygoing disposition invariably handles stress better than one with a nervous, overreactive temperament. But even highly reactive children can acquire resilience if they have a consistent, stabilizing element in their young lives—something like an attentive parent or mentor.

The most dramatic evidence on that score comes not from humans but from their more

---

There are sharp differences in the way children bear up under stress

3. CHILDHOOD: Social and Emotional Development

researchable cousins, the apes. In one five-year-long study, primate researcher Stephen Suomi has shown that by putting infant monkeys in the care of supportive mothers, he could virtually turn their lives around. Suomi, who heads the Laboratory of Comparative Ethology at the National Institute of Child Health and Human Development, has been comparing "vulnerable" and "invulnerable" monkeys to see if there are useful nurturing approaches to be learned. Differences of temperament can be spotted in monkeys before they're a week old. Like their human counterparts, vulnerable monkey infants show measurable increases in heart rate and stress-hormone production in response to threat situations. "You see a fairly consistent pattern of physiological arousal, and also major behavioral differences," says Suomi. "Parallel patterns have been found in human-developmental labs, so we feel we're looking at the same phenomena."

Left alone in a regular troop, these high-strung infants grow up to be marginal figures in their troops. But by putting them in the care of particularly loving, attentive foster mothers within their first four days of life, Suomi turns the timid monkeys into social lions. Within two months, they become bold and outgoing. Males in the species Suomi has been working with normally leave their native troop at puberty and eventually work their way into a new troop. The nervous, vulnerable individuals usually are the last to leave home. But after being "cross-fostered" to loving mothers, they develop enough confidence so that they're first to leave.

Once on their own, monkeys have complicated (but somehow familiar) patterns of alliances. Their status often depends on whom they know and to whom they're related. In squabbles, they quickly generate support among friends and family members. The cross-fostered monkeys grow very adept at recruiting that kind of support. It's a knack they somehow get through interaction with their foster mothers, in which they evidently pick up coping styles as well as information. "It's essentially a social-learning phenomenon," says Suomi. "I would argue that's what's going on at the human level, too. Evidently, you can learn styles in addition to specific information."

In the long run, the vulnerable infants not only were turned around to normality, they often rose to the top of their hierarchies; they became community leaders. Boyce notes there are significant "commonalities" between Suomi's findings and studies of vulnerable children. "The implications are that vulnerable children, if placed in the right social environment, might become extraordinarily productive and competent adult individuals," he says.

Children, of course, can't be fostered off to new parents or social conditions as readily as monkeys. Most resiliency research is based on children who have not had such interventions in their lives. Nevertheless, some of the findings are revealing. One of the definitive studies was conducted by Emmy E. Werner, a professor of human development at the University of California, Davis, and Ruth S. Smith, a clinical psychologist on the Hawaiian island of Kauai. Together,

they followed 698 children, all descendants of Kauaiian plantation workers, from their birth (in 1955) up to their early 30s. About half the children grew up in poverty; one in six had physical or intellectual handicaps diagnosed between birth and age 2. Of the 225 designated as high risk, two thirds had developed serious learning or behavior problems within their first decade of life. By 18 they had delinquency records, mental-health problems or teenage pregnancies. "Yet one out of three," Werner and Smith noted, "grew into competent young adults who loved well, worked well, played well and expected well."

Some of the protective factors the two psychologists identified underscore the nature-nurture connection. Like other researchers, they found that children who started out with robust, sunny personalities were often twice lucky: not only were they better equipped to cope with life to begin with, but their winning ways made them immediately lovable. In effect, the "nicer" the children, the more readily they won affection—both nature and nurture smiled upon them. There were also other important resiliency factors, including self-esteem and a strong sense of identity. Boyce says he encounters some children who even at 2 or 3 have a sense of "presence" and independence that seem to prefigure success. "It's as if these kids have had the 'Who am I' questions answered for them," he says.

One of the more intriguing findings of the Kauai research was that resilient children were likely to have characteristics of both sexes. Boys and girls in the study tended to be outgoing and autonomous, in the male fashion, but also nurturant and emotionally sensitive, like females. "It's a little similar to what we find in creative children," observes Werner. Some other key factors were inherent in the children's surroundings rather than their personalities. It helped to have a readily available support network of grandparents, neighbors or relatives. Others note that for children anywhere, it doesn't hurt at all to be born to well-off parents. "The advantage of middle-class life is there's a safety net," says Arnold Sameroff, a developmental psychologist at Brown University's Bradley Hospital. "If you screw up, there's someone to bail you out."

In most cases, resilient children have "clusters" of protective factors, not just one or two. But the sine qua non, according to Werner, is a "basic, trusting relationship" with an adult. In all the clusters in the Kauai study, "there is not one that didn't include that one good relationship, whether with a parent, grandparent, older sibling, teacher or mentor—someone consistent enough in that person's life to say, 'You count,' and that sort of begins to radiate other support in their lives." Even children of abusive or schizophrenic parents may prove resilient if they have had at least one caring adult looking out for them—someone, as Tom Boyce says, "who serves as a kind of beacon presence in their lives."

Such relationships do the most good when they are lasting. There is no lasting guarantee for resiliency itself, which is subject to change, de-

Researchers can spot differences of temperament in monkeys before they're a week old

pending on what sort of ups and downs people encounter. Children's ability to cope often improves naturally as they develop and gain experience, although it may decline after a setback in school or at home. Werner notes that around half the vulnerable children in the Kauai study had shaken off their previous problems by the time they reached their late 20s or early 30s. "In the long-term view, more people come through in spite of circumstances. There is an amazing amount of recovery, if you don't focus on one particular time when things are falling apart."

Ironically, this "self-righting" tendency has made the resiliency issue something of a political football. Conservatives have seized on the research to bolster their case against further social spending. "It's the politics of 'It's all within the kid'," says Lisbeth Schorr, a lecturer in social medicine at Harvard Medical School whose book, "Within Our Reach: Breaking the Cycle of Disadvantage," has had a wide impact in the field. "The conservative argument against interventions like Operation Head Start and family-support programs is that if these inner-city kids and families just showed a little grit they would pull themselves up by their own bootstraps. But people working on resilience are aware that when it comes to environments like the inner city, it really doesn't make a lot of sense to talk about what's intrinsic to the kids, because the environment is so overwhelming."

So overwhelming, indeed, that some researchers voice serious doubts over how much change can be brought about in multiple-risk children. Brown's Sameroff, who has been dealing with poor inner-city black and white families in Rochester, N.Y., says the experience has left him "more realistic" about what is possible. "Interventions are important if we can target one or two things wrong with a child. So you provide psychotherapy or extra help in the classroom, then there's a lot better chance." But the children he deals with usually have much more than that going against them—not only poverty but large families, absent fathers, drug-ridden neighborhoods and so on. "We find the more risk factors the worse the outcome," says Sameroff. "With eight or nine, *nobody* does well. For the majority of these children, it's going to involve changing the whole circumstance in which they are raised."

Others are expressing their own reservations, as the first rush of enthusiasm in resiliency research cools somewhat. "A lot of the early intervention procedures that don't follow through have been oversold," says Emmy Werner. "Not everyone benefited equally from such programs as Head Start." Yet, according to child-development specialists, only a third of high-risk children are able to pull through relatively unaided by such interventions. Says Werner: "At least the high-risk children should be guaranteed basic health and social programs."

Interestingly, when Suomi separates his vulnerable monkeys from their foster mothers at 7 months—around the same time that mothers in the wild go off to breed, leaving their young behind—the genes reassert themselves, and the monkeys revert to fearful behavior. According to Suomi, they do recover again when the mothers return and their new coping skills seem to stay with them. Yet their experience underscores the frailty of change. Boyce, an admirer of Suomi's work, acknowledges that the question of how lasting the effects of early interventions are remains open. But, he adds, programs like Head Start continue to reverberate as much as 15 years later, with reportedly higher school-completion rates and lower rates of delinquency and teen pregnancies.

Boyce recalls that years ago, when he was at the University of North Carolina, he dealt with an 8-year-old child from an impoverished, rural black family, who had been abandoned by his mother. The boy also had "prune-belly syndrome," an anomaly of the abdominal musculature that left him with significant kidney and urinary problems, requiring extensive surgery. But he also had two doting grandparents who had raised him from infancy. They showered him with love and unfailingly accompanied him on his hospital visits. Despite his physical problems and loss of a mother, the boy managed to perform "superbly" in school. By the age of 10, when Boyce last saw him, he was "thriving."

Children may not be as manageable or resilient as laboratory monkeys. If anything, they are more susceptible in the early years. But with the right help at the right time, they can overcome almost anything. "Extreme adversity can have devastating effects on development," says psychologist Ann Masten, who did some of the groundbreaking work in the resiliency field with her University of Minnesota colleague Norman Garmezy. "But our species has an enormous capacity for recovery. Children living in a hostile caregiving environment have great difficulty, but a lot of ability to recover to better functioning if they're given a chance. That's a very important message from the resiliency literature." Unfortunately, the message may not be getting through to the people who can provide that chance.

There are kids from adverse situations who do beautifully and seem to rise to their potential

PROFILE
ROBERT J. STERNBERG

Three Heads are Better than One

THE TRIARCHIC THEORY SAYS WE ARE GOVERNED BY THREE ASPECTS OF INTELLIGENCE AND SUGGESTS WAYS OF MAXIMIZING STRENGTHS AND MINIMIZING WEAKNESSES.

ROBERT J. TROTTER

Robert J. Trotter is a senior editor at Psychology Today.

I really stunk on IQ tests. I was just terrible," recalls Robert J. Sternberg. "In elementary school I had severe test anxiety. I'd hear other people starting to turn the page, and I'd still be on the second item. I'd utterly freeze."

Poor performances on IQ tests piqued Sternberg's interest, and from rather inauspicious beginnings he proceeded to build a career on the study of intelligence and intelligence testing. Sternberg, IBM Professor of Psychology and Education at Yale University, did his undergraduate work at Yale and then got his Ph.D. from Stanford University in 1975. Since then he has written hundreds of articles and several books on intelligence, received numerous fellowships and awards for his research and proposed a three-part theory of intelligence. He is now developing an intelligence test based on that theory.

Running through Sternberg's work is a core of common-sense practicality not always seen in studies of subjects as intangible as intelligence. This practical bent, which stems from his early attempts to understand his own trouble with IQ tests, is also seen in his current efforts to devise ways of teaching people to better understand and increase their intellectual skills.

Sternberg got over his test anxiety in sixth grade after doing so poorly on an IQ test that he was sent to retake it with the fifth-graders. "When you are in elementary school," he explains, "one year makes a big difference. It's one thing to take a test with sixth-graders, but if you're taking it with a bunch of babies, you don't have to worry." He did well on the test, and by seventh grade he was designing and administering his own test of mental abilities as part of a science project. In 10th grade he studied how distractions affect people taking mental-ability tests.

After graduating from high school, he worked summers as a research assistant, first at the Psychological Corporation in New York, then at the Educational Testing Service in Princeton, New Jersey. These jobs gave him hands-on experience with testing organizations, but he began to suspect that the intelligence field was not going anywhere. Most of the tests being used were pretty old, he says, and there seemed to be little good research going on.

This idea was reinforced when Sternberg took a graduate course at Stanford from Lee. J. Cronbach, a leader in the field of tests and measurements. Intelligence research is dead, Cronbach said; the psychometric approach—IQ testing—has run its course and people are waiting for something new. This left Sternberg at a loss. He knew he wanted to study intelligence, but he didn't know how to go about it.

About this time, an educational publishing firm (Barron's) asked Sternberg to write a book on how to prepare for the Miller Analogies Test. Since Sternberg had invented a scheme for classifying the items on the test when he worked for the Psychological Corporation, which publishes the test, he was an obvious choice to write the book. Being an impecunious graduate student, he jumped at the chance, but he had an ulterior motive. He wanted to study intelligence and thought that because analogies are a major part of most IQ tests, working on the book might help. This work eventually led to his dissertation and a book based on it.

At this stage, Sternberg was analyzing the cognitive, or mental, processes people use to solve IQ test items, such as analogies, syllogisms and series. His research gave a good account of what people did in their heads, he says, and also seemed to account for individual differences in IQ test performance. Sternberg extended this work in the 1970s and in 1980 published a paper setting forth what he called his "componential" theory of human intelligence.

"I really thought I had the whole bag here," he says. "I thought I knew what was going on, but that was just a delusion on my part." Psychology comes out of everyday experiences, Sternberg says. And his own experiences—teaching and working with graduate students at Yale—gave him the idea that there was much more to intelligence than what his componential theory was describing. He brings this idea to life with stories of three idealized graduate students—Alice, Barbara and Celia.

Alice, he says, is someone who looked very smart according to conventional theories of intelligence. She had almost a 4.0 average as an undergraduate, scored extremely high on the Graduate Record Exam (GRE) and was supported by excellent letters of recommendation. She had everything that smart graduate students are supposed to have and was admitted to Yale as a top pick.

"There was no doubt that this was Miss Real Smarto," Sternberg says, and she performed just the way the tests predicted she would. She did extremely well on multiple-choice tests and was great in class, especially at critiquing other people's work and analyzing arguments. "She was just fantastic," Sternberg says. "She was one of our top two students the first year, but it didn't stay that way. She wasn't even in the top half by the time she finished. It just didn't work out. So that made me suspicious, and I wanted to know what went wrong."

The GRE and other tests had accurately predicted Alice's performance for the first year or so but then got progressively less predictive. And what became clear, Sternberg says, is that although the tests did measure her critical thinking ability, they did not measure her ability to come up with good ideas. This is not unusual, he says. A lot of people are very good analytically, but they just don't have good ideas of their own.

Sternberg thinks he knows why people with high GRE scores don't always do well in graduate school. From elementary school to college, he explains, students are continuously reinforced for high test-smarts. The first year of graduate school is similar—lots of multiple-choice tests and papers that demand critical thinking. Then around the second year there is a transition, with more emphasis on creative, or synthetic, thinking and having good ideas. "That's a different skill," Sternberg says. "It's not that test taking and critical thinking all of a sudden become unimportant, it's just that other things become more important."

When people who have always done well on tests get to this transition point, instead of being continually reinforced, they are only intermittently reinforced. And that is the kind of reinforcement most likely to sustain a particular type of behavior. "Instead of helping people try to improve their performance in other areas, intermittent reinforcement encourages them to overcapitalize on test-smarts, and they try to use that kind of intelligence in situations in which it is not relevant.

"The irony is that people like Alice may have other abilities, but they never look for them," he says. "It's like psychologists who come up with a theory that's interesting and then try to expand it to everything under the sun. They just can't see its limitations. It's the same with mental abilities. Some are good in certain situations but not in others."

The second student, Barbara, had a very different kind of record. Her undergraduate grades were not great, and her GRE scores were really low by Yale standards. She did, however, have absolutely superlative letters of recommendation that said Barbara was extremely creative, had really good ideas and did exceptional research. Sternberg thought Barbara would continue to do creative work and wanted to accept her. When he was outvoted, he hired her as a research associate. "Academic smarts," Sternberg says, "are easy to find, but creativity is a rare and precious commodity."

Sternberg's prediction was correct. In addition to working full time as a

PSYCHOLOGY COMES OUT OF EVERYDAY EXPERIENCES; WORKING WITH GRADUATE STUDENTS SUGGESTED THAT HIS THEORY OF INTELLIGENCE WAS INCOMPLETE.

3. CHILDHOOD: Cognitive and Language Development

research associate she took graduate classes, and her work and ideas proved to be just as good as the letters said they would be. When the transition came, she was ready to go. "Some of the most important work I've done was in collaboration with her," Sternberg says.

Barbaresque talent, Sternberg emphasizes, is not limited to psychology graduate school. "I think the same principle applies to everything. Take business. You can get an MBA based on your academic smarts because graduate programs consist mostly of taking tests and analyzing cases. But when you actually go into business, you have to have creative ideas for products and for marketing. Some MBA's don't make the transition and never do well because they overcapitalize on academic smarts. And it's the same no matter what you do. If you're in writing, you have to have good ideas for stories. If you're in art, you have to have good ideas for artwork. If you're in law.... That's where Barbaresque talent comes in."

The third student was Celia. Her grades, letters of recommendation and GRE scores were good but not great. She was accepted into the program and the first year, Sternberg says, she did all right but not great. Surprisingly, however, she turned out to be the easiest student to place in a good job. And this surprised him. Celia lacked Alice's super analytic ability and Barbara's super synthetic, or creative, ability, yet she could get a good job while others were having trouble.

Celia, it turns out, had learned how to play the game. She made sure she did the kind of work that is valued in psychology. She submitted her papers to the right journals. In other words, Sternberg says, "she was a streetsmart psychologist, very street-smart. And that, again, is something that doesn't show up on IQ tests."

Sternberg points out that Alice, Barbara and Celia are not extreme cases. "Extremes are rare," he says, "but not good. You don't want someone who is incredibly analytically brilliant but never has a good idea or who is a total social boor." Like all of us, Alice, Barbara and Celia each had all three of the intellectual abilities he described, but each was especially good in one aspect.

After considering the special qualities of people such as Alice, Barbara and Celia, Sternberg concluded that his componential theory explained only one aspect of intelligence. It could account for Alice, but it was too narrow to explain Barbara and Celia. In an attempt to find out why, Sternberg began to look at prior theories of intelligence and found that they tried to do one of three things:

Some looked at the relation of intelligence to the internal world of the individual, what goes on inside people's heads when they think intelligently. "That's what IQ tests measure, that's what information processing tasks measure, that's the componential theory. It's what I had been doing," Sternberg says. "I'd take an IQ test problem and analyze the mental processes involved in solving it, but it's still the same damned problem. It's sort of like we never got away from the IQ test as a standard. It's not that I thought the componential work was wrong. It told me a lot about what made Alice smart, but there had to be more."

Other theories looked at the relation of intelligence to experience, with experience mediating between what's inside—the internal, mental world—and what's outside—the external world. These theories say you have to look at how experience affects a person's intelligence and how intelligence affects a person's experiences. In other words, more-intelligent people create different experiences. "And that," says Sternberg, "is where Barbara fits in. She is someone who has a certain way of coping with novelty that goes beyond the ordinary. She can see old problems in new ways, or she'll take a new problem and see how some old thing she knows applies to it."

A third kind of theory looks at intelligence in relation to the individual's external world. In other words, what makes people smart in their everyday context? How does the environment interact with being smart? And what you see, as with Celia, is that there are a lot of people who don't do particularly well on tests but who are just extremely practically intelligent. "Take Lee Iacocca," Sternberg says. "Maybe he doesn't have an IQ of 160 (or maybe he does, I don't know), but he is extremely effective. And there are plenty of people that are that way. And there are plenty of people going

Academic smarts are easy to find, but creativity is rare and precious.

around with high IQ's who don't do a damned thing. This Celiaesque kind of smartness—how you make it in the real world—is not reflected in IQ tests. So I decided to have a look at all three kinds of intelligence."

He did, and the result was the triarchic theory. A triarchy is government by three persons, and in his 1985 book, *Beyond IQ*, Sternberg suggests that we are all governed by three aspects of intelligence: componential, experiential and contextual. In the book, each aspect of intelligence is described in a subtheory. Though based in part on older theories, Sternberg's work differs from those theories in a number of ways. His componential subtheory, which describes Alice, for example, is closest to the views of cognitive psychologists and psychometricians. But Sternberg thinks that the other theories put too much emphasis on measuring speed and accuracy of performance components at the expense of what he calls "metacomponents," or executive processes.

"For example," he explains, "the really interesting part of solving analogies or syllogisms is deciding what to do in the first place. But that isn't isolated by looking at performance components, so I realized you need to look at metacomponents—how you plan it, how you monitor what you are doing, how you evaluate it after you are done. [See "Stalking the IQ Quark," *Psychology Today*, September 1979.]

"A big thing in psychometric theory," he continues, "is mental speed. Almost every group test is timed, so if you're not fast you're in trouble. But I came to the conclusion that we were really misguided on that. Almost ev-

18. Three Heads Are Better Than One

THE TRIARCHIC THEORY

Componential

Alice had high test scores and was a whiz at test-taking and analytical thinking. Her type of intelligence exemplifies the componential subtheory, which explains the mental components involved in analytical thinking.

Experiential

Barbara didn't have the best test scores, but she was a superbly creative thinker who could combine disparate experiences in insightful ways. She is an example of the experiential subtheory.

Contextual

Celia was street-smart. She learned how to play the game and how to manipulate the environment. Her test scores weren't tops, but she could come out on top in almost any context. She is Sternberg's example of contextual intelligence.

ILLUSTRATIONS BY JEAN TUTTLE

A BIG THING IN IQ TESTING IS SPEED, BUT ALMOST EVERYONE REGRETS SOME DECISION THAT WAS MADE TOO FAST.

eryone regrets some decision that was made too fast. Think of the guy who walks around with President Reagan carrying the black box. You don't want this guy to be real fast at pushing the button. So, instead of just testing speed, you want to measure a person's knowing when to be fast and when to be slow—time allocation—it's a metacomponent. And that's what the componential subtheory emphasizes."

The experiential subtheory, which describes Barbaresque talent, emphasizes insight. Sternberg and graduate student Janet E. Davidson, as part of a study of intellectual giftedness, concluded that what gifted people had in common was insight. "If you look at Hemingway in literature, Darwin in science or Rousseau in political theory, you see that they all seemed to be unusually insightful people," Sternberg explains. "But when we looked at the research, we found that nobody seemed to know what insight is."

Sternberg and Davidson analyzed how several major scientific insights came about and concluded that insight is really three things: selective encoding, selective combination and selective comparison. As an example of selective encoding they cite Sir Alexander Fleming's discovery of penicillin. One of Fleming's experiments was spoiled when mold contaminated and killed the bacteria he was studying. Sternberg says most people would have said, "I screwed up, I've got to throw this out and start over." But Fleming didn't. He realized that the mold that killed the bacteria was more important than the bacteria. This selective encoding insight—the ability to focus on the really critical information—led to the discovery of a substance in the mold that Fleming called "penicillin." "And this is not just something that famous scientists do," Sternberg explains. "Detectives have to decide what are the relevant clues, lawyers have to decide which facts have legal consequences and so on."

The second kind of insight is selective combination, which is putting the facts together to get the big picture, as in Charles Darwin's formulation of the theory of natural selection. The facts he needed to form the theory were already there; other people had them too. But Darwin saw how to put them together. Similarly, doctors have to put the symptoms together to figure out what the disease is. Lawyers have to put the facts together to figure out how to make the case. "My triarchic theory is another example of selective combination. It doesn't have that much in it that's different from what other people have said," Stern-

85

berg admits. "It's just putting it together that's a little different."

A third kind of insight is selective comparison. It's relating the old to the new analogically, says Sternberg. It involves being able to see an old thing in a new way or being able to see a new thing in an old way. An example is the discovery of the molecular structure of benzene by German chemist August Kekule, who had been struggling to find the structure for some time. Then one night he had a dream in which a snake was dancing around and biting its own tail. Kekule woke up and realized that he had solved the puzzle of benzene's structure. In essence, Sternberg explains, Kekule could see the relation between two very disparate elements—the circular image of the dancing snake and the hexagonal structure of the benzene molecule.

Sternberg and Davidson tested their theory of insight on fourth-, fifth- and sixth-graders who had been identified through IQ and creativity tests as either gifted or not so gifted. They used problems that require the three different kinds of insights. A selective-encoding problem, for example, is the old one about four brown socks and five blue socks in a drawer. How many do you have to pull out to make sure you'll having a matching pair? It's a selective-encoding problem because the solution depends on selecting and using the relevant information. (The information about the 4-to-5 ratio is irrelevant.)

As expected, the gifted children were better able to solve all three types of problems. The less gifted children, for example, tended to get hung up on the irrelevant ratio information in the socks problem, while the gifted children ignored it. When the researchers gave the less gifted children the information needed to solve the problems (by underlining what was relevant, for example), their performance improved significantly. Giving the gifted children this information had no such effect, Sternberg explains, because they tended to have the insights spontaneously.

Sternberg and Davidson also found that insight skills can be taught. In a five-week training program for both gifted and less gifted children, they greatly improved children's scores on insight problems, compared with children who had not received the training. Moreover, says Sternberg, the gains were durable and transferable. The skills were still there when the children were tested a year later and were being applied to kinds of insight problems that had never appeared in the training program.

Sternberg's contextual subtheory emphasizes adaptation. Almost everyone agrees that intelligence is the ability to adapt to the environment, but that doesn't seem to be what IQ tests measure, Sternberg says. So he and Richard K. Wagner, then a graduate student, now at Florida State University, tried to come up with a test of adaptive ability. They studied people in two occupations: academic psychologists, "because we think that's a really important job," and business executives, "because everyone else thinks that's an important job." They began by asking prominent, successful people what one needs to be practically intelligent in their fields. The psychologists and executives agreed on three things:

First, IQ isn't very important for success in these jobs. "And that makes sense because you already have a restricted range. You're talking about people with IQ's of 110 to 150. That's not to say that IQ doesn't count for anything," Sternberg says. "If you were talking about a range from 40 to 150, IQ might make a difference, but we're not. So IQ isn't that important with regard to practical intelligence."

They also agreed that graduate school isn't that important either. "This," says Sternberg, "was a little offensive. After all, here I was teaching and doing the study with one of my own graduate students, and these people were saying graduate training wasn't that helpful." But Sternberg remembered that graduate school had not fully prepared him for his first year on the job as an academic. "I really needed to know how to write a grant proposal; at Yale, if you can't get grants you're in trouble. You have to scrounge for paper clips, you can't get students to work with you, you can't get any research done. Five years later you get fired because you haven't done anything. Now, no one ever says you are being hired to write grants, but if you don't get them you're dead meat around here." Sternberg, who has had more than $5 million in grants in the past 10 years, says he'd be five years behind where he is now without great graduate students.

"What you need to know to be practically intelligent, to get on in an environment," Sternberg says, is tacit knowledge, the third area of agreement. "It's implied or indicated but not always expressed, or taught." Sternberg and Wagner constructed a test of such knowledge and gave it to senior and junior business executives and to senior and junior psychology professors. The results suggest that tacit knowledge is a result of learning from experience. It is not related to IQ but is related to success in the real world. Psychologists who scored high on the test, compared with those who had done poorly, had published more research, presented more papers at conventions and tended to be at the better universities. Business executives who scored high had better salaries, more merit raises and better performance ratings than those who scored low.

The tacit-knowledge test is a measure of how well people adapt to their environment, but practical knowledge also means knowing when not to adapt. "Suppose you join a computer software firm because you really want to work on educational software," Sternberg says, "but they put you in the firm's industrial espionage section and ask you to spy on Apple Computer. There are times when you have to select another environment, when you have to say 'It's time to quit. I don't want to adapt, I'm leaving.'"

There are, however, times when you can't quit and must stay put. In such

> THERE ARE PLENTY OF PEOPLE GOING AROUND WITH HIGH IQ'S WHO DON'T DO A DAMNED THING.

18. Three Heads Are Better Than One

IT'S REALLY IMPORTANT TO ME THAT MY WORK HAS AN EFFECT THAT GOES BEYOND THE PSYCHOLOGY JOURNALS, TO BRING INTELLIGENCE INTO THE REAL WORLD AND THE REAL WORLD INTO INTELLIGENCE.

situations, you can try to change the environment. That, says Sternberg, is the final aspect of contextual, or practical, intelligence—shaping the environment to suit your needs.

One way to do this is by capitalizing on your intellectual strengths and compensating for your weaknesses. "I don't think I'm at the top of the heap analytically," Sternberg explains. "I'm good, not the greatest, but I think I know what I'm good at and I try to make the most of it. And there are some things I stink at and I either try to make them unimportant or I find other people to do them. That's part of how I shape my environment. And that's what I think practical intelligence is about—capitalizing on your strengths and minimizing your weaknesses. It's sort of mental self-management.

"So basically what I've said is there are different ways to be smart, but ultimately what you want to do is take the components (Alice intelligence), apply them to your experience (Barbara) and use them to adapt to, select and shape your environment (Celia). That is the triarchic theory of intelligence."

What can you do with a new theory of intelligence? Sternberg, who seems to have a three-part answer for every question (and whose triangular theory of love will be the subject of a future *Psychology Today* article), says, "I view the situation as a triangle." The most important leg of the triangle, he says, is theory and research. "But it's not enough for me to spend my life coming up with theories," he says. "So I've gone in two further directions, the other two legs of the triangle—testing and training."

He is developing, with the Psychological Corporation, now in San Antonio, Texas, the Sternberg Multidimensional Abilities Test. It is based strictly on the triarchic theory and will measure intelligence in a much broader way than traditional IQ tests do. "Rather than giving you a number that's etched in stone," he says, "this test will be used as a basis for diagnosing your intellectual strengths and weaknesses."

Once you understand the kind of intelligence you have, the third leg of the triangle—the training of intellectual skills—comes into play. One of Sternberg's most recent books, *Intelligence Applied*, is a training program based on the theory. It is designed to help people capitalize on their strengths and improve where they are weak. "I'm very committed to all three aspects," Sternberg says. "It's really important to me that my work has an effect that goes beyond the psychology journals. I really think it's important to bring intelligence into the real world and the real world into intelligence."

How Kids Learn

Ages 5 through 8 are wonder years. That's when children begin learning to study, to reason, to cooperate. We can put them in desks and drill them all day. Or we can keep them moving, touching, exploring. The experts favor a hands-on approach, but changing the way schools teach isn't easy. The stakes are high and parents can help.

BARBARA KANTROWITZ & PAT WINGERT

With Howard Manly in Atlanta and bureau reports

It's time for number games in Janet Gill's kindergarten class at the Greenbrook School in South Brunswick, N.J. With hardly any prodding from their teacher, 23 five- and six-year-olds pull out geometric puzzles, playing cards and counting equipment from the shelves lining the room. At one round table, a group of youngsters fits together brightly colored wooden shapes. One little girl forms a hexagon out of triangles. The others, obviously impressed, gather round to count up how many parts are needed to make the whole.

After about half an hour, the children get ready for story time. They pack up their counting equipment and settle in a circle around Gill. She holds up a giant book about a zany character called Mrs. Wishy-washy who insists on giving farm animals a bath. The children recite the whimsical lines along with Gill, obviously enjoying one of their favorite tales. (The hallway is lined with drawings depicting the children's own interpretation of the book; they've taken a few literary liberties, like substituting unicorns and dinosaurs for cows and pigs.) After the first reading, Gill asks for volunteers to act out the various parts in the book. Lots of hands shoot up. Gill picks out four children and they play their parts enthusiastically. There isn't a bored face in the room.

This isn't reading, writing and arithmetic the way most people remember it. Like a growing number of public- and private-school educators, the principals and teachers in South Brunswick believe that children between the ages of 5 and 8 have to be taught differently from older children. They recognize that young children learn best through active, hands-on teaching methods like games and dramatic play. They know that children in this age group develop at varying rates and schools have to allow for these differences. They also believe that youngsters' social growth is as essential as their academic achievement. Says Joan Warren, a teacher consultant in South Brunswick: "Our programs are designed to fit the child instead of making the child fit the school."

Educators call this kind of teaching "developmentally appropriate practice"—a curriculum based on what scientists know about how young children learn. These ideas have been slowly emerging through research conducted over the last century, particularly in the past 30 years. Some of the tenets have appeared

The Lives and Times of Children

Each youngster proceeds at his own pace, but the learning curve of a child is fairly predictable. Their drive to learn is awesome, and careful adults can nourish it. The biggest mistake is pushing a child too hard, too soon.

● Infants and Toddlers
They're born to learn. The first important lesson is trust, and they learn that from their relationships with their parents or other caring adults. Later, babies will begin to explore the world around them and experiment with independence. As they mature, infants slowly develop gross motor (sitting, crawling, walking) and fine motor (picking up tiny objects) skills. Generally, they remain egocentric and are unable to share or wait their turn. New skills are perfected through repetition, such as the babbling that leads to speaking.

■ 18 months to 3 years
Usually toilet training becomes the prime learning activity. Children tend to concentrate on language development and large-muscle control through activities like climbing on jungle gyms. Attention spans lengthen enough to listen to uncomplicated stories and carry on conversations. Vocabulary expands to about 200 words. They enjoy playing with one other child, or a small group, for short periods, and learn that others have feelings too. They continue to look to parents for encouragement and protection, while beginning to accept limits on their behavior.

▲ 3-year-olds
Generally, they're interested in doing things for themselves and trying to keep up with older children. Their ability to quietly listen to stories and music remains limited. They begin telling stories and jokes. Physical growth slows, but large-muscle development continues as children run, jump and ride tricycles. They begin to deal with cause and effect; it's time to plant seeds and watch them grow.

● 4-year-olds
They develop better small motor skills, such as cutting with scissors, painting, working with puzzles and building things.
They can master colors, sizes and shapes. They should be read to and should be encouraged to watch others write; let them scribble on paper but try to keep them away from walls.

■ 5-year-olds
They begin to understand counting as a one-to-one correlation. Improved memories make it easier for them to recognize meaningful words, and with sharper fine motor skills, some children will be able to write their own names.

▲ Both 4s and 5s
Both groups learn best by interacting with people and concrete objects and by trying to solve real problems. They can learn from stories and books, but only in ways that relate to their own experience. Socially, these children are increasingly interested in activities outside their immediate family. They can play in groups for longer periods, learning lessons in cooperation and negotiation. Physically, large-muscle development continues, and skills such as balancing emerge.

● 6-year-olds
Interest in their peers continues to increase, and they become acutely aware of comparisons between themselves and others. It's a taste of adolescence: does the group accept them? Speech is usually well developed, and children are able to joke and tease. They have a strong sense of true and false and are eager for clear rules and definitions. However, they have a difficult time differentiating between minor and major infractions. Generally, children this age are more mature mentally than physically and unable to sit still for long periods. They learn better by firsthand experiences. Learning by doing also encourages children's "disposition" to use the knowledge and skills they're acquiring.

■ 7- to 8-year-olds
During this period, children begin developing the ability to think about and solve problems in their heads, but some will continue to rely on fingers and toes to help them find the right answer. Not until they're 11 are most kids capable of thinking purely symbolically; they still use real objects to give the symbols—such as numbers—meaning. At this stage they listen better and engage in give and take. Generally, physical growth continues to slow, while athletic abilities improve—children are able to hit a softball, skip rope or balance on a beam. Sitting for long periods is still more tiring than running and jumping.

---

under other names—progressivism in the 1920s, open education in the 1970s. But they've never been the norm. Now, educators say that may be about to change. "The entire early-childhood profession has amassed itself in unison behind these principles," says Yale education professor Sharon Lynn Kagan. In the last few years, many of the major education organizations in the country—including the National Association for the Education of Young Children and the National Association of State Boards of Education—have endorsed remarkably similar plans for revamping kindergarten through third grade.

3. CHILDHOOD: Cognitive and Language Development

Bolstered by opinions from the experts, individual states are beginning to take action. Both California and New York have appointed task forces to recommend changes for the earliest grades. And scores of individual school districts like South Brunswick, figuring that young minds are a terrible thing to waste, are pushing ahead on their own.

The evidence gathered from research in child development is so compelling that even groups like the Council for Basic Education, for years a major supporter of the traditional format, have revised their thinking. "The idea of putting small children in front of workbooks and asking them to sit at their desks all day is a nightmare vision," says Patte Barth, associate editor of Basic Education, the council's newsletter.

At this point, there's no way of knowing how soon change will come or how widespread it will be. However, there's a growing recognition of the importance of the early grades. For the past few years, most of the public's attention has focused on older children, especially teenagers. "That's a Band-Aid kind of approach," says Anne Dillman, a member of the New Jersey State Board of Education. "When the product doesn't come out right, you try and fix it at the end. But we really have to start at the beginning." Demographics have contributed to the sense of urgency. The baby boomlet has replaced the baby-bust generation of the 1970s. More kids in elementary school means more parents asking if there's a better way to teach. And researchers say there is a better way. "We've made remarkable breakthroughs in understanding the development of children, the development of learning and the climate that enhances that," says Ernest Boyer of The Carnegie Foundation for the Advancement of Teaching. But, he adds, too often, "what we know in theory and what we're doing in the classroom are very different."

The early grades pose special challenges because that's when children's attitudes toward school and learning are shaped, says Tufts University psychologist David Elkind. As youngsters move from home or preschool into the larger, more competitive world of elementary school, they begin to make judgments about their own abilities. If they feel inadequate, they may give up. Intellectually, they're also in transition, moving from the intensely physical exploration habits of infancy and toddlerhood to more abstract reasoning. Children are born wanting to learn. A baby can spend hours studying his hands; a toddler is fascinated by watching sand pour through a sieve. What looks like play to an adult is actually the work of childhood, developing an understanding of the world. Studies show that the most effective way to teach young kids is to capitalize on their natural inclination to learn through play.

But in the 1980s, many schools have tried to do just the opposite, pressure instead of challenge. The "back to basics" movement meant that teaching methods intended for high school students were imposed on first graders. The lesson of the day was more: more homework, more tests, more discipline. Children should be behind their desks, not roaming around the room. Teachers should be at the head of the classrooms, drilling knowledge into their charges. Much of this was a reaction against the trend toward open education in the '70s. Based on the British system, it allowed children to develop at their own pace within a highly structured classroom. But too many teachers and principals who tried open education thought that it meant simply tearing down classroom walls and letting children do whatever they wanted. The results were often disastrous. "Because it was done wrong, there was a backlash against it," says Sue Bredekamp of the National Association for the Education of Young Children.

At the same time, parents, too, were demanding more from their elementary schools. By the mid-1980s, the majority of 3- and 4-year-olds were attending some form of pre-school. And their parents expected these classroom veterans to be reading by the second semester of kindergarten. But the truth is that many 5-year-olds aren't ready for reading—or most of the other academic tasks that come easily to older children—no matter how many years of school they've completed. "We're confusing the numbers of years children have been in school with brain development," says Martha Denckla, a professor of neurology and pediatrics at Johns Hopkins University. "Just because a child goes to day care at age 3 doesn't mean the human brain mutates into an older brain. A 5-year-old's brain is still a 5-year-old's brain."

As part of the return to basics, parents and districts demanded hard evidence that their children were learning. And some communities took extreme measures. In 1985 Georgia became the first state to require 6-year-olds to pass a standardized test before entering first grade. More than two dozen other states proposed similar legislation. In the beginning Georgia's move was hailed as a "pioneering" effort to get kids off to a good start. Instead, concedes state school superintendent Werner Rogers, "We got off on the wrong foot." Five-year-olds who used to spend their days finger-painting or singing were hunched over ditto sheets, preparing for the big exam. "We would have to spend a month just teaching kids how to take the test," says Beth Hunnings, a kindergarten teacher in suburban Atlanta. This year Georgia altered the tests in favor of a more flexible evaluation; other states have changed their minds as well.

The intense, early pressure has taken an early toll. Kindergartners are struggling with homework. First graders are taking spelling tests before they even understand how to read. Second graders feel like failures. "During this critical period," says David Elkind in his book "Miseducation," "the child's bud-

19. How Kids Learn

In Japan, First Grade Isn't a Boot Camp

Japanese students have the highest math and science test scores in the world. More than 90 percent graduate from high school. Illiteracy is virtually nonexistent in Japan. Most Americans attribute this success to a rigid system that sets youngsters on a lock-step march from cradle to college. In fact, the early years of Japanese schooling are anything but a boot camp; the atmosphere is warm and nurturing. From kindergarten through third grade, the goal is not only academic but also social—teaching kids to be part of a group so they can be good citizens as well as good students. "Getting along with others is not just a means for keeping the peace in the classroom but something which is a valued end in itself," says American researcher Merry White, author of "The Japanese Educational Challenge."

Lessons in living and working together grow naturally out of the Japanese culture. Starting in kindergarten, youngsters learn to work in teams, with brighter students often helping slower ones. All children are told they can succeed if they persist and work hard. Japanese teachers are expected to be extremely patient with young children. They go over lessons step by step and repeat instructions as often as necessary. "The key is not to scold [children] for small mistakes," says Yukio Ueda, principal of Mita Elementary School in Tokyo. Instead, he says, teachers concentrate on praising and encouraging their young charges.

As a result, the classrooms are relaxed and cheerful, even when they're filled with rows of desks. On one recent afternoon a class of second graders at Ueda's school was working on an art project. Their assignment was to build a roof with poles made of rolled-up newspapers. The children worked in small groups, occasionally asking their teacher for help. The room was filled with the sound of eager youngsters chatting about how to get the job done. In another second-grade class, the subject was math. Maniko Inoue, the teacher, suggested a number game to practice multiplication. After a few minutes of playing it, one boy stood up and proposed changing the rules just a bit to make it more fun. Inoue listened carefully and then asked if the other students agreed. They cheered, "Yes, yes," and the game continued according to the new rules.

Academics are far from neglected in the early grades. The Education Ministry sets curriculum standards and goals for each school year. For example, third graders by the end of the year are supposed to be able to read and write 508 characters (out of some 2,000 considered essential to basic literacy). Teachers have time for play and lessons: Japanese children attend school for 240 days, compared with about 180 in the United States.

Mothers' role: Not all the teaching goes on in the classroom. Parents, especially mothers, play a key role in education. Although most kindergartens do not teach writing or numbers in any systematic way, more than 80 percent of Japanese children learn to read or write to some extent before they enter school. "It is as if mothers had their own built-in curriculum," says Shigefumi Nagano, a director of the National Institute for Educational Research. "The first game they teach is to count numbers up to 10."

For all their success in the early grades, the Japanese are worried they're not doing well enough. After a recent national curriculum review, officials were alarmed by what Education Minister Takeo Nishioka described as excessive "bullying and misconduct" among children—the result, according to some Japanese, of too much emphasis on material values. So three years from now, first and second graders will no longer be studying social studies and science. Instead, children will spend more time learning how to be good citizens. That's "back to basics"—Japanese style.

BARBARA KANTROWITZ *with* HIDEKO TAKAYAMA *in Tokyo*

---

ding sense of competence is frequently under attack, not only from inappropriate instructional practices . . . but also from the hundred and one feelings of hurt, frustration and rejection that mark a child's entrance into the world of schooling, competition and peer-group involvement." Adults under similar stress can rationalize setbacks or put them in perspective based on previous experiences; young children have none of these defenses. Schools that demand too much too soon are setting kids off on the road to failure.

It doesn't have to be this way. Most experts on child development and early-childhood education believe that young children learn much more readily if the teaching methods meet their special needs:

Differences in thinking: The most important ingredient of the nontraditional approach is hands-on learning. Research begun by Swiss psychologist Jean Piaget indicates that somewhere between the ages of 6 and 9, children begin to think abstractly instead of concretely. Younger children learn much more by touching and seeing and smelling and tasting than by just listening. In other words, 6-year-olds can easily understand addition and subtraction if they have actual objects to count instead of a series of numbers written on a blackboard. Lectures don't help. Kids learn to reason and communicate by engaging in conversation. Yet most teachers still talk at, not with, their pupils.

Physical activity: When they get to be 10 or 11, children can sit still for sustained periods. But until they are physically ready for long periods of inactivity, they need to be active in the classroom. "A young child has to make a conscious effort to sit still," says Denckla. "A large chunk of children can't do it for very long. It's a very energy-consuming activity for them." Small children actually get more tired if they have to sit still and listen to a teacher talk than if they're allowed to move around in the classroom. The frontal lobe, the part of the brain that applies the brakes to children's natural energy and curiosity, is still immature in 6- to 9-year-olds, Denckla says. As the lobe develops, so

3. CHILDHOOD: Cognitive and Language Development

does what Denckla describes as "boredom tolerance." Simply put, learning by doing is much less boring to young children.

Language development: In this age group, experts say language development should not be broken down into isolated skills—reading, writing and speaking. Children first learn to reason and to express themselves by talking. They can dictate stories to a teacher before they actually read or write. Later, their first attempts at composition do not need to be letter perfect; the important thing is that they learn to communicate ideas. But in many classrooms, grammar and spelling have become more important than content. While mastering the technical aspects of writing is essential as a child gets older, educators warn against emphasizing form over content in the early grades. Books should also be interesting to kids—not just words strung together solely for the purpose of pedagogy. Psychologist Katherine Nelson of the City University of New York says that her extensive laboratory and observational work indicates that kids can learn language—speaking, writing or reading—only if it is presented in a way that makes sense to them. But many teachers still use texts that are so boring they'd put anybody to sleep.

Socialization: A youngster's social development has a profound effect on his academic progress. Kids who have trouble getting along with their classmates can end up behind academically as well and have a higher incidence of dropping out. In the early grades especially, experts say youngsters should be encouraged to work in groups rather than individually so that teachers can spot children who may be having problems making friends. "When children work on a project," says University of Illinois education professor Lillian Katz, "they learn to work together, to disagree, to speculate, to take turns and de-escalate tensions. These skills can't be learned through lecture. We all know people who have wonderful technical skills but don't have any social skills. Relationships should be the first 'R'."

Feelings of competence and self-esteem: At this age, children are also learning to judge themselves in relation to others. For most children, school marks the first time that their goals are not set by an internal clock but by the outside world. Just as the 1-year-old struggles to walk, 6-year-olds are struggling to meet adult expectations. Young kids don't know how to distinguish between effort and ability, says Tynette Hills, coordinator of early-childhood education for the state of New Jersey. If they try hard to do something and fail, they may conclude that they will never be able to accomplish a particular task. The effects of obvious methods of comparison, such as posting grades, can be serious. Says Hills: "A child who has had his confidence really damaged needs a rescue operation."

Rates of growth: Between the ages of 5 and 9, there's a wide range of development for children of normal intelligence. "What's appropriate for one child may not be appropriate for another," says Dr. Perry Dyke, a member of the California State Board of Education. "We've got to have the teachers and the staff reach children at whatever level they may be at . . . That takes very sophisticated teaching." A child's pace is almost impossible to predict beforehand. Some kids learn to read on their own by kindergarten; others are still struggling to decode words two or three years later. But by the beginning of the fourth grade, children with very different histories often read on the same level. Sometimes, there's a sudden "spurt" of learning, much like a growth spurt, and a child who has been behind all year will catch up in just a few weeks. Ernest Boyer and others think that multigrade classrooms, where two or three grades are mixed, are a good solution to this problem—and a way to avoid the "tracking" that can hurt a child's self-esteem. In an ungraded classroom, for example, an older child who is having problems in a particular area can practice by tutoring younger kids.

Putting these principles into practice has never been easy. Forty years ago Milwaukee abolished report cards and started sending home ungraded evaluations for kindergarten through third grade. "If anything was developmentally appropriate, those ungraded classes were," says Millie Hoffman, a curriculum specialist with the Milwaukee schools. When the back-to-basics movement geared up nationally in the early 1980s, the city bowed to pressure. Parents started demanding letter grades on report cards. A traditional, direct-teaching approach was introduced into the school system after some students began getting low scores on standardized tests. The school board ordered basal readers with controlled vocabularies and contrived stories. Milwaukee kindergarten teachers were so up-

---

A Primer for Parents

When visiting a school, trust your eyes. What you see is what your child is going to get.

● Teachers should talk to small groups of children or individual youngsters; they shouldn't just lecture.

■ Children should be working on projects, active experiments and play; they shouldn't be at their desks all day filling in workbooks.

▲ Children should be dictating and writing their own stories or reading real books.

● The classroom layout should have reading and art areas and space for children to work in groups.

■ Children should create freehand artwork, not just color or paste together adult drawings.

▲ Most importantly, watch the children's faces. Are they intellectually engaged, eager and happy? If they look bored or scared, they probably are.

set by these changes that they convinced the board that their students didn't need most of the standardized tests and the workbooks that go along with the readers.

Some schools have been able to keep the progressive format. Olive School in Arlington Heights, Ill., has had a nontraditional curriculum for 22 years. "We've been able to do it because parents are involved, the teachers really care and the children do well," says principal Mary Stitt. "We feel confident that we know what's best for kids." Teachers say they spend a lot of time educating parents about the teaching methods. "Parents always think school should be the way it was for them," says first-grade teacher Cathy Sauer. "As if everything else can change and progress but education is supposed to stay the same. I find that parents want their children to like school, to get along with other children and to be good thinkers. When they see that happening, they become convinced."

Parental involvement is especially important when schools switch from a traditional to a new format. Four years ago, Anne Norford, principal of the Brownsville Elementary School in Albemarle County, Va., began to convert her school. Parents volunteer regularly and that helps. But the transition has not been completely smooth. Several teachers refused to switch over to the more active format. Most of them have since left the school, Norford says. There's no question that some teachers have trouble implementing the developmentally appropriate approach. "Our teachers are not all trained for it," says Yale's Kagan. "It takes a lot of savvy and skill." A successful child-centered classroom seems to function effortlessly as youngsters move from activity to activity. But there's a lot of planning behind it—and that's the responsibility of the individual teacher. "One of the biggest problems," says Norford, "is trying to come up with a program that every teacher can do—not just the cadre of single people who are willing to work 90 hours a week." Teachers also have to participate actively in classroom activities and give up the automatic mantle of authority that comes from standing at the blackboard.

Teachers do better when they're involved in the planning and decision making. When the South Brunswick, N.J., schools decided in the early 1980s to change to a new format, the district spent several years studying a variety of curricula. Teachers participated in that research. A laboratory school was set up in the summer so that teachers could test materials. "We had the support of the teachers because teachers were part of the process," says teacher consultant Joan Warren.

One residue of the back-to-basics movement is the demand for accountability. Children who are taught in nontraditional classrooms can score slightly lower on commonly used standardized tests. That's because most current tests are geared to the old ways. Children are usually quizzed on specific skills, such as vocabulary or addition, not on the concepts behind those skills. "The standardized tests usually call for one-word answers," says Carolyn Topping, principal of Mesa Elementary School in Boulder, Colo. "There may be three words in a row, two of which are misspelled and the child is asked to circle the correctly spelled word. But the tests never ask, 'Does the child know how to write a paragraph?'"

Even if the tests were revised to reflect different kinds of knowledge, there are serious questions about the reliability of tests on young children. The results can vary widely, depending on many factors—a child's mood, his ability to manipulate a pencil (a difficult skill for many kids), his reaction to the person administering the test. "I'm appalled at all the testing we're doing of small children," says Vanderbilt University professor Chester Finn, a former assistant secretary of education under the Reagan administration. He favors regular informal reviews and teacher evaluations to make sure a student understands an idea before moving on to the next level of difficulty.

Tests are the simplest method of judging the effectiveness of a classroom—if not always the most accurate. But there are other ways to tell if children are learning. If youngsters are excited by what they are doing, they're probably laughing and talking to one another and to their teacher. That communication is part of the learning process. "People think that school has to be either free play or all worksheets," says Illinois professor Katz. "The truth is that neither is enough. There has to be a balance between spontaneous play and teacher-directed work." And, she adds, "you have to have the other component. Your class has to have intellectual life."

Katz, author of "Engaging Children's Minds," describes two different elementary-school classes she visited recently. In one, children spent the entire morning making identical pictures of traffic lights. There was no attempt to relate the pictures to anything else the class was doing. In the other class, youngsters were investigating a school bus. They wrote to the district and asked if they could have a bus parked in their lot for a few days. They studied it, figured out what all the parts were for and talked about traffic rules. Then, in the classroom, they built their own bus out of cardboard. They had fun, but they also practiced writing, problem solving, even a little arithmetic. Says Katz: "When the class had their parents' night, the teacher was ready with reports on how each child was doing. But all the parents wanted to see was the bus because their children had been coming home and talking about it for weeks." That's the kind of education kids deserve. Anything less should get an "F."

Article 20

Now We're Talking!

Bernard Ohanian and Greta Vollmer

Bernard Ohanian wrote about circumcision in the June/July issue of Parenting Magazine. *Greta Vollmer is a language-development specialist at the International Studies Academy in San Francisco.*

AS MARY STEINBERG OF STORRS, Connecticut, remembers, it happened in the kitchen, when a little voice suddenly burst forth from the high chair with "cat." For Judy Henry of Potsdam, New York, it came when months of her daughter's not-quite-intelligible babbling suddenly melted into one clear "bye-bye." Time to dash to the baby book and record the hour and place; it's Baby's First Word, a milestone in our children's lives—and in ours.

With this first of what linguists call "one-word utterances," your child has entered into the kingdom of words, where genius and beauty, evil and banality have lived for centuries. And now that the first "bye-bye" has emerged, can the plaintive "Daddy, will you buy me this?" the mortified "Mom, how could you?" or the tentative "Can I borrow the car tonight?" be far behind?

Well, yes, actually. In the meantime, there's lots for your child to learn about language, and the learning process is sure to provide you with plenty of stories that—first-time parents be warned—will inevitably be funnier and cuter to you than to your friends and coworkers. It also is sure to give you lots to worry about: Is my kid talking early enough? clearly enough? correctly enough? Does she know enough words? What can I do to help?

The first thing you can do is relax. Linguists will tell you that almost all children are born with an innate ability to learn language, and while parents can certainly enrich that learning experience, there's very little they can do that will get in the way. The second thing you can do is have fun, and marvel at a process that even leading child-language specialists will admit remains a mystery. Experts in the field don't know exactly how it happens, but by about age five your child will know some 8,000 words. Moreover, she will have grasped a basic grammatical system that, no matter what the language, is as complex as some of the most sophisticated theorems taught in university-level mathematics courses.

YOUR CHILD MAY HAVE just entered the kingdom of words, but she has been living in the realm of language for some time. In fact, whether you realize it or not, you've been teaching her language skills since the day she was born. By responding to her post-feeding burps with an approving "That's a good baby!" for instance, you've been introducing her to conversational patterns: One person "says" something, another person responds. You've been teaching her intonation patterns as well, albeit higher-pitched, more singsong versions than adults and older children use. She, meanwhile, has communicated by gesturing, grabbing, and, of course, crying.

In the weeks leading up to your baby's first word, she will start to understand specific words and will begin to practice a kind of preverbal speech: using varied intonations, gestures, and sounds that approximate what she hears. This is the stage at which an older sibling is likely to say with frustration, "My baby brother is talking; he's just not saying anything!"

Then, finally, it comes: the word, a seemingly simple feat perhaps, but one that requires the human brain to operate at its synchronic best. The basic meaning of a word bubbles up from one part of the brain's left hemisphere, the Wernicke's area, and is fired off to another part of the brain known as the Broca's area. Once there, it is processed for speech and sent to the brain's motor cortex, which controls muscles in the mouth, lips, tongue, and voice box. Receiving their signals from the cortex, these muscles contract simultaneously, and in less than a millisecond—the time it takes the idea to travel from the recesses of the brain to the tip of the tongue—your child has become a talker.

For most kids, this neurological wizardry first takes place sometime between the ages of 10 and 14 months, but bear in mind that as with any childhood development, there is a great range of what is normal. Parents don't necessarily need to worry if their child doesn't follow this timetable. Girls will generally talk sooner than boys, and both first-born children and children born at least four years after their closest sibling will talk sooner than other kids. Even though, as Harvard University linguist Catherine Snow says, "our culture's notion of intelligence is tied up with talking," there is absolutely no evidence that early talkers are any smarter than later talkers. Nor will the process necessarily be a linear one: A child may learn a word, forget it for a while, and come back to it. "The first stage of language learning is not easy," says Snow, "Parents don't realize how hard those first ten words are for kids."

YOU CAN'T HELP A CHILD LEARN LANGUAGE ANY FASTER OR EARLIER, BUT YOU CAN HELP HER LEARN TO LOVE WORDS

Parents also may not realize how different those first ten words may be from child to child. In recent studies of English-speaking children in the United States, Katherine Nelson, a psychology professor at City University of New York, has identified two types of language learners: referential and expressive. "The first words of referential kids usually will be the names of objects," she explains, "while expressive children will often first use socially useful phrases like 'I want,' 'stop it,' or 'all gone.'" She emphasizes, however, that most kids who are learning to talk are part referential and part expressive—with varying degrees of each—and that it's not clear whether these differences reflect other personality tendencies.

What is clear is that the referential-versus-expressive model of language acquisition is not universal. Bambi Schieffelin, an anthropologist at New York University who studied child language acquisition in Papua New Guinea, says, "In other cultures children might first learn people's names, kin terms, how to greet and tease people, and how to ask for things correctly." Schieffelin points out, however, that children in New Guinea follow more or less the same time frame as children here: They too will master their language by the time they are five. Children in different cultures, she says, share a common neurological denominator; but once set in motion, neurology takes a back seat to social influences.

WHEN CHILDREN BEGIN TO TALK, they suddenly seem more human to us. And so great is our desire to communicate with our new conversationalists that we repeat back to them what we think they're saying, a practice linguists call modeling or shaping. We also simplify our language to make it easier for children to understand, using what specialists call motherese or child-directed speech and what the rest of us call baby talk.

And here, perhaps, is the greatest controversy about the language-acquisition process. Ask ten parents whether baby talk is healthy or harmful, and you'll get ten different answers; linguists

---

TAKING IT TO THE PROFESSIONALS

While their pronunciation, grammar, and choice of words may occasionally be so goofy that we can't help but laugh, most children do get the hang of their native language pretty quickly. A small minority, however, develop one or more of a variety of language problems and will need the help of a speech pathologist.

"Most parents I work with are afraid they've done something wrong," says Sandy Friel-Patti, an associate professor of communication disorders at the University of Texas at Dallas. "But it's important that parents understand that they are not to blame for their child's speech disorder." That's not to say that linguists necessarily know what *is* causing the problem. "We don't know what causes language impairment in children, except in cases where they've suffered neurological traumas or head injuries," says Friel-Patti.

What Makes a Speech Problem?

Speech and language problems are difficult to pin down because they differ depending on a child's age. "A child who is unintelligible at the age of two is not a problem," says speech pathologist Joan Kaderavek of Perrysburg, Ohio. "But a three-year-old who is unintelligible could be." Parents shouldn't be alarmed, Kaderavek says, if their child is a little later in talking than her playmates or siblings, but "if a child is not using words at 18 months, then the parent should check with a specialist."

Another reason to see a specialist may be if your child stutters, but again, a child's age will determine whether the stuttering warrants professional attention or not. Many children between the ages of two-and-a-half and four pass through periods of stuttering or nonfluency, and as Kaderavek points out, there are both normal and abnormal patterns of nonfluency. "If your child says 'Mommy, Mommy, Mommy, I want to go outside,' and she's completing all of her words, there shouldn't be any cause for concern," Kaderavek says. "If she seems frustrated by her inability to complete words, or changes her choice of words to consistently avoid a certain sound, it might be more than just a period of nonfluency."

Other children have developmental articulation disorders—that is, they have problems with certain sounds. Some youngsters, says Kaderavek, can't pronounce r sounds until they are about seven years old. But if your three-year-old still can't pronounce the basic consonant sounds—b, p, m, t, w, d—you might want to make an appointment with a speech pathologist.

Speech Specialists: What They Do

When you bring your child to a speech pathologist, he will probably run her through standard tests for hearing, comprehension, and the complexity of her grammar and vocabulary, as well as conduct a physical exam of her articulation mechanism. The pathologist will also watch your child at play, to see how communicative and cooperative she is. Based on his findings, he may recommend speech games or exercises to do at home or speech therapy with a specialist.

Kaderavek, whose own five-year-old son has a speech disorder, counsels patience to parents. "People say to me, 'You're a speech pathologist. Why can't you fix your son?' But all I can do is what I do with any child who has speech problems, which is to try to create an environment that encourages his language growth," she says.

If you suspect your child has a language problem, you can call the American Speech-Language-Hearing Association in Rockville, Maryland, for a recommendation to a certified speech pathologist in your area: (800) 638-8255.

3. CHILDHOOD: Cognitive and Language Development

are somewhat divided as well, although current thinking leans toward the belief that child-directed speech is not only good but is also, in our culture at least, almost unavoidable. For linguists, child-directed speech isn't simply a matter of truncating our words and pitching our voice an octave higher. Rather, it involves adjusting adult speech down to a child's level so that it can be understood, and then constantly revising the level upward as the child learns more and more words.

Specialists believe that most parents, if they talk and listen to their children, will do this almost automatically. But a problem can arise when parents are too slow to raise their level of speech. We all know such parents, those slightly embarrassing folks who still call a rabbit a "bunny-wunny" when talking to a four-year-old. Sandy Friel-Patti, an associate professor of communication disorders at the University of Texas at Dallas, says baby talk is linguistically appropriate at certain stages, "but sometimes parents hang on to it beyond when it's appropriate."

These parents may give baby talk a bad name, but they're probably in the minority. In fact, says Joan Kaderavek, a Perrysburg, Ohio, speech pathologist who favors simplified language but opposes the use of exaggerated baby talk, the main problem in the achievement-oriented eighties is not that parents use baby talk too often, but that they "use language with kids that's too adult."

Snow, who falls into the there's-nothing-wrong-with-baby-talk school, agrees. "Too many parents claim, 'My kid is so verbal, and it's because I didn't talk baby talk to him.'" Snow says that's a false assumption. The most important part of language acquisition, to her, is that kids can express what they feel and think and can understand others. That may mean parents' using baby talk, she says, because "it's not crazy to use words that kids have some hope of replicating."

So if you and your 18-month-old are playing with a toy truck and she can say "truck" but is not yet stringing several words together, Kaderavek says, "don't bombard her with 'Look at the fast truck racing around the track. Isn't it a nice red truck?' You are better off saying 'See the truck? Red truck. Truck can go. Truck can go fast.'"

Besides, you'll be able to pull out your long sentences about fast red trucks racing around tracks soon enough. Somewhere between 18 and 21 months, your child will advance to the two-word-utterance stage—although just as some children skip crawling altogether, later talkers may bypass the one-word stage and start directly with two- and three-word phrases. Now the child is generally commenting on the world, often on the absence or presence of someone or something: "water gone," "Daddy back," or "doggie big."

Then comes the explosion. "At 18 months a parent could probably tell you every word in her kid's vocabulary," says Friel-Patti. "At 30 months there's no way she could." During this rapid expansion, children start out learning 30 to 60 words per month and wind up with a vocabulary of 250 words or so by the end of the second year.

It is during this period that parents will think, fleetingly at least, that their child is a genius. "It's pretty amazing," says Snow, "when your kid uses some word that you don't remember having taught her." But hold your nominations for Baby Nobels; while it's true that some children are in fact more verbal than others, it's also true that the neighbor's three-year-old is probably learning language just as quickly as yours.

GIVEN THE DAZZLING SPEED with which children learn language, parents of suddenly eloquent preschoolers can't help but wonder whether there is something they can do to enhance the process even further. Within limits, there is. While you can't do anything to speed up the pace by which your child learns words, you can enrich her understanding of language—and more important, convey to her a delight in using language to communicate.

The best thing to do, linguists say, is also the most obvious: Talk to your child as much as you can. Talking directly to her is much more effective than trying to expose her to language indirectly by, say, plopping her down in front of the television or just talking with your spouse in front of her; according to Snow, two-year-olds absorb very little language from what they overhear. Nor do they pick up language from other kids. "It's a misconception that children learn to talk in a play group or daycare center," says Friel-Patti. "That may be where they learn to use their language socially, but they learn to talk from one-on-one conversations with adults."

As part of talking, of course, parents should listen carefully to their kids. "You should try to make guesses at what your children are attempting to say," says Eve Clark, a professor of linguistics at Stanford University. "The more you let them know that you understand, the more they'll produce. Don't dismiss what they're saying as meaningless; instead, try to supply the context they need verbally." And try to talk about what the child is interested in at that moment. As Kaderavek says, "You don't want to talk to her about her blocks if she's playing with an eggbeater."

Reading to children also enriches their language skills by teaching them words in an imaginative context and by developing their sense of narrative. Parents should take care, however, to choose books that will captivate their children's imagination. "Kids learn what makes sense to them," says Nelson. "They're not going to learn if you read them the encyclopedia." The experience will also be less fruitful if a parent turns a story into a test, flipping the pages and pointing to pictures while asking the child to name the objects in them. "Asking children to label," says Kaderavek, "is not teaching language."

"Children don't learn 15 words a day just by having them recited to them," says Snow. "They have to hear the words in context. So don't simply show your child a picture of a window; ask her to open the window."

While expanding their vocabulary, young children are also busy increasing their understanding of grammar. In fact, once children begin to figure out grammatical rules, they become more systematic about language than the language is itself. Many of a child's grammatical mistakes, for instance, stem from the expectation that language will be more regular than it really is: hence made-up words like "plantman" for gardener or "fixman" for mechanic; plurals like "tooths," "foots," and "mans"; and verb forms like "he goed."

As linguists point out, although parents may think a child's language is slipping backward if she switches from saying "he

20. Now We're Talking!

Word by Word The chart below indicates, by age, the age progression of language development in children. Don't panic if your child doesn't follow this schedule exactly. Many kids will fall outside of the normal timetable at one stage or another. Girls and first-born children, for instance, can be expected to reach a new stage a bit sooner than boys and younger siblings, and there can be other perfectly acceptable individual differences as well.

0–3 months
- Smiles, responds to voices
- Coos, makes vowel-like sounds

4–6 months
- Chuckles, giggles, babbles

6–9 months
- Tries to imitate sounds
- Engages in speechlike babbling
- Uses consonant sounds

10–12 months
- May say first word
- Uses intonation patterns that sound like sentences

12–18 months
- Uses one-word utterances to express a thought: "cat," "cookie," "bye-bye"

18–21 months
- Begins two-word utterances: "red truck," "mommy sick"
- Uses verbs
- Vocabulary explosion begins: May know 50 words

24–27 months
- Regularly uses two- and three-word utterances: "No go outside," "Kitty come back"
- Begins using pronouns, although not always correctly: "Me no want"
- Vocabulary explosion continues: May know 250 words

30–33 months
- Vocabulary explosion is in full swing
- Uses three- to four-word sentences
- Word order and phrase structures approximate adult speech much of the time

3 years and up
- Uses well-formed sentences that follow grammatical rules
- Speaks with approximately 90 percent comprehensibility; makes most sounds correctly, with possible exceptions of r, l, s, and th

went" to "he goed," it's actually a sign of progress, because she has begun to grasp an important tenet of English grammar: To form the past tense, tack on -d or -ed to the infinitive verb. "When children first use a word like *went*," says Stanford's Clark, "it's not at all clear to them that it's part of the verb *to go*." But because children are, in Clark's words, "great patternmakers," once they start to understand English's verb-conjugation system, they may say "wents," "wenting," and "goed." Then when they learn that there are some exceptions to the way we form the past tense, they'll settle back on *went* and discard *goed*—but usually only after a period of using the two interchangeably.

Tempting as it may be, parents shouldn't bother correcting their children when they make grammatical mistakes. Parents who do will soon learn what a futile exercise it is. "Kids under the age of five won't accept a change unless it belongs in their system. They'll say the word the way you want one time to please you, but the next time they'll go back to their way," says Friel-Patti. This back-and-forth is likely to continue until the child figures out how the correct version fits into her sense of the way language works, a discovery she will make at her own pace. Constant correction "tells children that you're more focused on how they're talking than on what they're telling you," Friel-Patti points out. "If that message comes across too strongly, they'll quit talking."

BY TALKING TO KIDS, listening to them, reading to them, and of course playing with them, parents can help them learn to love language in all its power and mystery. But linguists caution against trying to do too much: "Parents should be parents," advises Friel-Patti. "They shouldn't be teachers. Using flash cards and teaching two-year-olds to read is highly inappropriate."

Still, it's natural to worry about how our kids are going to do in school, and to wonder what language skills will serve them best. Snow says you may be able to help prepare your kids for school by having extended conversations with them on single topics, by discussing events that are taking place beyond their immediate and present world, and by asking them to recount things that happened when you weren't around. And she adds an important caveat for parents whose native language is not English: "Parents need to be talking to kids in a language they speak very, very well," she says. "A child is better prepared for school if you speak good Polish at home rather than bad English. She'll learn English quickly enough at school."

What a child needs most, then, is a sense that communicating is fun, that language is a great source of creativity. Your child may not talk as much as other kids, but as Friel-Patti says, "not all adults are blabby adults, and not all kids are blabby kids." If your child focuses more on expressive language than referential language, it may lead to concern on your part that she's not analytical enough. Don't worry, says Nelson: "The world doesn't necessarily belong to the analytical types. We need children who tell stories, as well as children who solve math problems." We need Faulkners as well as Einsteins, and Woolfs as well as Curies.

And we need children who are relaxed enough with language, and curious enough about language, that they will enjoy the fullest range of expression. So encourage your little genius to talk, to make up stories and poems, and to play with words in ways that you've forgotten how to. Talk with her about what you're doing and what you've done, and realize that when she says "My feets hurt," she's actually outsmarting the language. We often forget what a breathtaking place this kingdom of words is; with the help of our children, we can discover its magic all over again.

Though nearly a million children are regularly given drugs to control "hyperactivity," we know little about what the disorder is, or whether it is really a disorder at all

SUFFER THE RESTLESS CHILDREN

ALFIE KOHN

Alfie Kohn is the author of No Contest: The Case Against Competition *(1986) and* The Brighter Side of Human Nature *(1990).*

IN MARCH OF 1902 DR. GEORGE STILL STOOD BEFORE the Royal College of Physicians, in London, and described some children he had observed—mostly boys—who seemed to him restless, passionate, and apt to get into trouble. The children were suffering, he declared, from "an abnormal defect of moral control."

Despite his invocation of morality, Still's lecture is often billed as the first recorded discussion of hyperactivity. Thousands of articles on the subject have been published in professional journals since then, the great majority of them within the past two decades. A good proportion of these papers begin by citing the pervasiveness of the disorder. An opening sentence such as "Hyperactivity is the single most prevalent childhood behavioral problem" is usually regarded as sufficient, because the readers of these journals are already convinced that, on average, at least one hyperactive child sits in every elementary school classroom in the United States.

The actual estimates vary, however, and not by a little. If a psychiatrist says that about three percent or 10 percent or between one and five percent of elementary school children are hyperactive, this is simply a rough average of studies whose findings differ dramatically. One series of papers estimated the rate of hyperactivity at 10 to 20 percent. A California survey put it at precisely 1.19 percent. A nationally recognized expert says without hesitation that it is six percent. The one thing researchers generally agree on is that among children labeled hyperactive, boys outnumber girls by at least four to one.

The disparities can be explained to some extent by the varying stringency of the criteria that are applied, and the assumptions guiding those applying them. According to the latest guidelines for diagnosing what is now officially called attention-deficit hyperactivity disorder (ADHD), the problem must have been noticed before age seven, must persist for at least six months, and must include any eight of fourteen symptoms, among which are the following: the child is easily distracted by extraneous stimuli; has difficulty sustaining attention, following through on instructions, or waiting his or her turn in games; and often does such things as talk excessively, fidget with hands or feet, squirm while sitting down, lose things, and fail to listen to what is being said to him or her.

Most experts emphasize the importance of teachers' observations, which are often quantified on a rating scale that was developed by Keith Conners, a psychologist, in the late 1960s. When that score—or any single judgment—is the sole basis for diagnosis, 10 percent or more of all elementary school children may be labeled hyperactive. But if the observations of others—parents or pediatricians, for example—are also taken into account, then the prevalence of the disorder can be as low as one percent.

The experience of two Canadian researchers, Nancy J. Cohen and Klaus Minde, is illustrative. They had the teachers of 2,900 kindergartners in one community submit the names of children thought to be hyperactive. The researchers expected, because of the estimate offered by a widely used textbook, to find that four to 10 percent would be referred. At first their procedure yielded sixty-three names. But when they looked more closely, they found that most of these children had altogether different psychological problems or else seemed to be suffering from poor nutrition or too little sleep. Only twenty-three children—less than eight tenths of one percent—were left in the hyperactive category after this more rigorous screening.

The wildly divergent estimates of prevalence are disturbing enough in themselves, given that each percentage point stands for hundreds of thousands of children. But they also underscore the fact that different criteria for diagnosis produce different conclusions about whether a particular child will carry the ADHD label and, as a consequence, be required to swallow a drug every day. Most unsettling is a flicker of doubt about the integrity of the diagnosis itself. Can we in fact be confident that any child has a disorder called hyperactivity, or ADHD?

Overwhelmingly, child psychiatrists and psychologists answer in the affirmative. Just because the prevalence of hyperactivity is difficult to pin down, or because we can't be sure a particular child is afflicted with it, doesn't mean the phenomenon isn't real, according to most people in the field. "If you'd ever seen a hyperactive kid, you'd know it," the psychologist Susan Campbell says. "Something's there." But my review of more than a hundred journal articles and book chapters, and also conversations I had with many of the leading researchers in the field, suggests that this assessment may be too sanguine.

First, whether such a distinctive disorder exists is open to question, because each of the symptoms that are supposed to lead to a diagnosis of hyperactivity—restlessness, impulsiveness, and difficulty paying attention—occurs at least as often in children who have entirely different problems, as Cohen and Minde discovered.

Second, the key symptoms often do not appear together. Douglas G. Ullman, a psychologist, and his colleagues have found that children said to be hyperactive do not always turn out to have difficulty paying attention, and vice versa. One's ability to predict that a child will be inattentive because he is restless—or the other way around—"is not much better than if one tossed a coin to decide the matter," the authors concluded.

Third, the procedure for deciding which behaviors belong on that list of fourteen, and also the decision that eight of them (rather than seven or ten) will suffice for a diagnosis of hyperactivity, are arbitrary. These decisions are "made by committee," as Dennis Cantwell, a leading researcher in the field who was himself on such committees, admits.

A score on the Conners Teacher Rating Scale, or any of the other scales used in diagnosis, gives the appearance of scientific precision, as though it were, say, a white-cell count. In reality, the score is nothing more than a numerical value that sums up a particular teacher's subjective judgments about whether a child bounces around too much.

Some theorists have argued that ADHD is actually "heterogeneous"—that it is characteristic not of a single population of hyperactive children but of several distinct subgroups. This has a professional ring to it and seems plausible on its face, but it simply sets the problem back a step. What *are* the disorders mistakenly collected under the ADHD umbrella? How do we know that *they* are valid diagnostic categories? One might say that using the word "heterogeneous" tells more about what we don't know than about what we do.

This history of the diagnosis does nothing to allay one's doubts. For many years children with symptoms identical to those that are now considered to add up to hyperactivity were said to have "minimal brain damage." When researchers eventually acknowledged that they had no proof these children's brains were actually damaged, the label was changed to "minimal brain dysfunction." This, in turn, gave way to the diagnosis of "hyperkinetic reaction," which became "attention deficit disorder with (or without) hyperactivity," which became "attention deficit hyperactivity disorder."

These changes—and the latest in the series will surely not be the last—reflect something more than quibbling over labels. They suggest a fundamental disagreement about what, if anything, is behind the labels. "The whole notion has gone through so many metamorphoses as to suggest a catastrophe in terms of conceptual integrity," says Gerald Coles, the author of *The Learning Mystique*, a critical analysis of what are commonly called learning disabilities. "Rather than moving toward ever greater precision, they're constantly sweeping over the disasters of last year's conception."

A new diagnosis has appeared in every successive revision of the mental-health clinician's bible, the *Diagnostic and Statistical Manual of Mental Disorders*, or *DSM*. Whereas *DSM* II talked about "hyperkinetic reaction," the third edition, published in 1980, switched to "attention deficit disorder": now difficulty paying attention seemed to be the core of the disorder, with excessive activity merely an optional by-product. In 1987, with the publication of the latest revision (*DSM* III-R), the definers changed their minds again, deciding that insufficient data existed to support the emphasis on attention, and that hyperactivity really was the center of the problem after all.

The Problem May Be the Classroom

IN LIGHT OF ALL THIS, DISAGREEMENTS ABOUT whether a given child is "hyperactive" seem to signify something more than uncertainty about the applicability of an established diagnosis. Consider the rather obvious point made by Kenneth D. Gadow, a professor of special education at the State University of New York at Stony Brook: "What is diagnosed as hyperactivity by one physician may be considered emotional disturbance or 'spoiled child syndrome' by another."

When perceptions of that same child are compared across different environments, one is even less likely to find consensus about his status. A number of studies have by now shown "relatively low levels of agreement among parents, teachers, and clinicians on which children should be regarded as hyperkinetic," the psychiatrist Michael Rutter has written.

Why does the parent at home rate the child differently from the teacher at school? Simple subjectivity is not the answer, as it turns out. The fact is that children act differently in different places. Therefore, the idea of a unified disorder threatens to slip away completely.

Those in the field accept as common knowledge that symptoms of hyperactivity often vanish when a child is watching TV, engaged in free play, or doing something else he likes. Similarly, the way a child's environment is organized and the way tasks are presented can mean the difference between normal behavior and behavior called hyperactive, a finding that has been replicated again and again. This is particularly true for the symptoms related to paying attention.

Since the early 1970s, for example, researchers have known that children diagnosed as hyperactive do well at tasks that they can work on at their own pace, as opposed to tasks controlled by someone else. Many hyperactive children also seem virtually problem-free when they receive individual attention from a teacher or when the experimenter stays in the room with them. And their ability to concentrate on what they're doing picks right up when a reward is hanging in the balance (although the effect doesn't always last if the reward is withdrawn). This suggests that the problem may be more one of willingness to comply—especially in performing tasks that the children find boring—than one of a built-in deficit.

"The degree to which hyperactives are viewed as deviant depends on the demands of the environment in which they function," the veteran Canadian researchers Gabrielle Weiss and Lily Hechtman wrote in *Science* in 1979. One might even amend that to read, "The degree to which children are viewed as hyperactive in the first place depends. . . ." But rather than seriously questioning the legitimacy of the diagnosis, specialists have responded by fashioning a subcategory of the disorder called "situational hyperactivity." Keith Conners has written, "When data from parent and teacher conflict . . . . there may be a true 'situational' hyperactivity, a pattern of behavior which only emerges, say, in the school setting but not the home setting."

Of course, this approach cannot be proved wrong, just as it would not be technically inaccurate to say that a child who cries when her friend moves away is suffering from a syndrome called "situational depression." The question is, how is such labeling useful, and what sorts of inquiry does it serve to encourage or discourage?

Some of us remember things more accurately if we see them rather than hear them; some of us learn better if abstract ideas are represented spatially. Similarly, some children learn better and jump around less if they receive personal attention or get to design their own tasks. In 1978 the psychologists Charles E. Cunningham and Russell A. Barkley offered the heretical suggestion that "hyperactive behavior may be the *result* rather than the cause of the child's academic difficulties." This possibility raises the question of why these children fail—whether it has to do with how they are being taught.

A small study described in 1976 compared a group of hyperactives in a traditional classroom with a group in a classroom where instruction was individualized, children were relatively free to move around the room, and the teacher planned lessons in cooperation with the children. After a year the teacher's ratings showed almost no change for the first group, but the hyperactivity scores of those in the open classroom had dropped dramatically. A second study, which compared hyperactive children with a control group, found that the difference between them—as judged by the experimenters rather than the teachers—remained significant in a formal classroom but effectively disappeared in an open classroom. Although the studies are by no means conclusive, virtually no one has taken the trouble to investigate the question further.

If a teacher finds few hyperactive children in her class, that may be because she designs appropriate tasks for students who might otherwise squirm, or because she is less rigid in her demands than other teachers, and more tolerant of what educators refer to as "off-task behavior." (The use of this designation may say as much about the teacher as about the student.) "Hyperactivity," one researcher says, "typically comes to professional attention . . . when the child cannot conform to classroom rules." This invites questions about how reasonable the rules are.

But the psychiatrists who design the research, shape the diagnostic categories, and prescribe the drugs rarely explore how children are being taught, and even then the question tends to be treated as an aside. In 1986 the *Journal of Children in Contemporary Society* and *Psychiatric Annals* both devoted special issues to hyperactivity, and neither addressed so much as a paragraph to such matters as classroom organization and teachers' attitudes.

The Potency of Family Dynamics

IN ADDITION TO THE POSSIBILITY THAT SYMPTOMS ascribed to hyperactivity may result from unsuitable classroom environments or academic failure, a number of studies have found that warped family patterns often accompany hyperactivity. As Weiss and Hechtman have summarized the research, "Families of hyperactives tended to have more difficulties, mainly in the areas of mental health of family members, marital relationships, and, most particularly, the emotional climate of the home. . . . [and they] tended to use more punitive, authoritative approaches in child rearing than [other] families."

Particular styles of discipline and interaction, of course, may be the consequence of a parent's frustration with a child who is already hyperactive for other reasons. This is the view of Russell Barkley, who formerly headed the American Psychological Association's section on clinical child psychology. His own research shows that parents' reliance on commands and punishments drops significantly when their hyperactive children are put on medication. "The majority of the problem is the effects of the child's behavior on the parents, not the other way around," he asserts.

But this may be too much of a leap. "Knowing that the behavior changes when the child is on Ritalin doesn't tell you how the behavior got started in the first place," says Susan Campbell, who adds that no one knows why some children are more fidgety or impulsive than others.

L. Alan Sroufe, a professor of child psychology at the University of Minnesota, thinks that early parent-child dynamics may play a key role. In a study with Deborah Jacobvitz, Sroufe followed children from birth until age eight and discovered that those who were eventually diagnosed as hyperactive were more likely, during their infancy, to have care-givers who were rated as "intrusive." Rather than responding to the baby's needs, such a care-giver might, for example, push a bottle into its mouth even though it was trying to turn its head away.

Sroufe reasons that most of us, with our parents' help, learned quite early to control ourselves when circumstances demanded. However, some parents may overstimulate their children precisely when the children are already out of control. These children may well come to fit the ADHD pattern. Data to confirm this conclusion do not yet exist, Sroufe concedes, but then, few people have gone looking for them. No other researcher has ever tried to predict hyperactivity from observations of early care-giving, and neither has anyone helped overstimulating parents to modify their behavior in order to see if the children have fewer problems later on.

While they were investigating parent-child interactions, Sroufe and Jacobvitz looked back to infancy for differences that might have existed between hyperactive and other children, in case those mattered more. They came up virtually empty-handed. Hyperactivity doesn't seem to be connected to delivery complications or prematurity, to infant reflexes or distractibility, or to any of dozens of other measures. Indeed, Michael Rutter has reported in the *American Journal of Psychiatry*, "There is no indication of any biochemical feature that is specific to the hyperkinetic syndrome."

Theoretically, the behavior of some tiny subset of those children called hyperactive may be traceable to neurotransmitters, the brain's chemical messengers, or to genes or neurological damage. To date, though, no generally accepted evidence of an organic, or biological, cause of hyperactivity has been found.

This has not been for lack of trying. The medical journals are littered with the remains of discarded theories that purported to explain restlessness in children as a symptom of disease. For example, for quite some time stimulant drugs were believed to have a "paradoxical effect" on hyperactive children; the very idea that hyperactives—and only hyperactives—were quieted by this sort of medication was said to prove that their troubles were biochemical. But in 1978 the psychiatrist Judith Rapoport and her colleagues published a study showing that stimulants had precisely the same effects on the motor activity and attention span of normal children. Later studies showed that similar effects occur in normal adults and in children with entirely different problems.

The overwhelming majority of the research has shown that most hyperactive children have no discernible brain damage or neurological abnormalities; their EEG readings are not distinctive. For a while clinicians thought that the nervous systems of hyperactives were overaroused. Then they were believed to be underaroused. Neither of these theories has been proved, however. What is remarkable here is not the series of failures to find a biological cause but the tenacity with which this line of investigation continues to be pursued. For every study investigating the families of hyperactive children, hundreds search for neurological abnormalities. This is the sort of research that gets funded—not merely in the case of hyperactivity but in mental health more generally—possibly because this is how the investigators (and the grantors themselves) were trained. The humanistic psychologist Abraham Maslow once observed that if people are given only hammers, they will treat everything they come across as if it were a nail.

"People who don't have a high tolerance for ambiguity aren't going to look at family factors," Susan Campbell says. "In the biological sphere it seems as if one is on firmer ground." In any case, most physicians continue to assume that hyperactivity is biologically based, and when researchers are asked whether any evidence supports this assumption, a typical response is "Not yet."

The Effects of Drug Therapy

IF THE EMPHASIS AMONG RESEARCHERS ON BIOLOGY crowds out work on prevention, environmental causes, and alternative forms of therapy, a similar pattern occurs among clinicians. The most striking consequence of assuming that an unusually distractible or impulsive child is suffering from a disease is the tendency to turn to medication to solve the problem.

"Assumptions of organicity have often been used, in practice, as a justification for prescribing drugs," Jacobvitz and Sroufe have written. In an interview Sroufe adds, "The majority of hyperactive kids today are treated only with Ritalin—the *vast* majority."

3. CHILDHOOD: Developmental Problems

Ritalin is the brand name for methylphenidate, which was approved for use with children in 1961. Like Dexedrine (dextroamphetamine), another stimulant sometimes prescribed for hyperactivity, Ritalin is classified as a Schedule II drug, meaning that among substances with legitimate medical use it is regarded as having the highest potential for abuse, and its manufacture is regulated by the Drug Enforcement Administration. (Other drugs in that class include morphine and barbiturates.)

The best available figures on the use of drugs for ADHD come from a careful biennial survey of Baltimore County schools by the psychiatrist Daniel Safer. In 1987 among public elementary school students in that county 5.9 percent were taking stimulants. Extrapolating to the nation as a whole, and correcting for the fact that Maryland doctors are a bit freer with their prescription pads than their counterparts elsewhere, Safer estimates that three quarters of a million children nationwide are now receiving stimulants. "It's been increasing steadily since we first took a look, in 1971, and it'll go over one million in the 1990s if the present trend continues," he says.

Indeed, about four out of five children diagnosed as hyperactive are put on stimulants at some point, making drug therapy far and away the treatment of choice in the United States. (This does not seem to be true elsewhere; in most of Western Europe, for example, children rarely or never receive medication for hyperactivity.)

In the early 1970s, media coverage of Ritalin use seeded a storm of controversy, culminating in the publication of a widely read book by Peter Schrag and Diane Divoky: *The Myth of the Hyperactive Child*. This period also saw the publication of Benjamin Feingold's *Why Your Child Is Hyperactive*, which argued that drugs were unnecessary because hyperactivity could be cured by restricting the amounts of sugar and food additives in the child's diet. (Subsequent studies have been unable to demonstrate that diet can bring about any significant improvement in the great majority of hyperactive children.)

Lately the medication controversy has been heating up again, largely because an arm of the Church of Scientology, which calls itself the Citizens Commission on Human Rights, is picketing professional conferences and helping to sponsor a series of legal actions charging physicians with malpractice. But if the church group's claim that "psychiatry is making drug addicts out of America's school children" is, understandably, not taken seriously by those in the field, neither does the drug deliver the benefits claimed by some proponents.

While the idea that stimulants can have a quieting effect may seem peculiar, the fact is that these drugs don't so much slow down activity as redirect it. A child on Ritalin may move around just as much as a nonmedicated child over the course of a day, but he will be better able to sit still for tasks that require concentration. His activity is more goal-directed, less aimless, more likely to be "on-task" than it was before. Besides being less distractible and better able to sustain attention, the medicated child typically becomes less aggressive and less apt to get into trouble, less obnoxious to his peers, easier for his teachers to handle, and generally more compliant. Unsurprisingly, parents and teachers are often pleased with the change they see in a child who is put on Ritalin.

That's the good news. The bad news has to do with side effects—about which more in a moment—and with the drug's efficacy, which is probably the greater of the two problems. The evidence shows, first, that drugs do absolutely nothing for 25 to 40 percent (depending on whose estimate you trust) of hyperactive children. Kenneth Gadow, in his book-length contribution to a series called *Children on Medication*, reported, "Some youngsters even become worse on medication! Unfortunately, there is no way to tell whether medication is going to work other than to have the child take it."

Second, a large proportion of the children who do respond to Ritalin also improve on a placebo. After weeding out the nonresponders, the pediatrician Esther Sleator followed a group of medicated children for two years and then began slipping some of them sugar pills. Of twenty-eight subjects for whom definite data were available, eleven continued to behave as if they were getting the real thing. Russell Barkley's review of several hundred studies indicates that about 40 percent of children are rated as improved when they're on a placebo, although the magnitude of the improvement generally isn't as great as it is for children receiving Ritalin.

Third, even for children who respond well to stimulants, the effect is a temporary suppression of symptoms, not a cure. A child may have been taking Ritalin for years, but within hours of the last dose he will be indistinguishable from a hyperactive child who has never taken Ritalin. Or *almost* indistinguishable: in what is known as a rebound effect, when the drug wears off the child will briefly become a little worse than he was before.

Fourth, although some children on stimulants are able to do more work and thus receive better grades, drugs do nothing to enhance actual academic achievement. Beneficial effects on concentration had long been assumed to translate into achievement, but an analysis that Russell Barkley and Charles Cunningham made of seventeen studies in the late 1970s, and a subsequent analysis by Kenneth Gadow of another sixteen studies in 1985, were uniformly discouraging. "Certain behavioral interventions are clearly superior to stimulant medication in facilitating academic performance," Gadow concluded.

On reflection, this doesn't seem so strange. Drugs do not remedy cognitive deficits or create skills. And if hyperactivity is the result of learning problems rather than the cause, two psychologists pointed out in a 1988 article, "interventions directed toward suppressing [hyperactive] behaviors will have no long-term effects in reducing either the [hyperactivity] itself or learning difficulties unless the latter are specifically treated."

Some children's behavior seems to improve only at relatively high dosages, around one milligram per kilogram of body weight. This much medication tends to have a detrimental effect on thinking skills, thus forcing the careful physician to choose between reducing hyperactivity and optimizing cognitive performance. What's more, "the [dosage] where teachers perceive the most improved classroom behavior is also associated with side effects," Gadow has reported.

These findings suggest hard questions about just why children are put on Ritalin in the first place. Even assuming that drugs make a difference, should they be prescribed to help a third-grader learn better? How about to reduce fidgeting, which, Safer points out, is "neither a disruptive influence nor highly unusual"? Or to establish docility, so that children will follow the rules and not annoy adults? At best the drug "may have much greater relevance for stress reduction in caregivers than intrinsic value to the child," Gadow has written. Gabrielle Weiss and her colleagues found that "children on the whole preferred being without 'the pills,'" and in a follow-up study of adults who had been medicated as children, Weiss's group found that slightly more listed medication as a hindrance than listed it as a help.

This reaction may be due in part to the social stigma of having to take pills every day, but part of it clearly has to do with side effects. Overall, research on these effects does not support the extravagant claims of some critics, including the Church of Scientology. Extreme adverse reactions are very rare and crop up occasionally with other medications as well.

If Ritalin stunts growth, the effect seems to be temporary. (It does seem possible that someone who continued taking medication straight into adulthood would be permanently affected, but no one knows for sure.) Other concerns, including reports of elevated blood pressure, facial tics, insomnia, and weight loss, have led specialists to recommend that younger children not be given stimulants. For older children, most of these side effects turn out to be either uncommon or controllable by modifying the dosage or the medication schedule.

Such adjustments may not, however, eliminate all the behavioral side effects. According to some studies, children on Ritalin sometimes become withdrawn and stare off into space, a behavior that critics call the "zombie" effect. While stimulants make these children less likely to annoy their peers or pick fights, they are also less likely to interact with others at all. And some investigators suspect that medication leads children and their parents to attribute any improvement to the pills rather than to social causes or to factors within their control.

Barbara Henker and Carol Whalen, psychologists at the University of California, have found that when someone is told that a given child is on medication, he or she is more likely to believe that the child's problems are serious and due to "nervous system dysfunction" than if told the child is in a behavioral treatment program. (Hence the circle is completed: assumptions of a biological cause lead to drugs, and drugs lead to assumptions of a biological cause.)

These concerns seem to have prompted little hesitation about prescribing stimulants. The number of prescriptions continues to rise, and more and more psychiatrists are talking seriously about keeping, or putting, adolescents and adults on stimulants too.

Ask professionals to name the most important finding relevant to hyperactivity within the past decade and they will tell you it is the discovery that the disorder doesn't disappear at puberty. Some hyperactive children continue to have problems with school and work, to be antisocial and otherwise troubled, as they get older.

But a closer look at these data suggests that something else is going on. One of the diagnoses that overlaps to a considerable extent with ADHD is "conduct disorder," which refers to aggressive, disobedient, troublemaking behavior; perhaps two thirds of hyperactive children also qualify for that diagnosis. Children who become delinquent in later life are primarily from the conduct-disorder group, rather than being a random sample of those who fidget and can't pay attention. Some of the non-conduct-disorder group, not surprisingly, may have trouble finishing school and may continue to be more distractible than most people when they grow up, but they apparently don't become mentally ill or get in trouble any more than the rest of us. The major revelation in the field during the past ten years, then, turns out to be this: if you were aggressive and antisocial as a child, you may also be aggressive and antisocial as an adult.

Hyperactivity in Perspective

EVEN THOSE RESEARCHERS WHO ARE COMFORTable with both the diagnosis of hyperactivity and the use of Ritalin have urged that considerable care be taken in prescribing the drug and deciding who gets it. But virtually every commonsense recommendation offered in the professional journals is routinely ignored by physicians throughout the country.

Even though studies have shown that a child cannot be properly diagnosed on the basis of an office visit, a California survey of pediatricians revealed that the way children acted in front of them "seemed to be the most important characteristic in physician judgments." Doctors who seek further evidence to confirm their diagnosis may simply prescribe drugs and wait to see whether they work. According to a national survey done in 1987, three quarters of pediatricians continued to believe that a child's response to medication was helpful for purposes of diagnosis—this despite proof that many hyperactive children do not respond to stimulants, and that many non-hyperactive children do.

According to studies conducted in several states, teachers often play little or no role in diagnosis or treatment of hyperactivity, even though their observations

3. CHILDHOOD: Developmental Problems

are critical. Moreover, against the advice of specialists, clinicians often prescribe unnecessarily high doses of Ritalin, fail to recommend counseling and other nonmedical treatments in addition to stimulants, and fail to schedule periodic "holidays" from the drug, as they should.

Much as the public outcry over Ritalin in the 1970s may well have "spurred a wave of better designed studies," according to the psychiatrist Mina K. Dulcan, so the newly filed malpractice suits, Russell Barkley concedes, may "motivate practitioners to bring their practices a little more up to date." Barkley emphasizes, nonetheless, that he believes that the practitioners being sued are not, strictly speaking, negligent. Negligence is judged according to customary practice in the field, not by the standards suggested by research. Put bluntly, this means that if most clinicians are diagnosing casually and prescribing irresponsibly, bringing legal action against any one of them will be difficult.

Diagnoses of other psychological disorders may be similarly arbitrary and subjective, made by committee and poorly defined, insensible of social factors and conducive to unfounded assumptions about biological causes. On the one hand, this may serve to excuse what goes on with hyperactivity, or at least to place it in perspective. On the other hand, it may provoke larger, more disturbing questions about the theory and practice of mental health in the United States.

Tykes and Bytes

With a Boost From Computers, Kids With Disabilities Find the World's Filled With Possibilities

Bettijane Levine
Times Staff Writer

The first thing to know about the bright-eyed toddlers who zoom, lurch, plop, play, sing and go potty at UCLA's Intervention Program is that they are the advance guard of an army yet to mobilize.

Mostly, they act just like toddlers everywhere—but they're not.

With varying degrees of disability due to cerebral palsy, Down's syndrome and an array of what program director Dr. Judy Howard calls "fancy diagnoses," they are among the world's youngest computer whizzes.

Sure, some have poor motor skills and muscle tone, little or no speech, minimal vision—all sorts of knotty physical or mental problems. But at ages 18 months to 3 years, these toddler technocrats are already equipped with PC's, power pads, switches, speech synthesizers and other electronic gear designed to even the playing field between them and so-called normal children.

They are part of the first toddler generation whose disabilities can be mitigated by technology, who can be judged by their potential rather than by their limitations. They are the first to prepare from babyhood for a life that will be computer-friendly in the extreme, and, as a result, productive.

Jay Horrell, 2, has used his computer since he was 18 months old. "I don't know where we'd be without it," says his father, Michael, who explains that some Down's syndrome children, such as his son, are "able to receive a lot more information than they can give back. They know the answers and they know what's going on, but can't respond" as they'd like to.

The computer allows Jay to display and improve various skills. It talks to him and waits patiently for answers; it puts him on more even footing with classmates at the Intervention Program and with his brother in their North Hollywood home, where Jay's setup includes an Apple PC, an electronic touch pad (in place of a keyboard), a speech synthesizer that gives voice to the letters and pictures he calls up on screen—and as many software programs as his parents can find.

Jay will attend a regular preschool in September. Down the line, Horrell adds, "I believe the computer will allow him to be a productive member of society."

Gabriela Cellini was 17 months old, had cerebral palsy and lacked certain motor skills when her parents took her to the Computer Access Center in Santa Monica. There, a staff member explained what the toddler could do with a computer and special accessories suited to her needs.

"Gabriela took one look and was riveted to the screen," her mother, Harriet, recalls. "Her muscle tone increased. She was so motivated to play with it that she sat up straight all by herself for about a half-hour. She quickly understood the cause-and-effect principle of hitting the switch and activating games."

Now 2 1/2 and a student at the UCLA Intervention Program, Gabriela uses computers at her Pacific Palisades home and in class. "It's delightful to watch," her mother says. "This strange computer voice says 'Gabriela, stack the blocks.' Or 'Gabriela, build a face.' Or she shoots airplanes off a carrier, increasing her speed each time she scores a hit." Gabriela still needs some assistance in other areas, Cellini says, but she's her own person in front of the computer.

UCLA's Howard, a pediatrician who has headed the Intervention Program since 1974, began teaming disabled toddlers with computers in 1981. She found children of that age are "automatically computer friendly, which immediately sets up a positive response in adults. Suddenly, you see they have abilities, and you start to set expectations for them that you weren't able to

3. CHILDHOOD: Developmental Problems

set before. When you have children who cannot talk, who are visually handicapped, who for any reason cannot pick up a crayon and draw or play with dolls, puzzles and toys to show you what they can do," it is difficult to know what they are capable of, she explains.

The first step is to find a way for each child to access the computer. In the early 1980s, there were few devices commercially available to provide that access. Now there are dozens: large switches, oversized alternative keyboards, touch windows with built-in sensors that attach with Velcro to a computer screen. And there is growing body of knowledge about how to rig the devices so a child can work the computer by using whatever part of his body he controls best. Says Howard: "Every child can work one, even if he can only use one finger, his head or a toe. With appropriate software, they can solve puzzles, build with blocks, dress dolls. They can even play all the traditional favorite toddler games—two kids at the computer together—so they learn sharing, success and winning.

"Toddlers soon start to visually track on the screen because they're so highly motivated. They hold their little heads up and you see all the things that eventually lead to reading. That's the purpose of all this."

Kit Kehr, executive director of the UCLA program, says: "The younger you help these children, the better they'll do down the road. A kid who can't build with blocks or push cars around the way other kids can is missing essential play experiences." He also falls behind in language development and social skills, she says.

Rev Korman, a computer consultant in special education for the Los Angeles Unified school district, remembers such a child, named Kim. "She'd had a stroke before she was born. It affected her vocal cords, so she had no speech and the doctors told her parents she'd always be a vegetable. She was 3 when they rolled her into my office in her wheelchair. I set up a communication board, a speech synthesizer and the computer, so that it would speak for her. She took about 10 minutes to learn to push the pictures that communicated her needs and wants. 'I don't want to go to bed. I want a red balloon.'

"We then moved to a 24-picture board, which she mastered quickly. By using this setup, she was able to communicate for the first time in her life so that people could hear her. She spent the next 45 minutes using the Muppet keyboard, and by the end of her visit she was teaching herself the alphabet.

"Kim's parents went right out and bought the computer, the speech synthesizer, the electronic board. Now she's reading and the whole bit."

(Computer setups for children like Kim cost about $2,000, Korman says.)

Dr. Phillip Callison, head of special education for the Los Angeles Unified School District, has participated in the UCLA project from its beginning, and is credited with providing assistance and inspiration. He says he believes in computers for all children, especially those with disabilities. Right now, the school district can provide such equipment for severely handicapped students, he says.

(Six hundred toddlers with a variety of disabilities are using computers in 45 United Cerebral Palsy Assn. nursery school programs across the country. The projects are run by a coalition of the association, UCLA and Apple Computers.)

Many adults still know little about home computers and next to nothing about the rest of the exotic equipment needed to adapt it for use by children with disabilities. In fact, UCLA's Kehr says that even the salespeople in most computer stores "won't know what you're talking about" if you walk in and ask for a power pad, a speech synthesizer and a special switch to help you adapt a PC for your child.

Jackie and Steve Brand of Albany, Calif., found little help in 1983, when they realized their 6-year-old daughter, Shoshana, "needed technology in her life." Shoshana has multiple disabilities, including cerebral palsy and poor vision. "The standard teaching tools just weren't working," Jackie Brand says. Her husband took a one-year sabbatical from his teaching job, went to computer school and eventually put together a system Shoshana could use. It had a touch-sensitive keyboard with large keys and a synthesizer that gave voice to whatever she typed, so she could hear what she was doing rather than having to see it on the monitor.

"My daughter played for the first time in her life," when she got her computer, Brand says. "By that time she was 9, and we realized we needed to establish a program so others don't have to sacrifice years of their kids' lives."

The couple started the Alliance for Technology Access, which now has 43 chapters in 32 states. Each is a resource center where anyone with any disability—or parents of disabled children—can learn in informal, friendly surroundings what technology is available and how to customize it for their needs.

Nothing is sold at the centers, but they house an array of computers, access devices and information on companies that manufacture accessories not available in local stores but essential for those with special needs. The goal, Brand says, is simply to show people what they can do for themselves or their children, without making them go through "the usual hoops."

"Typically, you go to a doctor, evaluation center or clinic where they do an extensive evaluation of a child and prescribe what they think the right technology might be. In some situations that's needed." But doctors and clinics are not always on target when it comes to decisions about recreation and education for kids with handicaps, she says.

Sometimes they tell parents a child won't benefit from technology, Brand says. "But the bottom line is that each family knows their child's potential. They need to find out what their options are and try out different hardware and software to see what works best for the child. They need to be empowered to make their own decisions. That is what the Alliance helps them do.

"To people who say we must have realistic expectations for our children, I say I hope every family [of a handicapped child] has *unrealistic* expectations. That's the only way you will find your child's potential, so that your child can show you who he is and what he can do."

Brand's daughter will go into ninth grade in a regular public school in the fall. She uses a wheelchair, a computer for writing and a tape recorder with special levers for recording her notes, which she listens to at night. "This is a kid who nobody would have thought could function in a regular school program, and without technology, she couldn't. Yet she is doing phenomenally well. And she's not unique, she is typical," Brand says.

The Computer Access Center, which rents space from the John Adams Middle School in Santa Monica, is one of three Southern California chapters of the Alliance. The volunteer staff helps each visitor (by appointment) to understand how life for a disabled person can be enhanced through the magic of technology. And they are very up-to-date.

Last week, for example, the staff arranged a demonstration by Daniel Fortune and John Ortiz of Zofcom, a Palo Alto–based firm.

The two men have designed a device called the TongueTouch Keypad, which looks like an ordinary orthodontic retainer worn by most kids after their braces are off.

But this retainer is a wireless transmitter with built-in sensors. By touching different sensors with one's tongue, a user unable to move any other part of his body can gain almost total control of his environment.

To the astonishment of onlookers, Fortune answered the phone when it rang without seeming to move a muscle. He turned on and off the VCR, the TV, the fan. He used the computer and explained that there is an almost "unlimited array" of equipment that can be operated (including a page-turner) with the new device, which will probably be marketed at the end of the year.

Mary Ann Glicksman, a staff member at the center, was intrigued. Her son, John Duganne, just graduated from Santa Monica High School and starts college in the fall. He intends to make animated films and already works part time, creating computer graphics for a software firm.

Duganne drives his power wheelchair with his chin, she says, but that's about the extent of what his body can do. (He has cerebral palsy.) To work the computer on which he does his schoolwork and art, he uses a headset with an ultrasonic device and a bite switch in his mouth. If the TongueTouch could work for him, it would be an improvement, she said. The designers cautioned that it is meant primarily for people with spinal cord injuries, and that those with cerebral palsy might not have enough tongue control. But with optimism typical of Alliance members, Glicksman said she'd rather give her son a chance to find out if it works than take the inventor's word that it won't.

Two other Southern California branches of the Alliance for Technology Access, funded by Apple Computers, are: The Special Awareness Computer Center in Simi Valley and Team of Advocates for Special Kids in Anaheim.

Facts About Dyslexia

Developmental dyslexia is a specific learning disability characterized by difficulty in learning to read. Some dyslexics also may have difficulty in learning to write, to spell and, sometimes, to speak or to work with numbers. The exact cause of dyslexia is not known, but we do know that it affects children who are academically capable, physically and emotionally healthy and who come from good home environments. In fact, many children with dyslexia have the advantages of excellent schools, high mental ability and parents who are well-educated and value learning.

Children are subject to a broad range of reading problems, many of which have an identifiable cause. However, a small group of children have difficulty in learning to read for no apparent reason. These children are called dyslexic. It is estimated that as many as 15 percent of American students may be classified as dyslexic.

Defining Dyslexia

Over the years, the term dyslexia has been given a variety of definitions, and for this reason, many teachers have resisted using the term at all. Instead, they have used such terms as "reading disability" or "learning disability" to describe conditions more correctly designated as dyslexia. Although there is no universally recognized definition of dyslexia, the one presented by the World Federation of Neurology has won broad respect: "A disorder manifested by difficulty in learning to read despite conventional instruction, adequate intelligence and sociocultural opportunity."

Symptoms

Children with dyslexia are not all alike. The only trait they share is that they read at levels significantly lower than is typical for children of their age and intelligence. This reading lag usually is described in terms of grade level. For example, a 4th-grader who is reading at a second grade level is said to be two years behind in reading.

Referring to grade level as a measure of reading is convenient, but it can be misleading. A student who has a 2-year lag when he is in fourth grade has a much more serious problem than a 10th-grader with a 2-year lag. The 4th-grader has learned few of the reading skills that have been taught in the early grades, while the 10th-grader, by this measure, has mastered eight years, or 80 percent, of the skills needed to be a successful reader.

Samuel T. Orton, a neurologist who became interested in the problems of learning to read in the 1920s, was one of the first scientific investigators of dyslexia. In his work with students in Iowa and New York, he found that dyslexics commonly have one or more of the following problems:

- difficulty in learning and remembering printed words;
- letter reversal (b for d, p for q) and number reversals (6 for 9) and changed order of letters in words (tar for rat, quite for quiet) or numbers (12 for 21);
- leaving out or inserting words while reading;
- confusing vowel sounds or substituting one consonant for another;
- persistent spelling errors;
- difficulty in writing.

Orton noted that many dyslexics are lefthanded or ambidextrous and that they often have trouble telling left from right. Other symptoms he observed include delayed or inadequate speech; trouble with picking the right word to fit the meaning desired when speaking; problems with direction (up and down) and time (before and after, yesterday and tomorrow); and clumsiness, awkwardness in using hands, and illegible handwriting. Orton also found that more boys than girls show these symptoms and that dyslexia often runs in families. Fortunately, most dyslexics have only a few of these problems, but the presence of even one is sufficient to create unique educational needs.

Possible Causes

Most experts agree that a number of factors probably work in combination to produce the disorder. Possible causes of dyslexia may be grouped under three broad categories: educational, psychological and biological.

Educational Causes

Teaching Methods. Some experts believe that dyslexia is caused by the methods used to teach reading. In particular, they blame the whole-word (look-say) method that teaches children to recognize words as units rather than by sounding out letters. They think that the phonetic method, which teaches children the names of letters and their sounds first, provides a better foundation for reading. They claim that the child who learns to read by the phonetic method will be able to learn new words easily and to recognize words in print that are unfamiliar as well as to spell words in written form after hearing them pronounced. Other reading authorities believe that combining the whole-word and the phonetic approaches is the most effective way to teach reading. Using this method, children memorize many words as units, but they also learn to apply phonetic rules to new words.

Whatever method they support, experts who think that instructional practices may cause dyslexia agree that strengthening the beginning reading programs in all schools would significantly decrease the number and the severity of reading problems among school children.

Nature of the English Language. Many common English words do not follow phonetic principles, and learning to read and spell these words can be difficult, especially for the dyslexic. Words such as cough, was, where and laugh are typical of those words that must be memorized since they cannot be sounded out. While such words undoubtedly contribute to reading problems, they constitute only a small percent of words in English and so cannot be considered a primary cause of dyslexia.

Intelligence Tests. The commonly accepted definition of dyslexia as a reading disability affecting children of normal intelligence is based on the assumption that we can measure intelligence with a fair degree of accuracy. Intelligence test results, usually referred to as IQ scores, must be interpreted carefully. IQ scores may be affected by factors other than intelligence. Those IQ tests which require the child to read or write extensively pose special problems for the dyslexic. Scores from such tests may reflect poor language skills rather than actual intelligence. Even those IQ tests that are individually administered and demand little or no reading and writing may fail to give a fair measure of intelligence; dyslexics often develop negative attitudes toward all testing situations. In addition, conditions such as noise, fatigue or events immediately preceding the testing session may adversely affect test results. With such a range of possible influences on IQ scores, we must regard these scores as, at best, an estimate of the range of the child's scholastic aptitude and, at worst, a meaningless number that can unjustly label the student.

Psychological Causes

Some researchers attribute dyslexia to psychological or emotional disturbances resulting from inconsistent discipline, absence of a parent, frequent change of schools, poor relationships with teachers or other causes. Obviously, a child who is unhappy, angry or disappointed with his or her relations with parents or other children may have trouble learning. Sometimes such a child is labelled lazy or stupid by parents and friends—even by teachers and doctors. Emotional problems may result from rather than cause reading problems. Although emotional stress may not produce dyslexia, stress can aggravate any learning problem. Any effective method of treatment must deal with the emotional effects of dyslexia.

Biological Causes

A number of investigators believe that dyslexia results from alterations in the function of specific parts of the brain. They claim that certain brain areas in dyslexic children develop more slowly than is the case for normal children, and that dyslexia results from a simple lag in brain maturation. Others consider the high rate of lefthandedness in dyslexics as an indication of differences in brain function. This theory may have some validity. Another theory is that dyslexia is caused by disorders in the structure of the brain. Few researchers accepted this theory until very recently, when brains of dyslexics began to be subjected to post-mortem examination. These examinations have revealed characteristic disorders of brain development. It now seems likely that structural disorders may account for a significant number of cases of severe dyslexia.

Genetics probably play a role as well. Some studies have found that 50 percent or more of affected children come from families with histories of dyslexia or related disorders. The fact that more boys than girls are affected could mean that nongenetic biological factors as well as environmental/sociological factors could contribute to the problem.

Treatment

Educators and psychologists generally agree that the remedial focus should be on the specific learning problems of dyslexic children. Therefore, the usual treatment approach is to modify teaching methods and the educational environment. Just as no two children with dyslexia are exactly alike, the teaching methods used are likewise varied.

Children suspected of being dyslexic should be tested by trained educational specialists or psychologists. By using a variety of tests, the examiners are able to identify the types of mistakes the child commonly makes. The examiner is then able to diagnose the problem and, if the child is dyslexic, make specific recommendations for treatment such as tutoring, summer school, speech therapy or placement in special classes. The examiner may also recommend specific remedial ap-

3. CHILDHOOD: Developmental Problems

proaches. Since no method is equally effective for all children, remediation should be individually designed for each child. The child's educational strengths and weaknesses, estimated scholastic aptitude (IQ), behavior patterns and learning style, along with the suspected causes of the dyslexia, should all be considered when developing a treatment plan. The plan should spell out those skills the child is expected to master in a specific time period, and it should describe the methods and materials that will be used to help the child achieve those goals.

Treatment programs for dyslexic children fall into three general categories: developmental, corrective and remedial. Some programs combine elements from more than one category.

The *developmental* approach is sometimes described as a "more of the same" approach: Teachers use the methods that have been previously used believing that these methods are sound, but that the child needs extra time and attention. Small-group or tutorial sessions in which the teacher can work on reading with each child allow for individual attention. Some researchers and educators believe, however, that this method is not effective for many children.

The *corrective* reading approach also uses small groups in tutorial sessions, but it emphasizes the child's assets and interests. Those who use this method hope to encourage children to rely on their own special abilities to overcome their difficulties.

The third approach, called *remedial*, was developed primarily to deal with shortcomings of the first two methods. Proponents of this method try to resolve the specific educational and psychological problems that interfere with learning. The instructor recognizes a child's assets but directs teaching mainly at the child's deficiencies. Remedial teachers consider it essential to determine the skills that are the most difficult and then to apply individualized techniques in a structured, sequential way to remedy deficits in those skills. Material is organized logically and reflects the nature of the English language. Many educators advocate a multisensory approach, involving all the child's senses to reinforce learning: listening to the way a letter or word sounds; seeing the way a letter or word looks; and feeling the movement of hand or mouth muscles in producing a spoken or written letter, word or sound.

Prognosis

For dyslexic children, the prognosis is mixed. The disability affects such a range of children and presents such a diversity of symptoms and such degrees of severity that predictions are hard to make.

Parents of dyslexic children may be told such things as "the child will read when he is ready" or "she'll soon outgrow it." Comments like these fail to recognize the seriousness of the problem. Recent research shows that dyslexia does not go away, that it is not outgrown and that extra doses of traditional teaching have little impact.

Fortunately, educators are becoming more aware of the complexities of dyslexia, placing greater emphasis on choosing the most appropriate teaching method for each child. Teachers are more willing to provide remedial teaching over longer periods of time, whereas prior practice often has been to cut off services if observable changes fail to occur in a limited time. Some dyslexics improve quickly, others make steady but very slow progress, and still others are highly resistant to instruction. Many have persistent spelling problems. Some acquire a basic reading skill but cannot read fluently.

A child's ability to conquer dyslexia depends on many things. An appropriate remedial program is critical. However, environmental and social conditions can undermine any treatment program. The child's relationships with family, peers and teachers have a major effect on the outcome of instruction. In a supportive atmosphere, a child's chance of success is enhanced. Attitudes such as "expectancy," the degree to which a teacher expects a child to learn, are important. Children who sense that they are not expected to succeed seldom do. Since slight progress in reading ability can make an enormous difference in academic success and vocational pursuits, children need to know that they are expected to progress.

The earlier dyslexia is diagnosed and treatment started, the greater the chance that the child will acquire adequate language skills. Untreated problems are compounded by the time a child reaches the upper grades, making successful treatment more difficult. Older students may be less motivated because of repeated failure, adding another obstacle to the course of treatment. The time at which remediation is given also affects a dyslexic's chances. Often, remedial programs are offered only in the early grades even though they may be needed through high school and college. Remedial programs should be available as long as the student makes gains and is motivated to learn. Adults can make significant progress, too, although there are fewer programs for older students.

A dyslexic child's personality and motivation may influence the severity of the condition. Because success in reading is so vital to a child, dyslexia can affect his or her emotional adjustment. Repeated failure takes it toll. The child with dyslexia may react to repeated failure with anger, guilt, depression, resignation and even total loss of hope and ambition; he or she may require counseling to overcome these emotional consequences of dyslexia. With help a dyslexic child can make gains but the assistance must be timely and thorough, dealing with everything that affects progress. For the child whose dyslexia is identified early, with supportive family and friends, with a strong self-image, and with a proper remedial program of sufficient length, the prognosis is good.

NICHD Research Support

The National Institute of Child Health and Human Development (NICHD) supports many studies designed to determine how most children learn to read and what may interfere with or prevent some children from acquiring this important skill. Some investigators are attempting to develop language tests that can predict which 5- or 6-year-old children have the necessary skills to learn to read and those who are at risk for reading failure. If these investigators are successful it is likely that many cases of dyslexia can be prevented. Other scientists are attempting to identify children at risk for dyslexia through the use of modern neurological examination procedures, including electroencephalography and PET scans (Positron Emission Tomography, an imaging technique that measures brain activity).

Some scientists supported by the NICHD are studying the children from families that have a higher than normal incidence of dyslexia and related language disorders to determine a possible genetic cause of the reading disorder. Other investigators are concentrating their research efforts on the development of specific descriptions of various subtypes of dyslexia with the hope that more appropriate therapies can then be planned.

Although a number of important advances have been made through research, many unanswered questions remain about this developmental disorder of childhood. Our ultimate goal is the complete prevention of dyslexia as well as other specific learning disabilities. Intermediate to that goal is the early identification of all children who are at risk for dyslexia so that prompt and appropriate procedures can be administered which will preclude the manifestation of dyslexic symptoms, or minimize their effects on the child's intellectual, academic, psychological or social development.

Although impressive evidence exists concerning the specific behaviors and neurological characteristics of dyslexia, continued research is essential. Such medical and educational research, along with sound diagnostic techniques and individually-designed educational programs, can open the doors through which the dyslexic may enter into full participation in our literate society.

This article was written by staff members of the Orton Dyslexia Society and the National Institute of Child Health and Human Development (NICHD), NIH.

For Further Information on Dyslexia

American Academy of Pediatrics
P.O. Box 927
141 Northwest Point Rd.
Elk Grove Village, IL 60007
(800) 433-9016 or (312) 228-5005

American Speech-Language-Hearing Association (ASHA)
10801 Rockville Pike
Rockville, MD 20852
(301) 897-5700

Association for Children and Adults with Learning Disabilities (ACLD)
4156 Library Rd.
Pittsburgh, PA 15234
(412) 341-1515

Council for Exceptional Children (CEC)
1920 Association Dr.
Reston, VA 22091
(703) 620-3660

National Institute of Child Health and Human Development (NICHD)
Office of Research Reporting
Bldg. 31, Rm. 2A32
9000 Rockville Pike
Bethesda, MD 20205
(301) 496-5133

Orton Dyslexia Society
724 York Rd.
Baltimore, MD 21204
(301) 296-0232

Family, School, and Cultural Influences on Development

- **Parenting (Articles 24-26)**
- **Stress and Maltreatment (Articles 27-30)**
- **Cultural Influences (Articles 31-33)**
- **Education (Articles 34-36)**

In the early 1900s, behaviorists held that children should be reared strictly in order to correctly shape their behavior. Post–World War II advice leaned toward permissive forms of child-rearing. Today, child-rearing advice seems to have struck a middle road between the strict and permissive approaches. Parents are encouraged to provide their children with ample love, to cuddle their infants, to use reason as the major disciplinary technique, and to encourage verbal interaction—all in an environment where rules are clearly spelled out and enforced. Suggestion, persuasion, and explanation have become the preferred techniques of rule enforcement, rather than spanking or withdrawal of love. This view is echoed in the first *Parenting* subsection article "Dr. Spock Had It Right."

Many parents, with the best of intentions, overindulge their children and produce kids who learn they can get something for nothing. In "Positive Parenting," Bruce A. Baldwin describes what he calls "Cornucopia Kids"—kids who want it all, right now, without working for it. These children often grow into adults who cannot function in the "real world." Guidelines are suggested to help parents provide an environment in which their children can develop a realistic perspective and sound values.

Part of parenting is spending time with your children. However, as "Can Your Career Hurt Your Kids?" points out, with the decreasing number of mothers who stay home and the increasing number of single-parent families, parents are spending less time with their children. Many of these children are in day care, especially the younger ones, but as they get older, a large number become latchkey children. This article looks at the advantages and disadvantages of day care, examines possible effects of leaving adolescents on their own, and considers the consequences on family relationships of parents coming home too tired to interact with their children.

As we enter the last decade of the twentieth century, parents, as well as children, are perhaps subject to more sources of psychological stress than at any time in the recent past. Some of the chief sources of parental stress stem from single parenting, the decline of intergenerational families, and the increase in teenage mothers and fathers. These factors and others have strained the support services available for families in our society. Families that are stressed may turn to expedient rearing techniques even when they are aware of "expert" opinion. One such expediency is the use of physical punishment in order to "discipline" children. However, physical punishment is an ineffective form of discipline that teaches lack of self-control and promotes the use of aggression in order to control someone else's behavior.

The *Stress and Maltreatment* subsection points out that, taken to extremes, severe disciplinary techniques can spill over into physical abuse. What are the long-term consequences of physical, psychological, and sexual abuse? We often assume that abused children become maladjusted and abusing parents. However, as discussed in "The Lasting Effects of Child Maltreatment," the story is not that simple. Many, perhaps most, abused children become well-adjusted adults. Similarly, most abusing adults were not targets of abuse as children. Other factors, such as poverty, play a major role in determining the outcome of earlier experiences.

"Children of Violence" points out that children who grow up with violence all around them may become desensitized as their exposure to violence increases. Many of these children suffer post-traumatic stress disorder and may exhibit a number of different symptoms such as aggression, indifference, or inability to concentrate. In spite of this, many of these children show an amazing resiliency in their ability to survive. Experts believe that early intervention can be crucial in helping children recover from incidents of violence in which they have been an observer or a participant.

Children and parents are subject to high levels of stress in society today. Some of the major sources of childhood stress are examined in "Children Under Stress." Many young children are being pushed too hard and too fast academically; poor children suffer from peer pressures and are more likely to be abused and neglected; children experiment with drugs and alcohol at an earlier age. Divorce is one of the more common major sources of stress. Regardless of who wins custody, children do not have an easy time adjusting to the separation of their parents. Suggestions for helping children to cope often require a level of parental cooperation that may be unrealistic. The results of a longitudinal study of the effects of divorce on children, presented in "Children After Divorce," indicate that the psychological effects many be

Unit 4

longer lasting than previously thought. In spite of forces threatening to tear families apart, however, the influence of the family on the life of an individual remains strong and pervasive.

The articles in the *Cultural Influences* subsection profile two key influences on development—ethnicity and sex. In "Rumors of Inferiority," differences in performance between blacks and whites are linked to self-doubt, feelings of inferiority, and fear of intellectual abilities—by-products of a long-standing process of discrimination. In "Biology, Destiny, and All That," Paul Chance reviews evidence for sex differences, and concludes that many differences between the sexes are differences in how characteristics are expressed rather than absolute differences.

School expands a child's social network beyond the neighborhood peer group and often presents new social adjustment problems. In "Alienation and the Four Worlds of Childhood," Urie Bronfenbrenner draws attention to the increase in disorganized families and environment, which contribute to alienation from family, friends, school, and work.

In the subsection *Education*, the article "Tracked to Fail" notes that among those who are at risk to fail are children who acquire negative labels at a very early age based on criteria such as reading ability or developmental test scores. The article shows how labels tend to stick all through school, regardless of the child's ability. All too often, slow-track children are expected to do poorly, are given "watered-down education," and suffer lasting effects on self-esteem.

Additional aspects of contemporary education are addressed in two selections. "Master of Mastery" describes the views of the distinguished educator Benjamin Bloom, who stresses the urgent need for innovation in our educational strategies. "Not Just for Nerds" focuses on the need to lure students back to science and math programs. American students lag far behind their counterparts in many foreign countries, and this can have far-reaching effects on our nation's economy in the years to come.

Looking Ahead: Challenge Questions

Do you agree with the premise that sex role attitudes about marriage, parenting, and family relationships are fragile and correlate poorly with actual behavior? If not, what kind of evidence would you require in order to be convinced?

Do you think that most divorces provide an atmosphere in which parents can set aside their conflicts and animosities sufficiently to act in the best interests of their children?

It is relatively easy to blame teen pregnancy and teen substance abuse on the disorganization of the American family. It is far more difficult to suggest effective solutions to the problems. What changes do you believe should be instituted in American society to resolve such problems as divorce, child abuse, teen pregnancy, racism, sexism, and substance abuse?

Will you raise your children the way your parents raised you? If not, how would you do it differently and why?

There are some obvious advantages to early identification and tracking of children who have exceptional abilities as well as those who seem destined to fail. Do you think the advantages outweigh the disadvantages or vice versa?

Why is it important for the American public to be scientifically literate? What are the probable consequences of a public that does not understand or appreciate science?

Dr. Spock had it right

Studies suggest that kids thrive when parents set firm limits

Using the serious tone once reserved for childhood ailments such as diphtheria and measles, a recent article in a journal published by the American Academy of Pediatrics described a new illness sweeping the nation: The "spoiled-child syndrome." Children exhibiting symptoms of the disease are excessively "self-centered and immature," as a result of parental failure to enforce "consistent, age-appropriate limits." Often, the article goes on to suggest, spoiled children grow into spoiled adolescents and adults, never learning how to delay gratification or tolerate not getting their own way.

Whether or not the epidemic of spoiled brats is genuine, the wave of parental anxiety on the subject of discipline is undeniable. Today's parents are torn by conflicting instincts and conflicting theories on permissiveness vs. discipline. On the one hand are feelings of guilt, particularly on the part of two-income parents who feel they don't spend enough time with their children and want to make every available minute a pleasant one; on the other are growing fears over drugs and violence in society. A recent national survey by Louis Harris & Associates reveals that most people—64 percent of those polled—say parents just don't do a good job disciplining their children.

Burdens of the past. Behind the parental concerns over doing the "right" thing is a long-running scientific debate. Several recently completed studies that tracked more than 100 children for nearly 20 years have provided the first objective test of which disciplinary styles work best, and all point in the same direction. Parents who are not harshly punitive, but who set firm boundaries and stick to them, are significantly more likely to produce children who are high achievers and who get along well with others.

Over the ages, child-rearing theories have changed as faddishly as fashions, reflecting a continual shift between viewing children as innocents who need little adult intervention and as inherently evil, and in dire need of straightening out. "The normal child is healthy in every way. His manners need no correcting. ... So, when they cry or scream or are upset, we should understand that it means something is disturbing them, and we must try to discover what they need and give it to them," wrote Greek physician Galen in A.D. 175, perhaps the first doctor on record to advise demand feedings. In the 18th century, French philosopher Jean Jacques Rousseau envisioned the child as an unspoiled creature of nature, a noble savage best left unfettered by society's constraints, as did 17th-century English philosopher John Locke.

By the turn of the century in America, with the coming of the Industrial Age and the increasing faith placed in scientific experts, the idea of giving the young unbridled freedom fell severely out of favor. The prevailing view of eminent child experts, such as pediatrician L. Emmett Holt, author of *The Care and Feeding of Children*, called for strict, regimented conditioning to create good eating, sleeping and social habits in children. "Infants who are naturally nervous should be left much alone ... and should never be quieted with soothing sirups or the pacifier," Holt wrote. In 1928, John Watson, a leader of the behaviorist school of psychology, applied Holt's principles to the child's mind. The result, the immensely popular *Psychological Care of Infant and Child*, is horrifying by modern parenting standards. Watson advocated molding babies by scientific control, with strict 4-hour feeding and sleeping schedules (just tune out the wailing), toilet training at 6 months—before the infant can sit up—and no thumb sucking, hugging or kissing. "Mother love is a dangerous instrument" that could wreck a child's chance for future happiness, Watson warned. The only physical contact he sanctioned between parent and child was a brisk handshake each morning.

The theorizing, and the receptions various theories received, often seems to have had more to do with cultural attitudes than with science. Dr. Benjamin Spock, who popularized Freudian theory and research on child psychology in his famous 1945 work, *The Common Sense Book of Baby and Child Care*, was seized upon by war-weary Americans who were looking for ways to free themselves of old conventions and who interpreted his emphasis on nurturing, gentleness and following common-sense instincts as a sanction of permissiveness. Later, in the 1960s and '70s, Spock became a political target when critics such as minister Norman Vincent Peale and Spiro Agnew blamed his so-called permissiveness for drugs, student riots, promiscuity and other excesses of the counterculture. Both, in fact, garbled Spock's message; he consistently urged parents to assert their authority at home. Through four editions, he never wavers from his 1945 message that the child "needs to feel that his mother and father, however agreeable, still have their own rights, know how to be firm, won't

From *U.S. News & World Report*, August 7, 1989, pp. 49-51. Copyright 1989, U.S. News & World Report. Reprinted by permission.

24. Dr. Spock Had It Right

let him be unreasonable or rude. ... It trains him from the beginning to get along reasonably with other people. The spoiled child is not a happy creature even in his own home."

Although Spock didn't condone permissiveness, other influential theorists did, as recently as a generation ago. A case in point is Scottish educator A. S. Neill in *Summerhill: A Radical Approach to Child Rearing,* which became popular in the United States after its publication in 1960. Neill, who established the progressive Summerhill School in Suffolk, England, during the 1920s, practiced and preached a philosophy of noninterference. Summerhill students chose whether or not they wanted to attend class, do homework, wear clothes, smoke cigarettes and experiment with sex. During the 1960s, Neill's child-rearing theories played a part in the creation of alternative schools and communes across America.

Friendly persuasion. The recent studies tend to support Spock's view that the most effective disciplinary style emphasizes warmth, verbal give-and-take and exertion of control without a parent acting like the child's jailer. The best evidence comes from a set of longitudinal studies conducted by psychologist Diana Baumrind of the University of California at Berkeley's Institute of Human Development. In her studies of approximately 150 middle-class, well-educated parents and children over a 20-year period, Baumrind identified three disciplinary styles that produced markedly different behavior traits in pre-school-age children. "Authoritarian" parents ("do it because I'm the parent") were more likely to have discontented, withdrawn and distrustful children. "Permissive" parents ("do whatever you want") had children who were the least self-reliant and curious about the world, and who took the fewest risks. "Authoritative" parents ("do it for this reason") were more likely to have self-reliant, self-controlled, contented children.

Baumrind's research teams collected and analyzed data at three intervals, when the subjects of her study reached ages 5, 10 and 15. One of her most interesting findings is that during adolescence "authoritative" parents are as effective in preventing their children's experimentations with drugs as the more restrictive "authoritarian" parents.

Baumrind's research dovetails with the clinical experiences of pediatrician Dr. Glenn Austin, who has applied her theories to his young patients and their parents in Los Altos, Calif. Austin has found that parents whose actions fit into Baumrind's category of "authoritative" make the most effective disciplinarians. For example, Austin cites how certain parents effectively break their children of the habit of throwing tantrums. They tune out the child until the tantrum ceases, or if that becomes intolerable, carry the child to his room, without "scolding or fussing or attempting to control the child's actions."

In Austin's summary of one of Baumrind's studies, he found that 85 percent of children raised by authoritative parents were "fully competent," which meant they possessed a long list of positive attributes including a sense of identity, a willingness to pursue tasks alone and a healthy questioning of adult authority, compared with 30 percent of children raised by authoritarian parents and 10 percent of children raised by permissive parents. While both boys and girls raised by authoritative parents turned out to be fully or partially "competent," major differences were found between boys and girls raised by other types of parents. For instance, authoritarian parents appeared to dominate their boys more than their girls: Only 18 percent of the boys were fully competent, compared with 42 percent of the girls.

Other research supports the insight that the key to a successful disciplinary strategy is helping the child develop inner controls, a process Freud called "internalization." University of Minnesota Prof. L. Alan Sroufe, who has studied 500 children in the past 18 years, found that the limits and boundaries parents imposed, before the child could regulate himself, served as a safety net that helped the child operate and develop his own internal controls. Authoritative parents usually establish an early pattern of trust through responding to a child's needs, meaning what they say and then following through. "If parents stick to a few rules that they're clear about, then their kids will be more compliant within those limits," Sroufe says. When parental expectations are not met, the principle of "minimum sufficiency," or less is more, should guide the punishment, because if the punishment is too harsh, that's what the child will focus on, not the behavior the punishment is intended to change.

Adjusting to conditions. Another study supports the view that inner controls are more apt to kick in when parents adjust their disciplinary style to external conditions. Psychologists Arnold Sameroff of Brown University's Bradley Hospital and Al Baldwin and Clara Baldwin at the University of Rochester studied 150 families over 13 years and confirmed that the most effective parental discipline matches the actual dangers facing the child. For example, in high-crime, inner-city neighborhoods, children whose parents imposed restrictive curfews and household rules, but displayed warmth as well as strictness, performed better than average in school. Conversely, in low-risk, suburban neighborhoods, the children of restrictive parents performed worse than average in school.

Although these studies are as objective as anyone can get, they can't bestow upon parents any magic formulas for preventing brattiness or rearing angels. Perhaps the most striking theme to emerge from all the scientific data is that establishing a pattern of love and trust and acceptable limits within each family is what really counts, and not lots of technical details. The true aim of discipline, a word that has the same Latin root as *disciple,* is not to punish unruly children but to teach and guide them and help instill inner controls. For in the end, as every mother and father know, parenting is far more of an art than a science.

by Beth Brophy

Positive Parenting
How to Avoid Raising Cornucopia Kids

Dr. Bruce A. Baldwin

Dr. Bruce A. Baldwin is a practicing psychologist who heads Direction Dynamics, a consulting service specializing in promoting professional development and quality of life in achieving men and women. He responds each year to many requests for seminars on topics of interest to professional organizations and businesses. This article has been adapted from Baldwin's book on parenting, Beyond the Cornucopia Kids: How to Raise Healthy Achieving Children. *For busy achievers, he has written* It's All in Your Head: Lifestyle Management Strategies for Busy People. *Both are available in bookstores or from Direction Dynamics in Wilmington, North Carolina.*

A recent advertisement promoting one company's top-of-the-line product had a catchy headline: "'I have the simplest of tastes.... I am always satisfied with the best'—Oscar Wilde." This same headline unfortunately hits home with all too many frustrated parents these days. From coast to coast, in cities and hamlets, parents are deeply troubled about their children.

Their concern is focused on their children's expectations. Children today want it all—right now—and they don't particularly want to work for it. Parents are expected to keep their offspring up with the latest fashions and every fad item, and to provide lavish supplies of spending money. All unearned, of course.

From an objective point of view, these children of good families remain immature. Parents complain that they don't seem to be motivated to do anything but have a good time. More often than not, they are rude and self-centered. And even after they leave home, someone else is still expected to provide it all and to foot the bills. These are the Cornucopia Kids. They are becoming more common with each passing year.

Cornucopia Kids simply don't deal well with the "real world" because they've never been exposed to it. They are children who learn through years of direct experience in the home that the good life will always be available for the asking, without personal accountability or achievement motivation. These indulged children have lived their lives in an artificial environment created by naive and compliant parents. Life is always easy. They've never had to struggle; nor are they expected to give back in return for what they've been given. The real world, when at last they confront it, finds them unprepared and overwhelmed.

In other words, these are children who are indulged materially from an early age. They get their way with parents through persuasion or manipulation, and they escape the consequences of their actions. Their parents help out by taking over and doing their work for them. Lucky children? Far from it. These kids have everything—*except what they really need*. What they need is a sound value system and associated achievement-oriented skills that will enable them to confront the real world and succeed there. Only parents can provide these necessities.

While there is no question that Cornucopia Kids are on the increase these days, it is ironic that the parents who raise them usually do so with the best of intentions. These parents work hard. They are achievers, and through dedication and sacrifice they've attained financial success. Then, without thinking about what they are doing, they begin indiscriminately to share their largess with their children—and it's mighty easy to do.

These days there is more in the stores, and the kids want it all. There is more discretionary income in two-career families. Credit is easy and instantaneous. Sophisticated advertising links products with status and personal adequacy. Having what is necessary to be part of the "in crowd" is reinforced by the child's peers. The result is tremendous pressure on parents to give and to give in. Some parents simply buy this material-oriented lifestyle outright. Many others rationalize what they are doing because everyone else seems to be doing the same thing. Still others, busy and overwhelmed with responsibility, are simply worn down and just give up trying to do what is right for their kids.

It takes much more character to be a good parent now than it did several decades ago. Back in the late '40s and '50s, most middle-class parents made enough to get by. Income was allocated to necessities and there was not much left over. Back then, if the kids wanted something extra, there was only one way to get it, and that was to work for it. For some people these days all that has changed. The central issue facing parents now is different and more difficult: "If we have it to give and the kids know that we have it to give, are we strong enough to say no?"

It's well known that struggle builds confidence and confidence builds character. The life-styles of indulged children today make for a most difficult transition when they leave home and are on their own. On the other hand, it is entirely possible for parents to provide a home environment in which children learn healthy values that will help them to grow into adulthood as emotionally mature and responsible achievers. Further, parents need not apologize for their successes, but an important distinction must be clearly understood. Just because parents have made it economically, it does not follow that children must be indulged.

To help parents see through the superficiality of many of the negative values permeating our society and being adopted wholesale via the immaturity of children, there are several myths, each of which has to do with material indulgence, that must be directly confronted and resolved.

A child who has everything is advantaged and therefore more likely to succeed. Experience shows that quite the op-

25. Positive Parenting

posite often happens. A child who gets everything for nothing learns to expect everything for nothing. As an adult, this expectation may continue unabated, and that young adult may find no base of experience from which to deal effectively with the problems, sacrifices, struggles, setbacks and hard work required in the real world. They become marginal workers who continue to rely on their parents to provide them with the life-style to which they have become accustomed.

Having everything helps a child feel loved and secure. Hogwash and horsefeathers! In no way does indulging a child contribute to his or her self-esteem and personal security. When given everything, children never learn to deal effectively with the real world and its expectations. What is learned is dependence on others, particularly parents. As a result, personal confidence never develops. Loving a child entails providing a sound set of values and effective life skills. Love does not include giving in to everything a child wants. Indulged children feel helpless, hopeless, depressed and insecure when everything isn't being provided for them.

The child who is given everything is protected from the emotional pain and frustration of exclusion by peers. Children often complain vociferously about having to do chores, or about not having the absolute best of everything, or about being forced to endure what is perceived to be unfair parental discipline. This is especially true when their peers seem to have everything, without household responsibilities, and always seem to escape the consequences of their actions. From the immature perspective of a child, it is a totally unfair situation. From a developmental perspective, however, it's a wise parent who helps a child learn accountability and responsibility through positive experiences at home *regardless of what that child's peers do.* That child will then be able to fit into the real world and deal positively with its frustrations.

This is not to imply that our purpose is to deprive the youngsters of the good things in life, or of the joys and freedom of childhood. The goal is to provide children with sound values and solid skills throughout their developmental years so that they will be self-sufficient when the time comes to leave the family. Teaching these behaviors requires persistence, patience and knowledge on the part of parents, all applied consistently and positively through the growing years.

These adaptive responses are learned very slowly during the course of development. And they are learned with difficulty by children, who are all born completely impulsive, essentially self-centered and absolutely pleasure-oriented. But this seemingly insurmountable task is being done every day by wise parents who understand the big picture. "Parents must give their children what they need, not what they want" is one of the fundamentals of raising healthy achieving children these days. Here are some ideas to help you put that guideline into practice in your home.

The Benefits of Struggle

Learning to strive toward personal goals in healthy ways is an invaluable life skill; yet many parents do not really understand the positive psychological benefits of struggle. They perceive it only in terms of personal pain and deprivation. Certainly, most children resist working for what is wanted, and as a result, many parents unwittingly short-circuit their children's healthy striving through indiscriminate giving and giving in. Thus, immense potential for learning remains untapped.

Successfully meeting challenges is certainly not an easy lesson to learn during childhood. As an adult, however, these same skills are even more difficult, sometimes impossible, to acquire when there is no foundation of healthy childhood striving. Parents must, despite their material success in life, create a challenging environment for their children. The payoffs for setting goals, working hard to reach them and overcoming obstacles along the way are deep and lasting. Here is a baker's dozen of these payoffs just for starters.

Developing a clear perception of reality. A common childhood fantasy is one about growing up, quickly becoming rich and famous, and then living happily ever after. In a young child such a fantasy is fine. However, if during the course of development a child is not exposed to the realities of setting and achieving personal goals, then such fantasies may persist into adulthood. These adults become "dreamers" who have great ideas about what they're going to do but never seem to get around to doing it. This is a common pattern in adult Cornucopia Kids. Their dreams remain dreams because they don't know how to make them come true.

Defining personal limits and strengths. It is through confrontation with adversity that self-awareness develops. During the course of healthy striving to meet challenges, a process of self-evaluation takes place naturally. Supportive feedback from parents and significant others who see the broader perspective further helps a child to define positive and negative responses. Once weaknesses and strengths are clearly conceptualized, a child is in a position to maximize positive traits and find ways to strengthen weaknesses. Cornucopia Kids don't benefit from such experiences because they have never had to put themselves on the line. With their parents' help, these "soft" kids are sheltered and pampered instead.

Accepting failure as a part of learning. Henry Ford believed that "failure is the opportunity to begin again more intelligently." Children who are sheltered and indulged typically don't confront failure or learn to deal with it. Rather, they are rescued from the consequences of their actions or parents fix problems and otherwise make life so easy that no challenges or chances for experiencing failure are encountered. Such children, as adults, cannot deal with setbacks because they have no experience doing so. They are more likely to give up when the going gets difficult or try to find someone to "take over" for them as they did in childhood.

Learning to delay gratification. Young children expect immediate gratification. So do Cornucopia Kids, even as adults. They want it all right now. In children, this expectation is natural, but in Cornucopia Kids it is a significant defect. Learning to delay gratification—to wait for a payoff until later—is a lesson in perseverance. It's "hanging in there" until setbacks, obstacles or problems are overcome. Cornucopia Kids don't have to learn perseverance—they just get parents to gratify them "right now" instead.

Developing the capacity for self-discipline. Self-discipline always involves a choice. People consciously choose to do either what is right, what is easy or what is pleasurable. Self-discipline is the capacity to impose internal limits on one's own behavior rather than relying on external sources (parents, teachers, other authorities) to set behavioral boundaries. Cornucopia Kids, whose every whim is satisfied, are usually quite impulsive and are consistently allowed to escape taking responsibility for their actions. As a result, these immature persons never learn to anticipate the possible consequences of their actions or to set their own internal limits. One consequence of this is that they are often in trouble—but that's okay; their parents will bail them out.

Attaining long-term goals becomes possible. Parents often complain that today's kids seem to want everything *right now.* This is especially true of indulged children who are given everything *right now.* These disadvantaged children rarely have to struggle, sacrifice or deny themselves short-term rewards to get what they want. Yet success today requires just that ability. Learning to work for long-term goals requires self-sacrifice, and healthy achievers have practiced that skill all along at home. Cornucopia Kids haven't often had to do so.

4. FAMILY, SCHOOL, AND CULTURAL INFLUENCES: Parenting

Understanding the effort-reward relationship. An important asset in adult life is to have learned that there is a direct relationship between effort and reward. Studying harder for an exam brings a better grade. Doing those extras on the job brings a merit raise or even a promotion. But if a child is given everything merely because it is asked for, or practices manipulation to get it, this important effort-reward relationship has no chance to develop. In its stead grows the false perception that everything in life comes easily. For Cornucopia Kids that may be true at home, but the rest of the world doesn't work that way.

Building a reservoir of confidence. Many people naively believe that confidence comes through emotional support from others and "go get 'em" pep talks. While outside support is certainly helpful, the crucial element in personal confidence is a base of personal experience from which to confront challenges and succeed despite adversity. Those who constantly need massive infusions of external support have *not* developed a foundation of success experiences.

Experiencing the internal rewards of accomplishment. A feeling of pride and the satisfaction of accomplishment can only be experienced when effort has been expended toward attaining a personal goal. "It was hard but I've done it, and it makes me feel good" is emotionally reinforcing and accompanies any tangible rewards that may accrue from one's efforts.

Creating appreciation for personal property. A child who earns the money to buy a bicycle not only experiences pride in that accomplishment but also tends to take very good care of the bike. Why? Because the personal effort involved in obtaining it is attached to its value. A common characteristic of Cornucopia Kids is that they have lots of possessions but don't bother to take care of them. Abuse and neglect usually extend to the property of parents, siblings and friends. In their eyes, property can always be replaced by parents with little effort. With that certainty, why should a child take care of anything?

Gaining practice in making appropriate choices. Parents are fond of saying to their children, "You can't have your cake and eat it, too." From the Cornucopia Kids' point of view, the response is simple: "Wanna bet?" To the extent that they are consistently indulged, Cornucopia Kids *can* have it both ways. They can misbehave and escape the consequences. With parental help, they can enjoy rewards without effort. Kids who have it all continue to expect it all, even as adults. But healthy achievers set priorities and make healthy choices because that's what they have been taught to do all along at home.

Developing a sense of personal control. As a child successfully meets challenges and attains personal goals, he or she develops an internalized sense of control. "I know how to make good things happen" is part of the psychological makeup of a healthy achiever. It is this kind of child who, later as an adult, can meet the world on its own tough terms and succeed. By contrast, those who are materially indulged and allowed to escape the consequences of their behavior remain childlike and insecure. They must depend on others to "make good things happen." Those others are most often parents, and the process becomes increasingly expensive and exasperating as the years pass.

Nourishing healthy self-esteem. Self-direction, propelled by an awareness of internal control, is linked to a sense of personal adequacy and healthy self-esteem. Self-sufficiency is a major ingredient in feeling good about yourself. Cornucopia Kids are not self-sufficient and they know it. As a result, they are insecure, often frightened and emotionally dependent young men and women. "Take care of me" becomes the refrain in their relationship with their parents. With these characteristics it's practically impossible for them to feel good about themselves.

The Consequences of Neglect

Creating a challenging home environment where children can learn healthy life values is possible *regardless of income level*. It's the parents' knowledge (or the lack of it) that makes the difference. For example, it is not uncommon to find parents with modest incomes creating Cornucopia Kids when they cannot really afford it. Just as frequently, those who are extremely well off can be found raising healthy achieving children. The important question is whether parents allow negative material values and the demands of their children to subvert positive parenting.

Interestingly, parents who give too much and give in to their children often admit that they feel something is wrong with what they are doing. But, they argue, no one else seems to be doing what their intuition tells them is right. Up and down the street, neighbors strive to keep up with one another. Colleagues at work boast of how they give and give to their children. In some areas, it's almost a contest among status-minded adults to see who can give their children more and bigger and better. On the home front, the kids know exactly how to apply pressure, and our materialistic culture reinforces the "need" to acquire a continually renewed supply of expensive possessions to keep abreast of the latest fashions and fads.

It takes a strong parent with character and clear values to see through all of this pressure and do what is best—not what is easiest. To their credit, more parents are doing just that. These savvy parents, achievers themselves, are beginning to recognize the serious problems that result from giving too much materially and not giving enough of what children really need: healthy values and adaptive life skills. Conversely, the problems that accrue from an indulgent parental style can easily last a lifetime.

It is not to the benefit of parents, children or society to raise Cornucopia Kids. Yet many parents manage to do just that, without seeing the long-range effects of their day-to-day actions.

For parents who are still wavering in their commitment to do what is right, not what is easy, here's a bit of wisdom to keep in mind: Every person is born with an equal opportunity to become unequal. In a democracy, basic constitutional rights are given to each person, but there the guarantees end, and individuals must use their skills and expertise either to make their way in the world or to fail. Cornucopia Kids, pampered and indulged early in life, just don't have an equal chance.

CAN YOUR CAREER HURT YOUR KIDS?

Mommy often gets home from work too tired to talk. Daddy's almost never around. Says one expert: "We can only guess at the damage being done to the very young."

Kenneth Labich

BECAUSE CHILDREN are the future, America could be headed for bad bumps down the road. Some of the symptoms are familiar—rising teenage suicides and juvenile arrest rates, average SAT scores lower than 30 years ago. But what is the disease festering beneath that disturbing surface? Says Alice A. White, a clinical social worker who has been counseling troubled children in Chicago's prosperous North Shore suburbs for nearly two decades: "I'm seeing a lot more emptiness, a lack of ability to attach, no sense of real pleasure. I'm not sure a lot of these kids are going to be effective adults."

Not all children, or even most of them, are suffering from such a crisis of the spirit. In fact, some trends are headed in a promising direction. For example, drug use among young people has fallen sharply since the 1970s. But a certain malaise does seem to be spreading. Far more and far earlier than ever before, kids are pressured to take drugs, have sex, deal with violence. In a world ever more competitive and complex, the path to social and economic success was never more obscure.

And fewer traditional pathfinders are there to show the way. Divorce has robbed millions of kids of at least one full-time parent. With more and more women joining the work force, and many workaholic parents of both sexes, children are increasingly left in the care of others or allowed to fend for themselves. According to a University of Maryland study, in 1985 American parents spent on average just 17 hours a week with their children.

This parental neglect would be less damaging if better alternatives were widely available, but that is decidedly not the case. Families that can afford individual child care often get good value, but the luxury of

REPORTER ASSOCIATE *Jung Ah Pak*

a compassionate, full-time, $250-a-week nanny to watch over their pride and joy is beyond the reach of most American parents. They confront a patchwork system of informal home arrangements and more structured day care centers. In far too many cases, parents with infants or toddlers cannot feel secure about the care their children get. Says Edward Zigler, a professor of child development at Yale, who has spent much of his career fighting the abuses of child care: "We are cannibalizing children. Children are dying in this system, never mind achieving optimum development."

For older children with no parental overseer, the prospects can be equally bleak. **Studies are beginning to show that preteens and teenagers left alone after school, so-called latchkey children, may be far more prone than other kids to get involved with alcohol and illegal drugs.**

For some experts in the field, the answer to all this is to roll back the clock to an idyllic past. Mom, dressed in a frilly apron, is merrily stirring the stew when Dad gets home from work. Junior, an Eagle Scout, and Sally—they call her Muffin—greet him with radiant smiles. Everybody sits down for dinner to talk about schoolwork and Mom's canasta party.

For others more in touch with the economic temper of the times—especially the financial realities behind the rising number of working mothers—the solution lies in improving the choices available to parents. Government initiatives to provide some financial relief may help, but corporations could make an even greater difference by focusing on the needs of employees who happen to be parents. Such big companies as IBM and Johnson & Johnson have taken the lead in dealing with employee child care problems, and many progressive corporations are discovering the benefits of greater

flexibility with regard to family issues. At the same time, an array of professional child care organizations has sprung up to help big corporations meet their employees' demands.

Without doubt, helping improve child care is in the best interest of business—today's children, after all, are tomorrow's labor pool. Says Sandra Kessler Hamburg, director of education studies at the Committee for Economic Development, a New York research group that funnels corporate funds into education projects: "We can only guess at the damage being done to very young children right now. From the perspective of American business, that is very, very disturbing. As jobs get more and more technical, the U.S. work force is less and less prepared to handle them."

The state of America's children is a political mine field, and threading through the research entails a lot of gingerly probing as well as the occasional explosion. Much work in the field is contradictory, and many additional longer-range studies need to be done before anyone can say precisely what is happening.

Moreover, any researcher who dwells on the problems of child care—of infants in particular—risks being labeled antiprogressive by the liberal academic establishment. If the researcher happens to be male, his motives may seem suspect. If he says babies are at risk in some child care settings, he may be accused of harboring the wish that women leave the work force and return to the kitchen. Much valid research may be totally ignored because it has been deemed politically incorrect.

For example: Jay Belsky, a Penn State professor specializing in child development, set off a firestorm in 1986 with an article in *Zero to Three*, an influential journal that summarizes existing academic re-

4. FAMILY, SCHOOL, AND CULTURAL INFLUENCES: Parenting

search. His conclusions point to possible risks for very small children in day care outside the home. Though he scrupulously threw in a slew of caveats and even went so far as to confess a possible bias because his own wife stayed home with their two children, Belsky came under heavy attack. Feminist researchers called his scholarship into question. Says Belsky: "I was flabbergasted by the response. I felt like the messenger who got shot."

Belsky's critics charged, among other things, that he had ignored studies that document some more positive results from infant day care. Since then, for example, a study conducted by researchers at the University of Illinois and Trinity College in Hartford, Connecticut, found that a child's intellectual development may actually be helped during the second and third years of life if the mother works. The study, which tracked a nationwide sample of 874 children from ages 3 to 4, determined that the mental skills of infants in child care outside the home were lower than those of kids watched over by their mothers during the first year, but then picked up enough at ages 2 and 3 to balance out.

Whatever the merits of his critics' assault, Belsky presents a disturbing picture of the effects on infants of nonparental child care outside the home. In a 1980 study he cited in the article, involving low-income women in the Minneapolis-St. Paul area, infants in day care were disproportionately likely to avoid looking at or approaching their mothers after being separated from them for a brief period.

ANOTHER STUDY, conducted in 1974, concluded that 1-year-olds in day care cried more when separated from their mothers than those reared at home; still another, in 1981, found that day care infants threw more temper tantrums. To at least some extent, the observations seem to apply across socioeconomic boundaries. A 1985 University of Illinois study of infants from affluent Chicago families showed that babies in the care of full-time nannies avoided any sort of contact with their mothers more often than those raised by moms during their first year.

An infant's attachment, or lack of it, to the mother is especially crucial because it can portend later developmental problems. In the Minnesota study, toddlers who had been in day care early on displayed less enthusiasm when confronted with a challenging task. They were less likely to follow their mothers' instructions and less persistent in dealing with a difficult problem. Another study, which took a look at virtually all 2-year-olds on the island of Bermuda, found more poorly adjusted children among the early day care group regardless of race, IQ, or socioeconomic status.

Researchers in Connecticut investigating 8- to 10-year-old children in 1981 found higher levels of misbehavior and greater withdrawal from the company of others among those who had been in day care as infants, no matter what the educational level of their parents. In a study of kindergarten and first-grade children in North Carolina, the early day care kids were found more likely than others to hit, kick, push, threaten, curse, and argue with their peers.

What we should take away from the research, says an unrepentant Belsky, is this: "There is an accumulating body of evidence that children who were cared for by people other than their parents for 20 or more hours per week during their first year are at increased risk of having an insecure relationship with their parents at age 1— and at increased risk of being more aggressive and disobedient by age 3 to 8."

Belsky adds several "absolutely necessary caveats": First, the results of all these studies must be viewed in light of the added stress that many families experience when both parents work and of the fact that affordable high-quality day care is not always available. Belsky agrees with some of his academic opponents that the quality of day care matters. His second warning: The results of these studies are generalizations and do not apply to every single child. Third, he says, nobody really knows what causes underlie the findings.

Research on older children who spend at least part of the day on their own is far less controversial, though no less disturbing. A recent study by the American Academy of Pediatrics focused on substance abuse by nearly 5,000 eighth-graders around Los Angeles and San Diego. The sample cut across a wide range of ethnic and economic backgrounds and was split about half and half between boys and girls. The researchers concluded that 12- and 13-year-olds who were latchkey kids, taking care of themselves for 11 or more hours a week, were about twice as likely as supervised children to smoke, drink alcohol, and use marijuana. About 31% of the latchkey kids have two or more drinks at a time; only about 17% of the others do. Asked whether they expected to get drunk in the future, 27% of the latchkey kids and 15% of the others said yes.

Increasingly, isolation from parents is a problem even when the family is physically together. Beginning in infancy, children are highly attuned to their parents' moods. And when parents have little left to give to their offspring at the end of a stressful day, the kids' disappointment can be crushing. Says Eleanor Szanton, executive director of the National Center for Clinical Infant Programs, a nonprofit resource center in Virginia: "What happens between parents and children during the first hour they are reunited is as important as anything that happens all day. If the mother is too exhausted to be a mother, you've got a problem."

When children become adolescents and begin to test their wings by defying their parents' authority, stressed-out families may break down completely because no strong relationship between parents and children has developed over the years. In high-achiever families, says Chicago social worker Alice White, family life can become an ordeal where children must prove their worth to their parents in the limited time available. Conversation can be a series of "didjas"—Didja ace that test, win the election, score the touchdown? What's missing is the easygoing chatter, the long, relaxed conversations that allow parents and children to know each other.

White says that many of today's kids don't understand how the world works because they haven't spent enough time with their parents to understand how decisions are made, careers are pursued, personal relationships are formed. She finds herself spending more and more time acting as a surrogate mother for the seemingly privileged kids she counsels, advising them on everything from sexual issues to recipes for a small dinner party. Says White: "The parents serve as a model of success, but the kids are afraid they won't get there because nobody has shown them how."

JUST ABOUT EVERYONE in the child-development field agrees that all this adds up to a discouraging picture, but opinions vary wildly as to what ought to be done about it—and by whom. For a growing band of conservative social thinkers, the answer is simple: Mothers ought to stay home. These activists, working at private foundations and conservative college faculties, rail against what they see as the permissiveness of recent decades. They save their most lethal venom for organized child care, blaming it for everything from restraining kids' free will to contributing to major outbreaks of untold diseases. One conservative researcher, Bryce Christensen of the Rockford Institute in Illinois, has likened day care to the drug Thalidomide. Day care, he writes, is "a new threat to children that not only imperils the body, but also distorts and withers the spirit."

Gary L. Bauer, president of a conservative Washington research outfit called the Family Research Council, is among the most visible of these social activists. Bauer, a domestic-policy adviser in the Reagan White House, believes strongly that the entry of great numbers of women into the work force has harmed America's children. He says the importance of bonding between a mother and her children became clear to him and his wife one morning several years ago when they were dropping their 2-year-old daughter at a babysitter's home. The child went immediately to the sitter, calling her "Mommy." That was something of an epiphany for Bauer's wife, Carol: She quit her job as a government employment counselor soon after and has since stayed home to raise the couple's three children.

Still, the dual-career trend continues. No

26. Can Your Career Hurt Your Kids?

WHAT'S HAPPENING TO FAMILIES ...

Number of parents in the household (for children in U.S. under 18)

	1970	1990
Two parents	85%	72%
Mother only	11%	22%
Father only	1%	3%
Neither parent	3%	3%

Children with mothers in the labor force

	1970	1990
Age 5 and under	29%	53%
Ages 6–17	43%	66%

Who takes care of the working moms' preschoolers, 1987
- Relative 37%
- Commercial day care center 24%
- Neighborhood day care center 22%
- Other 17%

SOURCES: CHILD TRENDS INC., COLLEGE BOARD, MONITORING THE FUTURE, U.S. GOVERNMENT STATISTICS

... AND TO THE KIDS

High school seniors who read books, magazines, or newspapers (1976–'90, Female and Male, declining from ~60% to ~50%)

Average SAT scores (1963–'90, Math and Verbal, both declining)

Teen suicides, Ages 15–19 (1960–'88, rising from ~500 to over 2,000)

LINDA ECKSTEIN FOR FORTUNE

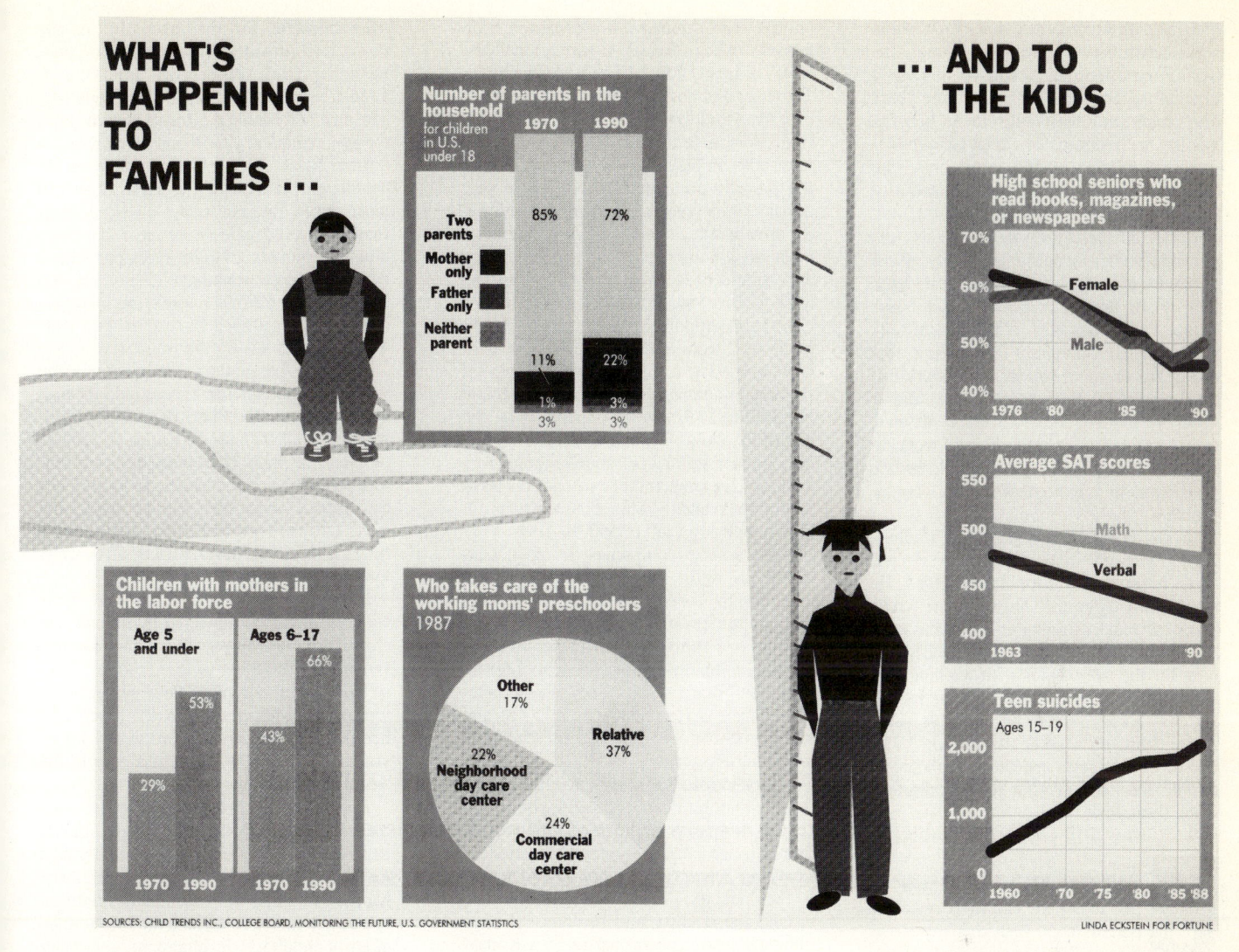

wonder. Most American families could not afford to forfeit a second income, a fact that renders the conservatives' yearning for a simpler past quixotic at best. Real weekly earnings for workers declined 13% from 1973 to 1990. So in most cases two paychecks are a necessity. Also, about a quarter of American children—and about half of black children—live in single-parent homes. Those parents are nearly all women. Though some receive child support or other income, their wages are usually their financial lifeblood. About half the mothers who have been awarded child support by the courts do not get full payments regularly.

From the standpoint of the national economy, a mass exodus of women from the work force would be a disaster: There simply won't be enough available males in the future. Women now make up over 45% of the labor force, and they are expected to fill about 60% of new jobs between now and the year 2000.

EVEN IF a child's welfare were the only consideration, in many cases full-time motherhood might not be the best answer. Children whose mothers are frustrated and angry about staying home might be better off in a high-quality day care center. And many kids may well benefit from the socializing and group activities available in day care. A mother and child alone together all day isn't necessarily a rich environment for the child.

In the end, the short supply of high-quality day care is the greatest obstacle to better prospects for America's children. Experts agree on what constitutes quality in child care—a well-paid and well-trained staff, a high staff-child ratio, a safe and suitable physical environment. They also generally concur that if those criteria are met, most children will not just muddle through but prosper. Says Barbara Reisman, executive director of the Child Care Action Campaign, a nonprofit educational and advocacy organization in New York: "Despite all the questions that have been raised, the bottom line is that if the quality is there, and the parents are comfortable with the situation, the kids are going to be fine."

But even tracking, much less improving, child care quality is a monumental task. Something like 60% of the approximately 11 million preschool kids whose mothers work are taken care of in private homes. That can be a wonderful experience: Grandma or a warm-hearted neighbor spends the day with the wee ones baking cookies and imparting folk wisdom. Or it can be hellish. Yale's Edward Zigler speaks in horror of the home where 54 kids in the care of a 16-year-old were found strapped into car seats all day. Low pay and the lack of status associated with organized day care centers make it tough to recruit and retain qualified workers. A study by a research group called the Child Care Employee Project in Oakland revealed an alarming 41% annual turnover rate at day care operations across the U.S. One big reason: an average hourly salary of $5.35. Says Zigler: "We pay these people less than we do zoo keepers—and then we expect them to do wonders."

The experts have floated various schemes to make more money available. The Bush Administration has offered $732 million in block grants to the states for child care and has also proposed increasing the modest tax credits (current maximum: $1,440) for lower-income parents using most kinds of day care. Another current idea: increase the personal exemption. If it had kept pace with inflation since 1950, it would be about $7,000 instead of $2,050. Zigler has proposed that couples be allowed to dip into

4. FAMILY, SCHOOL, AND CULTURAL INFLUENCES: Parenting

their Social Security accounts for a short while when their children are young. Under his plan families would be limited to tapping the accounts for up to three years per child, and their retirement benefits would be reduced or delayed proportionately. Zigler would also limit the amount of money withdrawn to some reasonable maximum.

Under all these proposals, parents—especially women—could more easily afford to pay for high-quality day care, scale back their work hours, or even stay at home longer with a newborn if they wished. A 1989 Cornell University study found that about two-thirds of mothers who work full time would cut their hours if they didn't need the extra income. In other surveys, even greater percentages of working mothers with infants say they would reduce their hours or stay home if money were no problem.

In his recent book, *Child Care Choices*, Zigler presents some innovative notions about improving care for older children. He would start organized classes in the public school system beginning at age 3, and keep schools open in the afternoon and early evening for the use of kids with working parents. School libraries, gyms, music rooms, and art rooms would be available. Says Zigler: "All you need is a traffic cop." The city of Los Angeles and the entire state of Hawaii have begun after-school programs along these lines.

BUSINESS has a crucial role to play in helping employees who are parents cope with their responsibilities. The U.S. Congress is currently considering legislation that would guarantee 12 weeks of unpaid leave in the event of a family illness or the birth of a child. According to a study conducted by economics professors at Cornell University and the University of Connecticut, the costs of allowing such leaves for most workers is less than letting them quit and hiring permanent replacements. Companies can provide big-time relief with smaller gestures as well: making a telephone available to assembly-line workers so they can check up on their latchkey kids, say, or letting office workers slip out for a parent-teacher conference without a hassle.

As the competition for good workers heats up, many companies will be forced to grapple with the problems working parents face, or risk losing desirable employees. Says Douglas Besharov, a resident scholar at the American Enterprise Institute in Washington: "If you need workers, you will do what has to be done."

Many corporations have already taken the plunge. IBM, among other companies, offers employees a free child-care referral service; IBM uses 250 different organizations. The company has also pledged $22 million over five years to improve the quality of day care available in the towns and cities where most of its employees live. Johnson & Johnson provides an array of goodies: one-year unpaid family care leaves, an extensive referral network, dependent care reimbursement accounts so that employees can pay child care expenses with pretax dollars, and up to $2,000 toward the cost of adoption.

J&J also supports an on-site day care center at its headquarters in New Brunswick, New Jersey. The company subsidizes part of the cost, but employees using the center still pay $110 to $130 a week depending on the age of the child. Depending on the region, average charges for a preschooler range from $67 to $115 a week. Infant care can cost up to $230 a week, and the affluent few with a full-time nanny pay $200 to $600.

Some smaller companies are paying attention as well. American Bankers Insurance Group, based in Miami, maintains a day care center for employees' children ages 6 weeks to 5 years. After that the child can attend a company-run private school for an additional three years. The school takes care of the child from 8 A.M. to 6:15 P.M. and keeps its doors open during school holidays and summer vacations.

The child care business is growing rapidly. Approximately 77,000 licensed child care centers now serve about four million children daily. Financially troubled Kinder-Care Learning Centers of Montgomery, Alabama, is the industry giant, with 1,257 centers around the country and revenues of $396 million a year. The runner-up is a Kansas City, Missouri, outfit called La Petite Academy Inc. It operates about 750 centers and did $201 million worth of business in 1990.

ServiceMaster, the widely diversified management company based near Chicago, jumped into the field last year and now runs three centers in suburban office parks, with four more under construction. The beauty part of the business is that ServiceMaster typically gets some form of financial help from a landlord or corporate client—reduced rent or lower occupancy costs—and then charges market rates for its services.

Parents working at companies such as Sears, Abbott Laboratories, Ameritech, and Mobil pay about $140 to $150 a week for infants and about $95 a week for 3- to 5-year-olds. ServiceMaster executives consider this a business with splendid growth prospects: They plan to open another half-dozen centers by the end of 1991 and then begin expanding beyond the Chicago area.

Child-development experts admire a relatively new and growing company, Bright Horizons Children's Centers of Cambridge, Massachusetts, for the innovative on-site day care it provides for major corporations including IBM, Prudential, and Dun & Bradstreet. Founded in 1986 by Linda Mason and her husband, Roger Brown, both former economic development workers in Africa, Bright Horizons now operates 38 centers up and down the East Coast. In most cases the corporate client donates space for child care in or near its office building, providing Bright Horizons with a handsome cost advantage. Fees can be steep—up to $225 per week for infants—but companies often subsidize the payments for employees low on the wage scale.

Because the centers are close to the workplace, parents are encouraged to drop in throughout the day. Bright Horizons' teaching staff members earn up to $20,000 per year, far more than most day care centers pay, plus a full benefits package. They can pursue a defined career track and move into management ranks if they qualify. As a result, turnover runs at a relatively low 16% to 24%.

Brown and Mason concede that many children might be damaged by second-rate child care, but they contend that parents rarely have anything to fear from the kind of high-quality attention their centers provide. Says Mason: "It's very much like health care. If you can afford to pay for it, you can receive the best child care in the world in this country."

For business, helping employees find their way through the child care thicket makes increasing sense—and not only as a method of keeping today's work force happy. More and more companies with an eye on the future recognize the importance of early childhood development, and many are alarmed by the discouraging signals they see in the upcoming generation of workers. At BellSouth in Atlanta, for instance, only about one job applicant in ten passes a battery of exams that test learning ability; ten years ago, twice as many did. Even those who make it through the tests often require extensive training and carry heavy personal baggage: A startling 70% of BellSouth's unmarried employees support at least one child.

More companies are finding that they have to help employees cope not just with their children but also with the gamut of life's vicissitudes. Says Roy Howard, BellSouth's senior vice president for corporate human resources: "Business used to feel that you ought to leave your personal problems at home. We can no longer afford to take that view." The psychic welfare of workers—and of their children—is increasingly a legitimate management concern, and companies that ignore it risk their employees' future as well as their own.

THE LASTING EFFECTS OF CHILD MALTREATMENT

Raymond H. Starr, Jr.

Raymond H. Starr, Jr., is a developmental psychologist on the faculty of the University of Maryland, Baltimore County. He has been conducting research with maltreated children and their families for more than sixteen years and was also a founder and first president of the National Down Syndrome Congress.

Every day, the media contain examples of increasingly extreme cases of child abuse and neglect and their consequences. The cases have a blurring sameness. Take, for example, the fourteen-year-old crack addict who lives on the streets by selling his body. A reporter befriends him and writes a vivid account of the beatings the boy received from his father. There is the pedophile who is on death row for mutilating and murdering a four-year-old girl. His record shows a sixth-grade teacher threatened to rape and kill him if he told anyone what the teacher had done to him. There is the fifteen-year-old girl who felt that her parents didn't love her. So she found love on the streets and had a baby she later abandoned in a trash barrel. And there are the prostitutes on a talk show who tell how the men their mothers had trusted sexually abused them as children. These and hundreds more examples assault us and lead us to believe that abused children become problem adolescents and adults.

Are these incidents the whole story? Case examples are dramatic, but have you ever wondered how such maltreatment changes the course of a child's life? In this sound-bite era, most of us rarely stop to think about this important question. We seldom ask why trauma should play such an important role in shaping the course of a child's life.

To examine these questions, we need to understand what psychologists know about the course of lives and how they study them—the subject of the field of life-span developmental psychology.

LIFE-SPAN DEVELOPMENT

Understanding why people behave the way they do is a complex topic that has puzzled philosophers, theologians, and scientists. The course of life is so complex that we tend to focus on critical incidents and key events. Most of us can remember a teacher who played an important role in our own development, but we have to consider that other teachers may have been important. If his seventh-grade civics teacher, Ms. Jones, is the person Bill says showed him the drama of the law, leading him to become a lawyer, does this mean that his sixth-grade English teacher, Ms. Hazelton, played no role in his career choice? An outside observer might say that Ms. Hazelton was the key person because she had a debate club and Bill was the most able debater in his class.

Case descriptions fascinate us, but it is hard to divine the reasons for life courses from such examples. It is for this reason that scientists studying human behavior prefer to use prospective studies. By following people from a certain age, we can obtain direct evidence about the life course and factors that influence it. However, most of our information comes from retrospective studies in which people are asked what has happened to them in the past and how it relates to their present functioning.

Life-span developmental theory seeks to explain the way life events have influenced individual development. Of necessity, such explanations are complex; lives themselves are complex. They are built on a biological foundation, shaped by genetic characteristics, structured by immediate events, and indirectly influenced by happenings that are external to the family. As if this were not complex enough, contemporary theory holds that our interpretation of each event is dependent on the prior interactions of all these factors.

Hank's reaction to the loss of his wife to cancer will differ from George's reaction to his wife's death from a similar cancer. Many factors can contribute to these differing reactions. Hank may have grown up with two parents who were loving and attentive, while George may never have known his father. He may have had a mother who was so depressed that from the time he was two, he had lived in a series of foster homes, never knowing a secure, loving, consistent parent.

MALTREATED CHILDREN AS ADULTS

Research has shown that there is a direct relation between a child's exposure to negative emotional, social, and environmental events and the presence of problems during adulthood. Psychiatrist Michael Rutter compared young women who were removed from strife-filled homes and who later came back to live with their parents to women from more harmonious homes.[1] The women from discordant homes were more likely to become pregnant as teens, were less skilled in parenting their children, and had unhappy marriages to men who also had psychological and social problems. Adversity begat adversity.

Do the above examples and theoretical views mean that abused and neglected children will, with great certainty, become adults with problems? Research on this issue has focused on three questions: First, do maltreated children grow up to maltreat their children? Second, are yesterday's maltreated children today's criminals? Third, are there more general effects of abuse and neglect on later psychological and social functioning? A number of research studies have examined these questions.

The cycle of maltreatment. It makes logical sense that we tend to raise our own children as we ourselves were raised. Different theoretical views of personality development suggest that this should be the case. Psychoanalytic theorists think that intergenerational transmission of parenting styles is unconscious. Others, such as learning theorists, agree that transmission occurs but differ about the mechanism. Learning parenting skills from our parents is the key mode by which child-rearing practices are transmitted from one generation to the next, according to members of the latter group of theorists.

Research suggests that the correspondence between being maltreated as a child and becoming a maltreating adult is far from the one-to-one relationship that has been proposed. Studies have focused on physical abuse; data are not available for either sexual abuse or neglect. In one recent review, the authors conclude that the rate of intergenerational transmission of physical abuse is between 25 percent and 35 percent.[2] Thus, it is far from certain that an abused child will grow up to be an abusive parent. Physical abuse should be seen as a risk factor for becoming an abusive adult, not as a certainty. Many abusive adults were never abused when they were children.

Researchers have also taken a broader approach by examining the cycle of family violence. Sociologist Murray Straus surveyed a randomly selected national sample of families about the extent of violence between family members.[3] Members of the

Research has shown that there is a direct relation between a child's exposure to negative emotional, social, and environmental events and the presence of problems during adulthood.

surveyed families were asked about experiences of violence when they were children and how much husband-wife and parent-child violence there had been in the family in the prior year.

Straus concluded that slightly fewer than 20 percent of parents whose mothers had been violent toward them more than once a year during childhood were abusive toward their own child. The child abuse rate for parents with less violent mothers was less than 12 percent. Having or not having a violent father was less strongly related to whether or not fathers grew up to be abusive toward their own children. Interestingly, the amount of intergenerational transmission was higher if a parent was physically punished by his or her opposite-sex parent.

Straus also found that the abusive adults in his study did not have to have been abused in childhood to become abusive adults. A violent home environment can lead a nonabused child to become an abusive adult. Boys who saw their fathers hit their mothers were 38 percent more likely to grow up to be abusive than were boys who never saw their father hit their mother (13.3 vs. 9.7 percent). Similarly, mothers who saw their mothers hit their fathers were 42 percent more likely to become abusive mothers (24.4 vs. 17.2 percent). Straus views seeing parents fight as a training ground for later child abuse.

To summarize, this evidence suggests that maltreatment during childhood is but one of many factors that lead to a person's becoming an abusive parent. Being abused as a child is a risk marker for later parenting problems and not a cause of such difficulties. It accounts for, at most, less than a third of all cases of physical abuse. Research suggests that a number of other factors, such as stress and social isolation, also play a role as causes of child abuse.[4]

Maltreatment and later criminality. Later criminal behavior is one of the most commonly discussed consequences of child abuse. Research on this subject has examined the consequences of both physical abuse and sexual abuse. Maltreatment has been linked to both juvenile delinquency and adult criminality.

It is difficult to do research on this topic. Furthermore, the results of studies must be carefully interpreted to avoid overstating the connection between maltreatment and criminality. For example, researchers often combine samples of abused and neglected children, making it hard to determine the exact effects of specific forms of maltreatment.

Two types of study have typically been done. Retrospective studies examine the family backgrounds of criminals and find the extent to which they were maltreated as children. It is obvious that the validity of the results of such studies may be compromised by the criminals' distortion of or lack of memory concerning childhood experiences. Prospective studies, in which a sample of children is selected and followed through childhood and into adolescence or adulthood, are generally seen as a more valid research strategy. Such studies are expensive and time-consuming to do.

One review of nine studies concluded that from 8 to 26 percent of delinquent youths studied retrospectively had been abused as children.[5] The rate for prospective studies was always found to be less than 20 percent. In one of the best studies, Joan McCord analyzed case records for more than 250 boys, almost 50 percent of whom had been abused by a parent.[6] Data were also collected when the men were in middle age. McCord found that 39 percent of the abused boys had been convicted of a crime as juveniles, adults, or at both ages, compared to 23 percent of a sample of 101 men who, as boys, had been classified as loved by their parents. The crime rate for both sets of boys is higher than would be expected because McCord's sample lived in deteriorated, urban areas where both crime and abuse are common.

Researchers have also examined the relationship between abuse and later violent criminality. Research results suggest that there is a weak relationship between abuse and later violence. For example, in one study, 16 percent of a group of abused children were later arrested —but not necessarily convicted—as suspects in violent criminal cases.[7] This was twice the arrest rate for nonabused adolescents and adults. Neglected children were also more likely to experience such arrests. These data are higher than would be the case in the general population because the samples contained a disproportionately high percentage of subjects from low-income backgrounds.

The connection between childhood sexual abuse and the commission of sex crimes in adolescence and adulthood is less clear. Most of the small number of studies that have been done have relied upon self-reports of childhood molestation made by convicted perpetrators. Their results show considerable variation in the frequency with which childhood victimization is reported. Incidence figures range from a low of 19 percent to a high of 57 percent. However, we should look at such data with suspicion. In an interesting study, perpetrators of sex crimes against children were much less likely to report that they had been sexually abused during their own childhood when they knew that the truthfulness of their answers would be validated by a polygraph examination and that lies were likely to result in being sent to jail.[8] Thus, people arrested for child sexual abuse commonly lie, claiming that they were abusing children because they themselves had been victims of sexual abuse as children.

To summarize, there is a link between childhood abuse and later criminality. Although some studies lead to a conclusion that this relationship is simple, others suggest that it is really quite complex. The latter view is probably correct.

The case of neglect is an example of this complexity. Widom, in her study discussed above, found that 12 percent of adolescents and adults arrested for violent offenses were neglected as children and 7 percent experienced both abuse and neglect (compared to 8 percent of her nonmaltreated control adolescents and adults).

These data raise an interesting question: Why is neglect, typically considered to be a nonviolent offense, linked to later criminality? Poverty seems to be the mediating factor. Neglect is more common among impoverished families. Poor families experience high levels of frustration, known to be a common cause of aggression. Similarly, we know that lower-class families are, in general, more violent.[9] For these reasons, all the forms of maltreatment we have considered make it somewhat more likely that a maltreated child will grow up to commit criminal acts.

Maltreatment in context. Research suggests that maltreatment during childhood has far-reaching consequences. These are best seen as the results of a failure to meet the emotional needs of the developing child. Indeed, in many cases, the trust the child places in the parent is betrayed by the parent.

This betrayal has been linked to many and varied consequences. The greatest amount of research has focused on the long-term effects of sexual abuse. Studies have looked at samples that are representative of the normal population and also at groups of adults who are seeking psychotherapy because of emotional problems. The most valid findings come from the former type of study. One review of research concluded that almost 90 percent of studies found some lasting effect of sexual abuse.[10]

Sexual abuse has been linked to

Psychoanalytic theorists think that intergenerational transmission of parenting styles is unconscious.

a wide variety of psychological disturbances. These include depression, low self-esteem, psychosis, anxiety, sleep problems, alcohol and drug abuse, and sexual dysfunction (including a predisposition to revictimization during adulthood). As was true for the research reviewed in the preceding two sections of this article, any particular problem is present in only a minority of adult survivors of childhood sexual victimization.

We know less about the long-term effects of physical abuse. Most of the limited amount of available research has used data obtained from clinical samples. Such studies have two problems. First, they rely on retrospective adult reports concerning events that happened during childhood. Second, the use of such samples results in an overestimate of the extent to which physical abuse has long-term consequences. Compared with a random sample of the general population, clinical samples contain individuals who are already identified as having emotional difficulties, regardless of whether or not they have been abused.

Researchers in one study found that more than 40 percent of inpatients being treated in a psychiatric hospital had been sexually or physically abused as children, usually by a family member.[11] Also, the abuse was typically chronic rather than a onetime occurrence. The abused patients were almost 50 percent more likely to have tried to commit suicide, were 25 percent more likely to have been violent toward others, and were 15 percent more likely to have had some involvement with the criminal justice system than were other patients at the same hospital who had not experienced childhood maltreatment.

Much research remains to be done in this area. We know little about the long-term consequences of particular forms of

4. FAMILY, SCHOOL, AND CULTURAL INFLUENCES: Stress and Maltreatment

abuse. The best that we can say is that many victims of physical and sexual abuse experience psychological trauma lasting into adulthood.

The lack of universal consequences. The above analysis suggests that many victims of childhood maltreatment do *not* have significant problems functioning as adults. Researchers are only beginning to ask why many adult victims apparently have escaped unsullied. Factors that mediate and soften the influence of abuse and neglect are called buffers.

The search for buffers is a difficult one. Many of the negative outcomes that have been discussed in the preceding sections may be the result of a number of factors other than maltreatment itself. For example, abused children commonly have behavior problems that are similar to those that have been reported in children raised by drug addicts or adults suffering from major psychological disturbances. Abused children do not exhibit any problems that can be attributed only to abuse. A given behavior problem can have many causes.

One view of the way in which buffers act to limit the extent to which physical abuse is perpetuated across succeeding generations has been proposed by David Wolfe.[12] He believes that there is a three-part process involving the parent, the child, and the relationships between the two. In the first stage, factors predisposing a parent to child abuse (including stress and a willingness to be aggressive toward the child) are buffered by such factors as social support and an income adequate for the purchase of child-care services. Next, Wolfe notes that children often do things that annoy parents and create crises that may lead to abuse because the parent is unprepared to handle the child's provocative behavior. Ameliorating factors that work at this level include normal developmental changes in child behavior, parental attendance at child management classes, and the development of parental ability to cope with the child's escalating annoying actions. Finally, additional compensatory factors work to limit the ongoing use of aggression as a solution to parenting problems.

The amount of intergenerational transmission was higher if a parent was physically punished by his or her opposite-sex parent.

Parents may realize that researchers are indeed correct when they say that physical punishment is an ineffective way of changing child behavior. In addition, children may respond positively to parental use of nonaggressive disciplinary procedures and, at a broader level, society or individuals in the parents' circle of friends may inhibit the use of physical punishment by making their disapproval known. Parents who were abused as children are therefore less likely to abuse their own children if any or all of these mediating factors are present.

Research suggests that the factors mentioned by Wolfe and other influences all can work to buffer the adult effects of childhood maltreatment. These include knowing a nurturing, loving adult who provides social support, intellectually restructuring the maltreatment so that it is not seen so negatively, being altruistic and giving to others what one did not get as a child, having good skills for coping with stressful events, and getting psychotherapy.

One study compared parents who broke the cycle of abuse to those who did not.[13] Mothers who were not abusive had larger, more supportive social networks. Support included help with child care and financial assistance during times of crisis. Mothers who did not continue the abusive cycle also were more in touch with their own abuse as children and expressed doubts about their parenting ability. This awareness made them more able to relive and discuss their own negative childhood experiences.

To summarize, investigators have gone beyond just looking at the negative consequences of childhood maltreatment. They are devoting increasing attention to determining what factors in a child's environment may inoculate the child against the effects of maltreatment. While research is starting to provide us with information concerning some of these mediating influences, much more work needs to be done before we can specify the most important mediators and know how they exert their influences.

CONCLUSIONS

We know much about the intergenerational transmission of childhood physical and sexual abuse. Research suggests that abused children are (1) at an increased risk of either repeating the acts they experienced with their own children or, in the case of sexual abuse, with both their own and with unrelated children; (2) more likely to be involved with the criminal justice system as adolescents or adults; and (3) likely to suffer long-lasting emotional effects of abuse even if they do not abuse their own children or commit criminal acts.

This does not mean that abused chil-

Physical abuse should be seen as a risk factor for becoming an abusive adult, not as a certainty.

> *People arrested for child sexual abuse commonly lie, claiming that they were abusing children because they themselves had been victims of sexual abuse as children.*

dren invariably grow up to be adults with problems. Many adults escape the negative legacy of abuse. They grow up to be normal, contributing members of society. Their escape from maltreatment is usually related to the presence of factors that buffer the effects of the physical blows and verbal barbs.

The knowledge base underlying these conclusions is of varied quality. We know more about the relationship of physical and sexual abuse to adult abusiveness and criminality, less about long-term psychological problems and buffering factors, and almost nothing about the relationship of neglect to any of these outcomes. Almost no research has been done on neglect, a situation leading to a discussion of the reasons behind our "neglect of neglect."[14] Our ignorance is all the more surprising when we consider that neglect is the most common form of reported maltreatment.

The issues involved are complex. We can no longer see the development of children from a view examining such simple cause-effect relationships as exemplified by the proposal that abused children grow up to be abusive adults. Contemporary developmental psychology recognizes that many interacting forces work together to shape development. Children exist in a context that contains their own status as biological beings, their parents and the background they bring to the task of child-rearing, the many and varied environments such as work and school that exert both direct and indirect influences on family members, and the overall societal acceptance of violence.

Advances in research methods allow us to evaluate the interrelationships of all the above factors to arrive at a coherent view of the course of development. Appropriate studies are difficult to plan and expensive to conduct. Without such research, the best that we can do is to continue performing small studies that give us glimpses of particular elements of the picture that we call the life course.

Research is necessary if we are to develop and evaluate the effectiveness of child maltreatment prevention and treatment programs. Our existing knowledge base provides hints that are used by program planners and psychotherapists to find families where there is a high risk of maltreatment and to intervene early. But when such hints are all we have to guide us in working to break the cycle of maltreatment, there continues to be risk of intergenerational perpetuation.

1. Michael Rutter, "Intergenerational Continuities and Discontinuities in Serious Parenting Difficulties," in *Child Maltreatment: Theory and Research on the Causes and Consequences of Child Abuse and Neglect*, ed., Dante Cicchetti and Vicki Carlson (New York: Cambridge University Press, 1989), 317–348.

2. Joan Kaufman and Edward Zigler, "Do Abused Children Become Abusive Adults?" *American Journal of Orthopsychiatry* 57 (April 1987): 186–192.

3. Murray A. Straus, "Family Patterns and Child Abuse in a Nationally Representative American Sample," *Child Abuse and Neglect* 3 (1979): 213–225.

4. Raymond H. Starr, Jr., "Physical Abuse of Children," in *Handbook of Family Violence* ed. Vincent B. Van Hasselt, et al. (New York: Plenum Press, 1988): 119–155.

5. Cathy Spatz Widom, "Does Violence Beget Violence? A Critical Examination of the Literature," *Psychological Bulletin* 106 (1989): 3–28.

6. Joan McCord, "A Forty-year Perspective on Effects of Child Abuse and Neglect," *Child Abuse and Neglect* 7 (1983): 265–270. Joan McCord, "Parental Aggressiveness and Physical Punishment in Long-term Perspective," in *Family Abuse and Its Consequences*, ed. Gerald T. Hotaling, et al. (Newbury Park, Calif.: Sage Publishing, 1988): 91–98.

7. Cathy Spatz Widom, "The Cycle of Violence," *Science*, 14 April 1989.

8. Jan Hindman, "Research Disputes Assumptions about Child Molesters," *National District Attorneys' Association Bulletin* 7 (July/August 1988): 1.

9. Murray A. Straus, Richard J. Gelles, and Suzanne K. Steinmetz, *Behind Closed Doors: Violence in the American Family* (New York: Anchor Press, 1980).

10. David Finkelhor and Angela Browne, "Assessing the Long-term Impact of Child Sexual Abuse: A Review and Conceptualization," in *Family Abuse and Its Consequences*, ed. Gerald T. Hotaling, et al.: 270–284.

11. Elaine (Hilberman) Carmen, Patricia Perri Rieker, and Trudy Mills, "Victims of Violence and Psychiatric Illness," *American Journal of Psychiatry* 141 (March 1984): 378–383.

12. David A. Wolfe, *Child Abuse: Implications for Child Development and Psychopathology* (Newbury Park, Calif.: Sage Publishing, 1987).

13. Rosemary S. Hunter and Nancy Kilstrom, "Breaking the Cycle in Abusive Families," 136 (1979): 1320–22.

14. Isabel Wolock and Bernard Horowitz, "Child Maltreatment as a Social Problem: The Neglect of Neglect," *American Journal of Orthopsychiatry* 54 (1984); 530–543.

CHILDREN AFTER DIVORCE

WOUNDS THAT DON'T HEAL

Judith S. Wallerstein

Judith S. Wallerstein is a psychologist and author of "Second Chances: Men, Women & Children a Decade After Divorce," published by Ticknor & Fields. This article, adapted from the book, was written with the book's co-author, Sandra Blakeslee, who is a regular contributor to The New York Times.

As recently as the 1970's, when the American divorce rate began to soar, divorce was thought to be a brief crisis that soon resolved itself. Young children might have difficulty falling asleep and older children might have trouble at school. Men and women might become depressed or frenetic, throwing themselves into sexual affairs or immersing themselves in work.

But after a year or two, it was expected, most would get their lives back on track, at least outwardly. Parents and children would get on with new routines, new friends and new schools, taking full opportunity of the second chances that divorce brings in its wake.

These views, I have come to realize, were wishful thinking. In 1971, working with a small group of colleagues and with funding from San Francisco's Zellerbach Family Fund, I began a study of the effects of divorce on middle-class people who continue to function despite the stress of a marriage breakup.

That is, we chose families in which, despite the failing marriage, the children were doing well at school and the parents were not in clinical treatment for psychiatric disorders. Half of the families attended church or synagogue. Most of the parents were college educated. This was, in other words, divorce under the best circumstances.

Our study, which would become the first ever made over an extended period of time, eventually tracked 60 families, most of them white, with a total of 131 children, for 10, and in some cases 15, years after divorce. We found that although some divorces work well—some adults are happier in the long run, and some children do better than they would have been expected to in an unhappy intact family—more often than not divorce is a wrenching, long-lasting experience for at least one of the former partners. Perhaps most important, we found that for virtually all the children, it exerts powerful and wholly unanticipated effects.

Our study began with modest aspirations. With a colleague, Joan Berlin Kelly—who headed a community mental-health program in the San Francisco area—I planned to examine the short-term effects of divorce on these middle-class families.

We spent many hours with each member of each of our 60 families—hearing their first-hand reports from the battleground of divorce. At the core of our research was the case study, which has been the main source of the fundamental insights of clinical psychology and of psychoanalysis. Many important changes, especially in the long run, would be neither directly observable nor easily measured. They would become accessible only through case studies: by examining the way each of these people processed, responded to and integrated the events and relationships that divorce brings in its wake.

We planned to interview families at the time of decisive separation and filing for divorce, and again 12 to 18 months later, expecting to chart recoveries among men and women and to look at how the children were mastering troubling family events.

We were stunned when, at the second series of visits, we found family after family still in crisis, their wounds wide open. Turmoil and distress had not noticeably subsided. Many adults were angry, and felt humiliated and rejected, and most had not gotten their lives back together. An unexpectedly large number of children were on a downward course. Their symptoms were worse than they had been immediately after the divorce. Our findings were absolutely contradictory to our expectations.

Dismayed, we asked the Zellerbach Fund to support a follow-up study in the fifth year after divorce. To our surprise, interviewing 56 of the 60 families in our original study, we found that although half the men and two-thirds of the women (even many of those suffering economically) said they were more content with their lives, only 34 percent of the children were clearly doing well.

Another 37 percent were depressed, could not concentrate in school, had trouble making friends and suffered a wide range of other behavior problems. While able to function on a daily basis, these children were not recovering, as everyone thought they would. Indeed most of them were on a downward course. This is a powerful statistic, considering that these were children who were functioning well five years before. It would be hard to find any other group of children—except, perhaps, the victims of a natural disaster—who suffered such a rate of sudden serious psychological problems.

The remaining children showed a mixed picture of good achievement in some areas and faltering achievement in others; it was hard to know which way they would eventually tilt.

The psychological condition of these children and adolescents, we found, was related in large part to the overall quality of life in the post-divorce family, to what the adults had been able to build in place of the failed marriage. Children tended to do well if their mothers and fathers, whether or not they remarried, resumed their parenting roles, managed to put their differences aside, and allowed the children a continuing relationship with both parents. Only a handful of kids had all these advantages.

We went back to these families again in 1980 and 1981 to conduct a 10-year follow-up. Many of those we had first interviewed as children were now adults. Overall, 45 percent were doing well; they had emerged as competent, compassionate and courageous people. But 41 percent were doing poorly; they were entering adulthood as worried, underachieving, self-deprecating and sometimes angry young men and women. The rest were strikingly uneven in how they adjusted to the world; it is too soon to say how they will turn out.

At around this time, I founded the Center for the Family in Transition, in Marin County, near San Francisco, which provides counseling to people who are separating, divorcing or remarrying. Over the years, my colleagues and I have seen more than 2,000 families—an experience that has amplified my concern about divorce. Through our work at the center and in the study, we have come to see divorce not as a single circumscribed event but as a continuum of changing family relationships—as a process that begins during the failing marriage and extends over many years. Things are not getting better, and divorce is not getting easier. It's too soon to call our conclusions definitive, but they point to an urgent need to learn more.

It was only at the 10-year point that two of our most unexpected findings became apparent. The first of these is something we call the sleeper effect.

28. Children After Divorce

A divorce-prone society is producing its first generation of young adults, men and women so anxious about attachment and love that their ability to create enduring families is imperiled.

The first youngster in our study to be interviewed at the 10-year mark was one who had always been a favorite of mine. As I waited for her to arrive for this interview, I remembered her innocence at age 16, when we had last met. It was she who alerted us to the fact that many young women experience a delayed effect of divorce.

As she entered my office, she greeted me warmly. With a flourishing sweep of one arm, she said, "You called me at just the right time. I just turned 21!" Then she startled me by turning immediately serious. She was in pain, she said.

She was the one child in our study who we all thought was a prime candidate for full recovery. She had denied some of her feelings at the time of divorce, I felt, but she had much going for her, including high intelligence, many friends, supportive parents, plenty of money.

As she told her story, I found myself drawn into unexpected intricacies of her life. Her trouble began, typically, in her late teens. After graduating from high school with honors, she was admitted to a respected university and did very well her freshman year. Then she fell apart. As she told it, "I met my first true love."

The young man, her age, so captivated her that she decided it was time to have a fully committed love affair. But on her way to spend summer vacation with him, her courage failed. "I went to New York instead. I hitchhiked across the country. I didn't know what I was looking for. I thought I was just passing time. I didn't stop and ponder. I just kept going, recklessly, all the time waiting for some word from my parents. I guess I was testing them. But no one—not my dad, not my mom—ever asked me what I was doing there on the road alone."

She also revealed that her weight dropped to 94 pounds from 128 and that she had not menstruated for a year and a half.

"I began to get angry," she said. "I'm angry at my parents for not facing up to the emotions, to the feelings in their lives, and for not helping me face up to the feelings in mine. I have a hard time forgiving them."

I asked if I should have pushed her to express her anger earlier.

She smiled patiently and said, "I don't think so. That was exactly the point. All those years I denied feelings. I thought I could live without love, without sorrow, without anger, without pain. That's how I coped with the unhappiness in my parents' marriage. Only when I met my boyfriend did I become aware of how much

4. FAMILY, SCHOOL, AND CULTURAL INFLUENCES: Stress and Maltreatment

feeling I was sitting on all those years. I'm afraid I'll lose him."

It was no coincidence that her acute depression and anorexia occurred just as she was on her way to consummate her first love affair, as she was entering the kind of relationship in which her parents failed. For the first time, she confronted the fears, anxieties, guilt and concerns that she had suppressed over the years.

Sometimes with the sleeper effect the fear is of betrayal rather than commitment. I was shocked when another young woman—at the age of 24, sophisticated, warm and friendly—told me she worried if her boyfriend was even 30 minutes late, wondering who he was with and if he was having an affair with another woman. This fear of betrayal occurs at a frequency that far exceeds what one might expect from a group of people randomly selected from the population. They suffer minute to minute, even though their partners may be faithful.

In these two girls we saw a pattern that we documented in 66 percent of the young women in our study between the ages of 19 and 23; half of them were seriously derailed by it. The sleeper effect occurs at a time when these young women are making decisions with long-term implications for their lives. Faced with issues of commitment, love and sex in an adult context, they are aware that the game is serious. If they tie in with the wrong man, have children too soon, or choose harmful life-styles, the effects can be tragic. Overcome by fears and anxieties, they begin to make connections between these feelings and their parents' divorce:

"I'm so afraid I'll marry someone like my dad."

"How can you believe in commitment when anyone can change his mind anytime?"

"I am in awe of people who stay together."

We can no longer say—as most experts have held in recent years—that girls are generally less troubled by the divorce experience than boys. Our study strongly indicates, for the first time, that girls experience serious effects of divorce at the time they are entering young adulthood. Perhaps the risk for girls and boys is equalized over the long term.

When a marriage breaks down, men and women alike often experience a diminished capacity to parent. They may give less time, provide less discipline and be less sensitive to their children, since they are themselves caught up in the maelstrom of divorce and its aftermath. Many researchers and clinicians find that parents are temporarily unable to separate their children's needs from their own.

In a second major unexpected finding of our 10-year study, we found that fully a quarter of the mothers and a fifth of the fathers had not gotten their lives back on track a decade after divorce. The diminished parenting continued, permanently disrupting the child-rearing functions of the family. These parents were chronically disorganized and, unable to meet the challenges of being a parent, often leaned heavily on their children. The child's role became one of warding off the serious depression that threatened the parents' psychological functioning. The divorce itself may not be solely to blame but, rather, may aggravate emotional difficulties that had been masked in the marriage. Some studies have found that emotionally disturbed parents within a marriage produce similar kinds of problems in children.

These new roles played by the children of divorce are complex and unfamiliar. They are not simple role reversals, as some have claimed, because the child's role becomes one of holding the parent together psychologically. It is more than a caretaking role. This phenomenon merits our careful attention, for it affected 15 percent of the children in our study, which means many youngsters in our society. I propose that we identify as a distinct psychological syndrome the "overburdened child," in the hope that people will begin to recognize the problems and take steps to help these children, just as they help battered and abused children.

One of our subjects, in whom we saw this syndrome, was a sweet 5-year-old girl who clearly felt that she was her father's favorite. Indeed, she was the only person in the family he never hit. Preoccupied with being good and helping to calm both parents, she opposed the divorce because she knew it would take her father away from her. As it turned out, she also lost her mother who, soon after the divorce, turned to liquor and sex, a combination that left little time for mothering.

A year after the divorce, at the age of 6, she was getting herself dressed, making her own meals and putting herself to bed. A teacher noticed the dark circles under her eyes, and asked why she looked so tired. "We have a new baby at home," the girl explained. The teacher, worried, visited the house and discovered there was no baby. The girl's story was designed to explain her fatigue but also enabled her to fantasize endlessly about a caring loving mother.

Shortly after this episode, her father moved to another state. He wrote to her once or twice a year, and when we saw her at the five-year follow-up she pulled out a packet of letters from him. She explained how worried she was that he might get into trouble, as if she were the parent and he the child who had left home.

"I always knew he was O.K. if he drew pictures on the letters," she said. "The last two really worried me because he stopped drawing."

Now 15, she has taken care of her mother for the past 10 years. "I felt it was my responsibility to make sure that Mom was O.K.," she says. "I stayed home with her instead of playing or going to school. When she got mad, I'd let her take it out on me."

I asked what her mother would do when she was angry.

"She'd hit me or scream. It scared me more when she screamed. I'd rather be hit. She always seemed so much bigger when she screamed. Once Mom got drunk and passed out on the street. I called my brothers, but they hung up. So I did it. I've done a lot of things I've never told anyone. There were many times she was so upset I was sure she would take her own life. Sometimes I held both her hands and talked to her for hours I was so afraid."

In truth, few children can rescue a troubled parent. Many become angry at being trapped by the parents' demands, at being robbed of their separate identity and denied their childhood. And they are saddened, sometimes beyond repair, at seeing so few of their own needs gratified.

Since this is a newly identified condition that is just being described, we cannot know its true incidence. I suspect that the number of overburdened children runs much higher than the 15 percent we saw in our study, and that we will begin to see rising reports in the next few years—just as the reported incidence of child abuse has risen since it was first identified as a syndrome in 1962.

The sleeper effect and the overburdened-child syndrome were but two of many findings in our study. Perhaps most important, overall, was our finding that divorce has a lasting psychological effect on many children, one that, in fact, may turn out to be permanent.

Children of divorce have vivid memories about their parents' separation. The details are etched firmly in their minds, more so than those of any other experiences in their lives. They refer to themselves as children of divorce, as if they share an experience that sets them apart from all others. Although many have come to agree that their parents were wise to part company, they nevertheless feel that they suffered from their parents' mistakes. In many instances, conditions in the post-divorce family were more stressful and less supportive to the child than conditions in the failing marriage.

If the finding that 66 percent of the 19- to 23-year-old young women experienced the sleeper effect was most unexpected, others were no less dramatic. Boys, too, were found to suffer unforeseen long-lasting effects. Forty percent of the 19- to 23-year-old young men in our study, 10 years after divorce, still had no set goals, a limited education and a sense of having little control over their lives.

In comparing the post-divorce lives of former husbands and wives, we saw that 50 percent of the women and 30 percent of the men were still intensely angry at their former spouses a decade after divorce. For women over 40 at divorce, life was lonely throughout the decade; not one in our study remarried or sustained a loving relationship. Half the men over 40 had the same problem.

In the decade after divorce, three in five children felt rejected by one of their parents, usually the father—whether or not it was true. The frequency and duration of visiting made no difference. Children longed for their fathers, and the need increased during adolescence. Thirty-four percent of the youngsters went to live with their fathers during adolescence for at least a year. Half returned to the mother's home disappointed with what they had found. Only one in seven saw both mother and father happily remarried after 10 years. One in two saw their mother or their father undergo a second divorce. One in four suffered a severe and enduring drop in the family's standard of living and went on to observe a lasting discrepancy between their parents' standards of living.

We found that the children who were best adjusted 10 years later were those who showed the most distress at the time of the divorce—the youngest. In general, pre-schoolers are the most frightened and show the most dramatic symptoms when marriages break up. Many are afraid that they will be abandoned by both parents and they have trouble sleeping or staying by themselves. It is therefore surprising to find that the same children 10 years later seem better adjusted than their older siblings. Now in early and mid-adolescence, they were rated better on a wide range of psychological dimensions than the older children. Sixty-eight percent were doing well, compared with less than 40 percent of older children. But whether having been young at the time of divorce will continue to protect them as they enter young adulthood is an open question.

Our study shows that adolescence is a period of particularly grave risk for children in divorced families. Through rigorous analysis, statistical and otherwise, we were able to see clearly that we weren't dealing simply with the routine angst of young people going through transition but rather that, for most of them, divorce was the single most important cause of enduring pain and anomie in their lives. The young people told us time and again how much they needed a family structure, how much they wanted to be protected, and how much they yearned for clear guidelines for moral behavior. An alarming number of teenagers felt abandoned, physically and emotionally.

For children, divorce occurs during the formative years. What they see and experience becomes a part of their inner world, influencing their own relationships 10 and 15 years later, especially when they have witnessed violence between the parents. It is then, as these young men and women face the developmental task of establishing love and intimacy, that they most feel the lack of a template for a loving relationship between a man and a woman. It is here that their

4. FAMILY, SCHOOL, AND CULTURAL INFLUENCES: Stress and Maltreatment

anxiety threatens their ability to create new, enduring families of their own.

As these anxieties peak in the children of divorce throughout our society, the full legacy of the rising divorce rate is beginning to hit home. The new families being formed today by these children as they reach adulthood appear particularly vulnerable.

Because our study was such an early inquiry, we did not set out to compare children of divorce with children from intact families. Lacking fundamental knowledge about life after the breakup of a marriage, we could not know on what basis to build a comparison or control group. Was the central issue one of economics, age, sex, a happy intact marriage—or would any intact marriage do? We began, therefore, with a question—What is the nature of the divorce experience?—and in answering it we would generate hypotheses that could be tested in subsequent studies.

This has indeed been the case. Numerous studies have been conducted in different regions of the country, using control groups, that have further explored and validated our findings as they have emerged over the years. For example, one national study of 699 elementary school children carefully compared children six years after their parents' divorce with children from intact families. It found—as we did—that elementary-age boys from divorced families show marked discrepancies in peer relationships, school achievement and social adjustment. Girls in this group, as expected, were hardly distinguishable based on the experience of divorce, but, as we later found out, this would not always hold up. Moreover, our findings are supported by a litany of modern-day statistics. Although one in three children are from divorced families, they account for an inordinately high proportion of children in mental-health treatment, in special-education classes, or referred by teachers to school psychologists. Children of divorce make up an estimated 60 percent of child patients in clinical treatment and 80 percent—in some cases, 100 percent—of adolescents in inpatient mental hospital settings. While no one would claim that a cause and effect relationship has been established in all of these cases, no one would deny that the role of divorce is so persuasively suggested that it is time to sound the alarm.

All studies have limitations in what they can accomplish. Longitudinal studies, designed to establish the impact of a major event or series of events on the course of a subsequent life, must always allow for the influence of many interrelated factors. They must deal with chance and the uncontrolled factors that so often modify the sequences being followed. This is particularly true of children, whose lives are influenced by developmental changes, only some of which are predictable, and by the problem of individual differences, about which we know so little.

Our sample, besides being quite small, was also drawn from a particular population slice—predominately white, middle class and relatively privileged suburbanites.

Despite these limitations, our data have generated working hypotheses about the effects of divorce that can now be tested with more precise methods, including appropriate control groups. Future research should be aimed at testing, correcting or modifying our initial findings, with larger and more diverse segments of the population. For example, we found that children—especially boys and young men—continued to need their fathers after divorce and suffered feelings of rejection even when they were visited regularly. I would like to see a study comparing boys and girls in sole and joint custody, spanning different developmental stages, to see if greater access to both parents counteracts these feelings of rejection. Or, does joint custody lead to a different sense of rejection—of feeling peripheral in both homes?

It is time to take a long, hard look at divorce in America. Divorce is not an event that stands alone in childrens' or adults' experience. It is a continuum that begins in the unhappy marriage and extends through the separation, divorce and any remarriages and second divorces. Divorce is not necessarily the sole culprit. It may be no more than one of the many experiences that occur in this broad continuum.

Profound changes in the family can only mean profound changes in society as a whole. All children in today's world feel less protected. They sense that the institution of the family is weaker than it has ever been before. Even those children raised in happy, intact families worry that their families may come undone. The task for society in its true and proper perspective is to strengthen the family—all families.

A biblical phrase I have not thought of for many years has recently kept running through my head: "Watchman, what of the night?" We are not I'm afraid, doing very well on our watch—at least for our children. We are allowing them to bear the psychological, economic and moral brunt of divorce.

And they recognize the burdens. When one 6-year-old boy came to our center shortly after his parents' divorce, he would not answer questions; he played games instead. First he hunted all over the playroom for the sturdy Swedish-designed dolls that we use in therapy. When he found a good number of them, he stood the baby dolls firmly on their feet and placed the miniature tables, chairs, beds and, eventually, all the playhouse furniture on top of them. He looked at me, satisfied. The babies were supporting a great deal. Then, wordlessly, he placed all the mother and father dolls in precarious positions on the steep roof of the doll house. As a father doll slid off the roof, the boy caught him and, looking up at me, said, "He might die." Soon, all the mother and father dolls began sliding off the roof. He caught them gently, one by one.

"The babies are holding up the world," he said.

Although our overall findings are troubling and serious, we should not point the finger of blame at divorce per se. Indeed, divorce is often the only rational solution to a bad marriage. When people ask whether they should stay married for the sake of the children, I have to say, "Of course not." All our evidence shows that children exposed to open conflict, where parents terrorize or strike one another, turn out less well-adjusted than do children from divorced families. And although we lack systematic studies comparing children in divorced families with those in unhappy intact families, I am convinced that it is not useful to provide children with a model of adult behavior that avoids problem-solving and that stresses martyrdom, violence or apathy. A divorce undertaken thoughtfully and realistically can teach children how to confront serious life problems with compassion, wisdom and appropriate action.

Our findings do not support those who would turn back the clock. As family issues are flung to the center of our political arena, nostalgic voices from the right argue for a return to a time when divorce was more difficult to obtain. But they do not offer solutions to the wretchedness and humiliation within many marriages.

Still we need to understand that divorce has consequences—we need to go into the experience with our eyes open. We need to know that many children will suffer for many years. As a society, we need to take steps to preserve for the children as much as possible of the social, economic and emotional security that existed while their parents' marriage was intact.

Like it or not, we are witnessing family changes which are an integral part of the wider changes in our society. We are on a wholly new course, one that gives us unprecedented opportunities for creating better relationships and stronger families—but one that also brings unprecedented dangers for society, especially for our children.

CHILDREN of VIOLENCE

What Happens to Kids Who Learn as Babies to Dodge Bullets and Step Over Corpses on the Way to School?

LOIS TIMNICK

Lois Timnick is a Times staff writer. Lilia Beebe contributed to this report.

THE morning after a 19-year-old gang member was gunned down at a phone box at 103rd and Grape streets in Watts, his lifeless body lay in a pool of blood on the sidewalk as hundreds of children walked by, lunch boxes and school bags in hand, on their way to the 102nd Street Elementary School. A few months later, during recess, kindergartners at the school dropped to the ground as five shots were fired rapidly nearby, claiming another victim. On still another occasion, an outdoor school assembly was disrupted by the crackle of gunshots and wailing sirens as students watched a neighborhood man scuffle with police officers.

Terrifying occurrences such as these have brought together six youngsters, ages 6 through 11, who sit in a circle around a box of Kleenex in a colorful classroom. The children are a bit fidgety and shy at first, as a psychiatric social worker asks if anyone would like to share a recent event that made them sad. With hesitation, then with the words spilling out, each tells his story—pausing frequently to grab a tissue to wipe away the tears.

"They shoot somebody every day," begins Lester Ford, who is 9 and lives with his mother and brother in the vast Jordan Downs housing project across from the school. When he's playing outside and hears gunshots, the solemn child says softly, "I go in and get under the bed and come out after the shooting stops."

He says he has lost seven relatives. "My daddy got knifed when he got out of jail," Lester explains, and suddenly tears begin streaming down his face. "My uncle got shot in a fight—there was a bucket of his blood. And I had two aunties killed—one of them was pushed off the freeway and there were maggots on her."

Sitting next to Lester, 11-year-old Trevor Dixon, whose mother and father died of natural causes, puts a comforting arm around his friend. "We don't come outside a lot now," he says of himself and his twin sister. "It's like the violence is coming down a little closer."

When it's her turn, 8-year-old Danielle Glover peers through thick glasses and says matter-of-factly: "Just three people [in my family] died." At night their ghosts haunt her, she says. "I been seein' two of them."

This is grief class at the 102nd Street Elementary School, and it is one of the front lines in the battle against violence in South-Central Los Angeles and other urban war zones.

29. Children of Violence

Experts and mental health professionals are just beginning to learn what happens to children like Lester, Trevor and Danielle as they grow into adulthood: **Even if these children of violence survive the drugs, the gangs and the shootings, they might not survive the psychological effects of the constant barrage.**

Though therapists are finding encouraging signs of resiliency, they believe that no child who is victimized, witnesses violent crime or simply grows up in its maelstrom escapes unscathed. Despite a fragmented and sometimes underfunded approach, these researchers are developing therapies to address the problem.

Two years ago, 102nd Street school principal Melba Coleman, the school guidance counselor and the school psychologist had seen some children regress to bed-wetting, others become overly withdrawn or hostile and good students struggle to concentrate. They called on the Los Angeles Unified School District's mental health center, and social worker Deborah Johnson, to develop a way to help the kids overcome their experiences.

So far, 30 children have participated in the weekly hour-long class, which is thought to be the first regular grief and loss program for elementary school students in the nation. They are encouraged to talk about life, death and ways to keep safe in an unsafe world. The hope is that, by sharing their thoughts and emotions with others, the children will come to terms with their feelings of loss, anger and confusion before long-term, irreversible problems develop. And it seems to be working.

Says Kentral Brim, 10, whose two older brothers were killed and who barely escaped injury himself during a gang fight that broke out at Martin Luther King Jr./Drew Medical Center: "I was getting mad and fighting." With the group's help, the neatly dressed, polite young boy says, "I settled down."

SETTLING DOWN IS hard in South-Central Los Angeles, in the shadow of the famous Watts Towers and within a few blocks of four squalid public housing projects. Sleepless nights are punctuated by gunshots, sirens and hovering police helicopters. Liquor stores are routinely robbed. Children as young as 6 are recruited as drug-runners. Some babies' first words and gestures are the names and hand signs of their parents' gangs. The color of a T-shirt can determine whether someone lives or dies, and the most important lesson of childhood is that survival depends on hitting the ground when the inevitable shooting starts. Some families are so fearful that whenever gang warfare flares up, they live behind closed curtains with the lights off, sleeping and eating on the floor to avoid stray bullets.

The scope of violence in South-Central Los Angeles is horrifying. In the first seven months of this year, there have been 237 homicides, 413 rapes, 5,864 robberies and 9,068 aggravated assaults in one of the most turbulent areas of the city. Thirty-five of the homicide victims were under 18. A study headed last year by Dr. Gary Ordog at King/Drew Medical Center found that 34 children under 10 years old were treated there for gunshot wounds between 1980 and 1987. Records showed none in earlier years.

Perhaps most telling of all is the fact that 90% of children taken to the psychiatric clinic at the hospital have witnessed some act of violence, a recent UCLA survey found.

"How can children see all that [violence] and *not* be affected?" asks Gwen Bozart, a third-grade teacher at Compton Avenue Elementary School. Especially when such violence takes place against a landscape of deprivation and failure. Standardized school test scores in South-Central Los Angeles are far below the average for the rest of the city and state. The dropout rate in some high schools is nearly twice that of the rest of the district. More than half the adult population is unemployed. And mental health professionals say depression and suicide attempts are disproportionately high among a despairing population that is barely surviving.

"There's just so much stress that an individual can take before he is completely overwhelmed," notes UCLA child psychiatrist Gloria Johnson Powell, who grew up in the tough Roxbury section of Boston and this fall will establish a center at Harvard University to focus on the special needs of minority children. Often, Powell points out, inner-city children witness violence at home as well as at school and in the streets. They frequently endure abusive family relationships and watch endless hours of romanticized TV violence. "[Many of] these children have daily stress from the time they wake up until they go to bed," she says.

Researchers have found that youngsters growing up in a war-zone environment such as South-Central Los Angeles are likely to become anxious or depressed. Youngsters who have been direct victims or who have witnessed, say, the brutal murder of a parent, are most likely to suffer post-traumatic stress disorder, today's term for a cluster of symptoms recognized in soldiers and others for many years but given labels such as "shellshock" or "combat neurosis."

In children, post-traumatic stress disorder takes the form of reliving the violent experience repeatedly in play, nightmares and sudden memories that intrude during class or other activities. Kids with the disorder can be easily startled, apathetic, hopeless or possessed by a fear of death. Many children regress to early childhood behaviors such as clinginess, become extremely irritable and develop stomachaches and headaches that have no organic cause.

They play differently, too, going beyond the roughhousing that is part of normal child development. Such traumatized children tend to be more aggressive and more willing to take risks—wrestling to hurt companions, for example, or jumping from high places. Others can become inhibited, forsaking sports they used to enjoy; still others might re-enact a gruesome event in play. (The popular childhood game of ring-around-the-rosy with its "ashes, ashes, all fall down" is thought to have originated as a response to the bubonic plague, when children watched people die and saw streets filled with corpses.) A psychiatrist tells of a child who, after having witnessed the stabbing of her mother, painted her hands red with her paintbrush.

Most devastating perhaps for school-age children, post-**traumatic stress disorder** reduces the ability to concentrate and remember, resulting in poor school performance. But it doesn't stop there. Principal Coleman says the teachers at the 102nd Street school identified about 10% of the school's

4. FAMILY, SCHOOL, AND CULTURAL INFLUENCES: Stress and Maltreatment

more than 1,200 youngsters as showing "high risk" behaviors and notes that these children also interfere with the others' ability to learn. "High risk" behaviors, which include repeatedly cursing at and hitting adults, are thought to be early signs that a child could become a social misfit.

Older children sometimes cope by affecting an indifference to the violence around them, experts say. They exhibit an emotional denial that can cripple all of their relationships. The seeming indifference is evident in random conversations with South-Central Los Angeles children over the last year. One group spoke dispassionately of finding a murdered woman's mutilated body—"Her eyeball was in her shoe," a child said. Another group, when told about a fatal shooting, appeared less interested in the victim than in the details of his shiny new truck and its equipment. And, when asked for class field-trip ideas, an 11-year-old boy suggested, "How about the cemetery?"

Other conversations reveal further emotional distancing: Several 8-year-olds discussing whether the murder of a lifeguard was justified because he had ordered someone to get out of the pool agreed that, as one child explained, "Yeah, he [the killer] shudda done it." To them, killing seemed a reasonable response to a perceived insult. A 16-year-old girl at Jordan High School, ticking off the

'The more kids are exposed to violence, the more desensitized they become.'

names of at least nine people who had been shot recently, put it this way: "The ones [bodies] I see in the street that are killed, that don't mean nothin' to me anymore."

Such callous talk among children alerts mental health professionals to underlying emotional difficulties. "You're not going to be very trusting [as an adult] if you observe or have close to you violent behavior," says Santa Monica therapist Ruth Bettelheim, a former consultant to Head Start and daughter of noted child psychologist Bruno Bettelheim. "You tend to keep your distance psychologically—making close and intimate relationships difficult. And that isolation fuels depression, which is already there because of previous losses."

But, Bettelheim says, some children are able to overcome these problems if the adults and children around them can offer support. "It's old psychology wisdom that the same fire that melts butter hardens steel," she adds, so the very violence that spells destruction for one child might make a strong survivor out of another.

Psychiatrist William Arroyo, acting director of the Los Angeles County/USC child-adolescent psychiatric clinic, has studied refugee children from Central America and children from inner-city areas as well. He acknowledges that supportive parents, schools and neighborhoods can serve as buffers against stressful environments, then shakes his head sadly at the fact that South-Central Los Angeles offers so little: Many households are chaotic and headed by uneducated welfare mothers, nutrition and prenatal care are poor, mental health programs are scarce, and Watts doesn't even have a YMCA, where kids can participate in structured activities instead of hanging out and getting into trouble.

"The more they are exposed to violence, the more desensitized they become until it's no longer horrifying but merely an occurrence in daily living," Arroyo says. "We are *very* concerned about those who already have psychiatric disorders and those who have poor impulse control, who then witness violence either in real life or on television. If youngsters learn that the way of succeeding in everyday tasks includes maiming or killing community members, stealing and generally engaging in sorts of behavior that larger society calls criminal, we'll see a larger population of these types."

THE CALENDAR SAYS it happened more than five years ago. But Ana Anaya Gonzalez, now nearly 16, still pictures it clearly. Even when she tries to forget, her scarred body and recurring nightmares have been constant reminders.

It was a winter Friday afternoon in South-Central Los Angeles, just as children were being dismissed from the 49th Street Elementary School. A deranged neighborhood resident fired 57 times at the playground from his second-story window across the street, killing two people and injuring 13 others.

Ana, a fifth-grader at the time, remembers that she was playing on the monkey bars while her sister, Rosa, went back inside to get her sweater.

"I thought it sounded like gunfire, but a friend said, 'No, it's firecrackers.' Then a bullet hit the ground near me, and everybody dropped. I fell on my knees. I couldn't feel my legs, but I didn't know I was hit."

Inside the school, Rosa's teacher pushed her to the floor and fell on top of her to protect her. A passing jogger spotted the sniper, shouted to him to stop and, seeing that Ana was moving, threw himself over her. He was hit by the next round of gunfire and died two months later.

"I was wearing a pink dress," Ana says, "and when I saw it was full of blood, I tried to scream to a teacher in the doorway to help me. He looked at me and then turned around and closed the door."

Ana spent five months in the hospital. She lost a kidney, still suffers leg pain, cannot bend backward and experiences stomach discomfort when she eats. A single bullet wound scars her back; surgical incisions mark her abdomen.

She also suffers from post-traumatic stress disorder, according to Dr. Quinton C. James, the psychiatrist who treated her and several others injured in the sniper attack. "Whether you're talking about violence in Belfast, Beirut or South-Central Los Angeles, I think it all has an impact, although you never know how it will manifest itself in a particular child. Up to a point, one child may adapt with no [apparent] impairment, while for another it may impair [his] ability to function adequately, socially and academically," says James, the former chief of child/adolescent services at the Augustus Hawkins Mental Health Center (a part of King/Drew Medical Center) who now works with the School Mental Health Center and the Centinela Child Guidance Clinic in Inglewood.

Immediately after the attack, mental health professionals from the school district, the Los Angeles County Mental

29. Children of Violence

Art Therapy: Drawing From Experience

EVEN WHEN children can't verbalize the effect that violence has on them, they sometimes express it by drawing pictures full of blood, guns and knives, says Dr. Spencer Eth, acting chief of psychiatry of the West Los Angeles Veteran's Administration Medical Center and medical director of a trauma unit associated with Cedars-Sinai Medical Center.

Eth worked with the crisis team that responded to the shooting at the 49th Street Elementary School. He says that when children too traumatized to talk are told, "Just draw about anything you want," their pictures reveal much about what's on their minds. This enables therapists to ask children to tell a story, which usually has some connection with the trauma they have suffered.

For example, after a boy threatened to jump off the roof at an elementary school as his horrified classmates watched, children spontaneously made drawings that depicted the incident and the hospital where they imagined the boy was taken.

Says Eth: "Drawing is one of the most effective techniques we have for getting a child to open up and confront difficult feelings—the first step in healing." —*L.T.*

Pictures from several schools, clockwise from top, depict a classmate's suicide; a man killing his baby; the "cemetery where baby is"; sadness about a classmate who drowned.

Health Department, County-USC and UCLA medical centers and the Cedars-Sinai Psychological Trauma Center and various university and private consultants volunteered to help students, as well as parents and teachers. Over the next several weeks, using art therapy (see box above) and in-depth discussions of the shooting, the experts helped those who had witnessed the attack deal with its shattering effects. Social workers continued to work with the children several days a week for the next year.

A month after the shooting, a team headed by Dr. Robert Pynoos, an associate professor of psychiatry at the UCLA School of Medicine, returned to see how the children were doing. Those most severely affected reported feeling stressed, upset and afraid just from thinking about the shooting, and fearful that it might happen again. They complained of jumpiness, nightmares, loss of interest in activities, difficulty paying attention in school and other disturbances. Those who had experienced other violence, an unexpected death or physical injury during the preceding year described having renewed thoughts and images of that event—even if they were not directly exposed to the playground shooting. Many were depressed or grieving a year later, the team found.

"You can cope with it," James says he told Ana at the start of her psychotherapy. "There are things that happen in life, but you don't have to be defeated by them. You'll have physical scars, emotional scars, but you have to accept that it happened

4. FAMILY, SCHOOL, AND CULTURAL INFLUENCES: Stress and Maltreatment

and that we don't know why. . . . The thing is you're still alive . . . and there is something you can do. We'll find out together; I'll help you, and so will your family and other relatives, people at school and other agencies."

With her determination and the help of intense psychotherapy, Ana's condition improved. After seeing James every day for the first month she was hospitalized and five days a week for the next four months, she was able to return to finish sixth grade at the 49th Street school, although she never again ventured onto the playground.

Ana continues to progress, but memories of the shooting plague her, James says. As recently as last February—on the anniversary of the incident—she told James: "I still feel the same way. I get scared about things. . . . I get nervous and start crying about the past."

Though she knows the sniper killed himself, Ana occasionally feels as if he is stalking her. She is sometimes afraid to be alone and sleeps in a bed with Rosa. Until recently, she has had a recurring nightmare: A man is chasing her and shoots her. "I wake up when he shoots me and can't go back to sleep."

Formerly an excellent student, Ana has had trouble concentrating, has required a tutor and is working at about average academic level—although her grades are improving at Jefferson High School, where she is now a junior.

Other members of Ana's family are dealing with the shooting as well. Rosa, now 14, appears withdrawn and more traumatized than Ana, perhaps because she has received less therapy, James says. She was reluctant to return to school, avoids discussing the attack and remains fearful and anxious.

Their mother, Esperanza "Blanca" Gonzalez, a waitress who has four other children, suffered a nervous breakdown and had to be hospitalized briefly.

"I feel sick for Ana," Gonzalez says. "She's very nervous and restless in the classroom. She's lost a lot of her spirit." Gonzalez knows that her daughter "cries every night . . . and is sad much of the time," but she says the family cannot afford psychotherapy.

The Gonzalezes live in the same neighborhood, still frightened by the gunshots they hear in the night. "I still get scared," Ana says. "When I hear the shots outside, sometimes I feel like they are shooting at me."

Ana's mother says she remains bitter about the police department's failure to respond to previous complaints about the sniper's brandishing and firing guns. "Until we see blood, we can't do anything about him," she says officers told neighbors.

On the outside, Ana is a pretty, dark-haired teen-ager who appears bubbly, caught up in plans with her girlfriends for her upcoming Sweet 16 birthday party. She landed a summer job selling theater tickets and says she goes to dances and parties as much as possible.

She is talkative—but not about the shooting. "It's not going to change anything to talk about it," she says, then adds, "but I would like to go back to therapy because I like to draw the pictures about what happened. I get too nervous now, and I think talking with the doctor helps."

James adds that Ana is one of the lucky ones, a child whose outlook is much brighter for having a "very supportive network." Without caring family, friends and school personnel, she might have become another trauma victim unable to envision a future. But Ana has career hopes, "like maybe becoming a doctor. I liked how they worked when I was in the hospital."

WHILE EDUCATORS and mental health professionals work closely with youngsters in the classroom, researchers continue studying—and in some cases, debating—how violence affects the young.

The first attempts to evaluate scientifically the phenomenon investigated children during wartime. Studies by Anna Freud and others after World War II suggested that children are minimally affected by war and sometimes find it exciting. This work has now been largely debunked. After studying the survivors of Belfast, Cambodia and Beirut, most experts contend instead that the effects of trauma can be masked, delayed or minimized—but never eliminated.

Some experts such as UCLA's Pynoos have found that exposure to violence can cause physiological changes in a child's developing brain stem, altering the brain's chemistry and causing personality changes—such as reduced impulse control, an attraction to danger or a debilitating sense of fear.

But whatever the theory, almost all the experts speak with awe of the emotional strength children possess, even those youngsters from the bleakest backgrounds.

One of the few studies exploring the roots of resilience in young children followed nearly 700 Hawaiian children over a 30-year period, ending in 1985. The study, conducted by Emmy Werner, a child psychologist at the University of California, Davis, found that one out of every four children classified as "high risk" infants had developed into a competent, confident and caring young adult. Some seemed to have a natural strength, but for others, the scales tipped from vulnerability to resilience because the children found strong emotional support at home, school, work or church.

Raiford Woods, manager of the Jordan High Student Health Clinic, sees examples of resiliency every day in the heart of Los Angeles' most violent neighborhood. With or without outside support, some of "these kids are marvelous and have the psychological strength to survive," he says. "You compare a kid from Watts to one from Orange County or Westwood; he can handle twice as much pressure."

Studies and observations like this form the basis for the widely held belief among experts that early intervention is essential—that anti-gang programs must begin in junior high, that grade-school children need support to get them off to a good start and make them less vulnerable, that "drug babies" and preschoolers need special care. And a growing number of programs seek to apply this premise in young lives.

"Some [older children] are lost causes," psychiatrist James says. "We have to focus our attention on those coming along. I met with some youngsters who all told me they'd already been in jail. 'Doc, you're wasting your time with us,' one said. 'You work with our little brothers and sisters.'"

29. Children of Violence

A few Los Angeles programs focus on prevention, others with helping youngsters cope after the damage has been done. The Los Angeles Unified School District, for instance, offers a kindergarten intervention project in which children who are identified as having social problems often stemming from exposure to violence, are assigned a volunteer "special friend" to act as a companion and confidant.

Preschoolers known to have been prenatally exposed to drugs are the focus of a new program at the Salvin Special Education Center, where early childhood specialists try to interrupt behaviors, often violent, that are forerunners of school failure.

At Jordan High School, alone in the district, all ninth-graders are required to take a violence prevention course. Among other things, the course stresses how to avoid fights, how to be manly without being macho and how to deal positively with anger.

UCLA's Program in Trauma, Violence and Sudden Bereavement responds to requests for assistance from cities across the United States where extreme acts of violence—such as sniper attacks, hostage-taking and shootings—have occurred. The Cleveland Elementary School in Stockton asked for assistance after the mass shooting in January. (While these incidents are obviously traumatic, experts say, they differ significantly from the chronic violence experienced in the inner city.) The program also trains mental health professionals, provides counseling and studies children who have witnessed violence in the home or community. Likewise, the Psychological Trauma Center affiliated with Cedars-Sinai Medical Center provides psychological assistance to schools where tragedy has struck.

'These children have stress from the time they wake up until they go to bed.'

Effective ways of coping with a reality that can't be erased are not likely to lie with any single approach, the experts say. Nor can therapy ignore the web of problems that make dealing with violence even worse. Success lies in the cooperative efforts of the police, mental health and health services, schools, churches and concerned parents, and in solutions that address violence as well as drug abuse, poverty and single-parent homes.

Early-prevention programs are important, such as camp programs that give children an opportunity to see the world outside their everyday existence and television. Parents, especially mothers, need exposure to alternatives, with programs that give them a break from the draining responsibility of caring for several children around the clock, on limited resources and without male support. And ways to bring fathers into the system also need to be found.

For two years, the 102nd Street school program has focused mainly on grief and loss. Run by the school staff, the class is based on the theory that, with the support of their peers, troubled children can learn to deal with feelings they might otherwise suppress or act out at school or at home.

The program's pilot group began with social worker Johnson reading a story about a young boy who flew kites with his uncle. The uncle dies—"and there was not a dry eye in the room," Johnson remembers—but at the end, the boy goes out alone with his kite and remembers the good times they had shared. "That set the stage for the rest," she says.

"Our focus is on recognizing and expressing feelings, getting them to come to grips with the fact that they've experienced a loss and leading toward an acceptance that loss is something we all experience," says Johnson, who has spent most of the past 15 years working with disadvantaged children and their families and responding to violent crises at various schools.

Activities in the grief class include using a "feeling board," on which children draw or write whatever they want. Children play a game in which they make faces in a mirror to reflect different feelings, and they perform relaxation exercises such as deep breathing and stretching. They also listen to soft music while they visualize a place that makes them feel good. And they plant small gardens, which helps instill in them a sense of responsibility while symbolizing the beginning, growth and end of all life.

The program has not been scientifically evaluated. Some participants remain deeply troubled and have required referrals for outside counseling, and a few families have moved in hopes that memories will fade faster in a new setting. But teachers say the attitude, behavior and academic performance of most of the youngsters have improved markedly.

It is those little ones who are maturing and progressing amid daily bloodshed who give hope to Johnson and her colleagues.

"Even though all these horrible things are happening, there is a resilience there. These children and their families do respond to interventions. They have strengths even though life circumstances don't allow them to live outside this war zone," she says. "They've seen a lot, but for some some reason they're still children, still trying to walk the tightrope between the craziness of the adult world and a carefree kind of kid world."

Article 30

Broken homes: 3 in 5 born today will live with a single parent by age 18.

Child care: 2 of 4 children age 13 and under live with parents who both work.

Drugs: 1 child in 6 has tried marijuana and 1 in 3 alcohol before 9th grade.

Sex: The share of girls under 15 who have had sex has tripled in 2 decades.

Suicide: The rate for youths under 15 has tripled since 1960.

CHILDREN UNDER STRESS

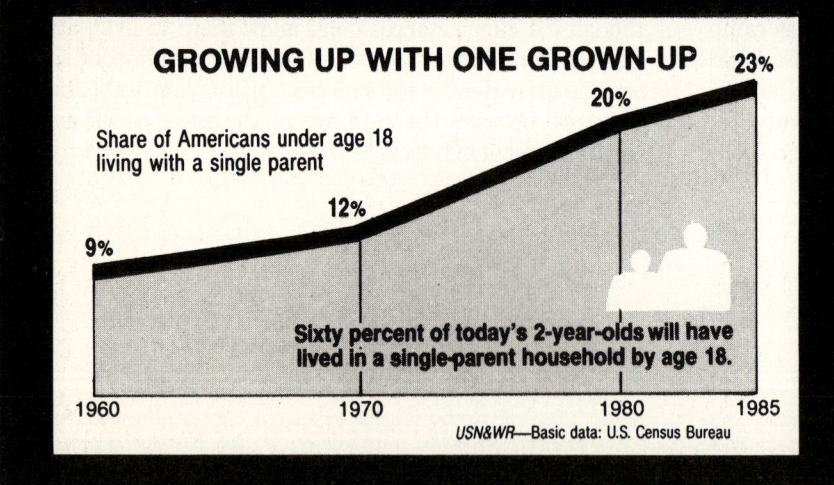

GROWING UP WITH ONE GROWN-UP

Share of Americans under age 18 living with a single parent

1960: 9%
1970: 12%
1980: 20%
1985: 23%

Sixty percent of today's 2-year-olds will have lived in a single-parent household by age 18.

USN&WR—Basic data: U.S. Census Bureau

■ Whatever happened to childhood?

Gone are the worries about dying from the flu, toiling in sweatshops or 5-mile treks to the country school. But for the children of the '80s, a host of psychological pressures have superseded the physical stresses previous generations had to bear.

What the late child psychologist Selma Fraiberg called "the magic years" don't seem so magical any more. Whether they come from cities, suburbs or farms, stable families or broken homes, kids today must cope with a world in which both parents work, sex and drugs cloud even the grade-school yard and violence is only as far away as the living-room television screen.

"The age of protection for children has ended," says Dr. Lawrence Brain, director of child and adolescent services at the Psychiatric Institute of Washington, D.C. "Today's children are increasingly thrust into independence and self-reliance before they have the skills and ability to cope." Laments Ann Gannon, 47, a Bethesda, Md., mother of 13 children, ages 2 to 23: "We're not letting them stop and smell the roses. We don't play sports for fun any more. Children today are constantly being pushed."

Most of the 48 million Americans under age 14 aren't candidates for therapy. But mental-health professionals and child experts are alarmed by the rising proportion of troubled children. "There is a growing awareness and growing incidence of psychological stress on children," says David Elkind, child psychologist and author of *The Hurried Child: Growing Up Too Fast Too Soon*.

The polls seem to bear Elkind out. According to a recent survey by Louis Harris & Associates, 3 out of 4 adults said they believed that the problems facing today's children are more severe than those they faced as children. Fewer than half believed that the nation's children are basically happy. One in 8 said their child had mental or emotional problems, and 1 in 20 admitted their child has had drug problems.

High cost of child rearing

"Today, preadolescence is more stressful than it used to be," says pediatrician and best-selling author T. Berry Brazelton. In his practice, Brazelton reports, there is a greater incidence of such psychosomatic ailments as wheezing, dizziness, chest pains and stomach problems among patients ages 10 to 12 than there was 10 years ago. In fact, up to 35 percent of American kids suffer stress-related health problems at some point, from pulling out one's hair to headaches, says psychologist Nicholas Zill, executive director of Child Trends, a research group in Washington, D.C.

From *U.S. News & World Report*, October 27, 1986, pp. 58-64. Copyright © 1986, U.S. News & World Report.

30. Children Under Stress

The stress on children begins at home. The "typical" American family—with dad at work, mom baking cakes and kids engaging only in innocent mischief—is mainly the stuff of television reruns. Families of the '80s bear little resemblance to the Cleaver clan. Today, only 1 in 5 families with children is composed of a father as breadwinner, mother at home and children under 18, compared with 41 percent 10 years ago, according to the Bureau of Labor Statistics. Spurred largely by economic pressures, women with young children have joined men in 9-to-5 jobs in record numbers over the last decade.

Each family member is going a separate way, as if "he is a corporation in a different business, and the family residence functions like the holding company," suggests Carl Thoresen, professor of education and psychology at Stanford University. "It's as if nobody's home any more."

It's no wonder families need the extra income. The average cost of raising a child born in 1984 to age 18 jumped to $140,927, according to a report by the Conference Board, a business-research group. Thirty years ago, it cost a third as much. The annual bill for one year of child rearing consumed 29 percent of the median family's budget in 1984, compared with an estimated 11 percent in 1966. "I used to think I would be able to stay home with my children," says Gale Whitfield, a Chicago nurse. But Whitfield and her husband David, a lab technician and part-time real-estate agent, found that their combined income of $55,000 last year was barely enough to support the eight children they share from their current and previous marriages.

Fewer children per family

The Whitfields are an unusually large family, by any norms. But parents today are having fewer children per family, which can have the effect of stepping up the pressure on each child. Twenty years ago, 31 percent of all families with children had only one child under 18; now, 42 percent have only one child under 18. And far fewer families—6 percent vs. 20 percent two decades ago—have four or more children. "I've never seen such a high level of caring [from parents]," says Susie Bond, a nursery-school teacher in suburban Washington, D.C. "But, you can lose perspective on what's good for your kid. Kids don't have time to play, to do nothing."

Some of the current stresses are self-inflicted by a generation of status-conscious parents. "There are cults about strollers," says Marian Blum, educational director of the Child Study Center at Wellesley College. "You've got to have the right stroller or you're not a good parent. You've got to dress the kid in certain clothes or you're not a good parent. If your kid doesn't go to a certain school, you're not a good parent."

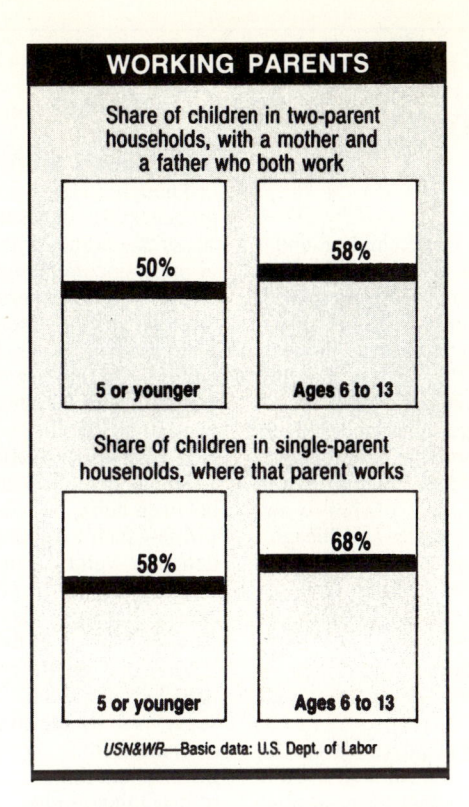

Indeed, most of the pressures on kids stem from America's new lifestyles. Take day care, for example. At least 7 million children nationwide are cared for in family day-care homes or in child-care centers. Baby-sitting a child in someone else's home for a fee accounts for about three fourths of all child care outside the home. The quality of family day care can vary greatly: Seventy-five to 90 percent of family day-care facilities are unlicensed or unregistered, says Representative George Miller (D-Calif.), chairman of the Select Committee on Children, Youth and Families.

Generally, the consensus is that sending children under age 3 to day care is more stressful for the kids than staying at home with a parent or a familiar baby-sitter. While research studies indicate that good-quality day care is not harmful to the very young, the stress of separation from the parents plus the pressures of a competitive program can exacerbate tensions, says Tufts University's Elkind. Warns Dr. Arnold Samuels, a child expert at Chicago's Institute for Psychoanalysis: "The child must comply with the environment; the program doesn't always respond to the child."

With more mothers and fathers pushing papers than strollers, toddlers barely out of diapers are being shipped to preschool. Half of the children under age 6 who live with both parents have working mothers and fathers, compared with 28 percent in 1970. Last year, 39 percent of all 3-and-4-year-olds were enrolled in preschool; 11 percent in 1965. "It seems that kids are enrolled in nursery school as soon as the mother gets pregnant," notes Dr. Judy Howard, a pediatrician at the University of California at Los Angeles.

The kindergarten hustle

In New York City, Mayor Edward Koch has pledged that by 1989 every 4-year-old will be eligible for enrollment in a publicly funded, voluntary prekindergarten program, making New York one of the first major cities to open classes to children that young. "This isn't a prekindergarten boot camp," insists Marian Schwarz, Koch's coordinator of youth services. Twenty-odd other states have plans for similar programs, and some educators are worried about how these programs will be implemented. "What used to be the second-grade curriculum is now in first grade, and the first-grade curriculum is being brought into kindergarten," says Polly Greenberg, publications director of the National Association for the Education of Young Children. Many kids, she argues, are being pushed too hard academically, too fast.

Partly because more families have both parents in the work force, all-day (6-hour) kindergarten programs are increasing. More than a third of all kindergarten children go a full day, compared with a fifth in 1973. In Stamford, Conn., for instance, all 11 elementary schools have offered full-day kindergarten since 1980. Moreover, many nursery schools offer "extended day" programs for families where both parents work outside the home.

But sending children to kindergarten before they are ready can harm them, according to some studies. Research by James Uphoff, an Ohio education professor, and June Gilmore, an Ohio psychologist, found that children who were enrolled in kindergarten below the age of 5 years and 3 months or in first grade before 6 years and 3 months don't do as well in subsequent schooling as those who began school later.

Uphoff's study of 278 pupils in the Hebron, Nebr., Elementary School suggests that parents should think twice before they push kids to start earlier than their classmates. "Summer children"—those born during the summer months and younger than most of their classmates—accounted for 75 percent of the school's academic failures.

The evidence is by no means one-sided. Even prekindergarten programs can have major benefits—as long as

---

WORKING PARENTS

Share of children in two-parent households, with a mother and a father who both work

5 or younger	Ages 6 to 13
50%	58%

Share of children in single-parent households, where that parent works

5 or younger	Ages 6 to 13
58%	68%

USN&WR—Basic data: U.S. Dept. of Labor

4. FAMILY, SCHOOL, AND CULTURAL INFLUENCES: Stress and Maltreatment

For many kids, the community net is gone

Growing up poor

The numbers tell a sorry story: 7 million white children and 4 million black children under age 15 in the U.S. live below the poverty line. One out of five children in America is poor, with children nearly twice as likely to be poor today as adults or the elderly. "We have become the first nation in history," laments Senator Daniel Patrick Moynihan (D-N.Y.), "where the children are the poorest segment of the population."

While divorce and separation are no longer unusual in higher-income families, half of all poor children grow up in households headed by women, and a quarter of poor black children live in homes where Mom has never married. "My mother *is* my father," says Reggie Minnis, an eighth grader at Chicago's Albert Einstein Elementary, a South Side school where 90 percent of the students are from single-parent families. Children in such homes often "become the boss of the family because a parent is always away," notes guidance counselor Annie Stephens of Houston's E. O. Smith Middle School. "When they try to become the boss of the school, they wreak havoc among the other kids."

In the 1950s, malnutrition, ill health and poverty were far more common among children, yet most poor children lived in rural communities that provided leadership that today is often missing. Marian Wright Edelman, head of the Washington, D.C.-based Children's Defense Fund, recalls her childhood in Bennettsville, S.C., as a time when "the churches, schools and your family all reinforced the message that you could make it, even if you were poor. We all expected to go to college and didn't have time to get into trouble." Poor children today, particularly in the inner city, are more isolated from others who have climbed out of the slums. "That community net, the role models—the doctor, the police officer and the minister," she says, "are not there now."

Peer pressure fills the vacuum of authority created by absent fathers and other missing role models, often with destructive results. Many poor children come to feel as though they've done something wrong if they refrain from sex, drugs, or joining a gang. "My buddies were delighted that I did well in school," says James Comer, a professor at Yale University's Child Study Center who grew up in East Chicago. "Poor kids today, particularly black ones, are considered turncoats or sellouts if they do well."

More abuse, more neglect

Though today's poor kids differ in some ways from their predecessors, they share the experience of scraping by—and the stress that comes with it. Compared with the children of middle or upper-income families, poor children are seven to eight times more likely to experience child abuse and neglect, and two to three times more likely to grow up in families headed by parents who describe their health as only fair or poor. Preschool children in mother-only families are also estimated to use mental-health facilites four times more often than children from two-parent families.

Not by bread alone

Social scientists are increasingly concerned about a link between being poor as a child and experiencing handicaps later in life—such as giving birth out of wedlock, becoming dependent on welfare, and dropping out of school. "There was 'nothing broke' in poor people who grew up in earlier years except that many had been subjected to terrible racism," says Charles Murray of the Manhattan Institute for Policy Research. "When they got the chance, they took it; it's a lot tougher now to help kids who don't know how to go to work or deal with a supervisor by the time they're 20."

Some critics say too much emphasis has been given to government aid to poor children, while too little has been paid to providing guidance. Education Under Secretary Gary Bauer, who heads an administration task force on families, argues that "parents and teachers have mistakenly tried to be morally neutral to the nth degree in recent years. What we learned out of the antipoverty efforts of the 1960s is that children, just like man, cannot live by bread alone."

Others stress that the civil-rights movement and such antipoverty programs as Head Start opened doors to many poor children. Maintains Robert Coles, Harvard psychiatrist and author of the five-volume study *Children of Crisis*, "There is more hope and expectation among poor children today than there was 30 years ago." At the same time, he notes, government efforts to reach out to the excluded are considerably more circumscribed today than they were during the 1960s and 1970s. Says Coles: "The time where society tells poor kids 'come and join in' is fading."

---

they emphasize play, not academic skills. Findings from a 20-year study by the High/Scope Educational Research Foundation show that children who attend preschool are less likely to require special educational services, are more likely to graduate from high school and get better-paying jobs and are less likely to receive welfare in adulthood.

There's little doubt that harried parents beget harried children. "Children have a greater concern about time than is normal," says Principal Rudolph White of Carderock Springs Elementary School in Bethesda, Md. One result of all the hurrying is that "the child's sense of achievement is compressed." Unlike previous generations, children's daily routines are likely to be dictated by Mom and Dad's workload. "Their schedules for meals, bedtime and baths are not predictable the way kids' schedules used to be," says Blum. Even so-called quality time—the special time parents devote to their children—cannot compensate for a lack of quantity time, argues Deborah Fallows, author of *A Mother's Work*. "It's hard to program quality-time moments. They just happen."

Quality-time squeeze

A recent study by the University of Michigan Institute for Social Research found that working mothers spend an average of 11 minutes daily of quality time (defined as exclusive playing or teaching) with their kids during weekdays and about 30 minutes per day on the weekends. Fathers spend about 8 minutes of quality time with their kids on weekdays and 14 minutes on weekends. Nonworking mothers spend 13 minutes per day of quality time with their children.

It's far from clear that children with working mothers are being disadvantaged. A recent Kent State University study of 573 elementary-school students in 38 states found that, on average, children of working mothers scored higher on IQ and reading tests, had better communication skills, were absent fewer days and were more self-reliant than children of nonworking mothers. But not all children of working mothers fared equally well. Children did better if their mothers worked part time rather than full time, were married rather than divorced and worked at high-status jobs.

Even the most self-reliant children face a dangerous world. Everything from the milk carton at breakfast—plastered with pictures of missing children—to the evening news can be a chilling reminder of this fact. There's extra stress on kids who feel there are

30. Children Under Stress

situations where they have to take care of themselves. Says Blum of Wellesley: "What children need to know is that the grown-ups in their world—parents, baby-sitters, neighbors—are there to take care of them and keep them safe." For Marie La Vere, 14, of Marine City, Mich., the fright is real. "I'm scared," she says. "You hear about kids getting killed all the time. You hear about people disappearing."

A stack of research points to the conclusion that TV is one of the culprits. "Violence on television does lead to aggressive behavior by children and teenagers who watch the programs. Children who watch a lot of violence on television come to accept violence as normal behavior," says a 1982 report by the National Institute of Mental Health. Others argue that just because "heavy viewers" may be more violent or don't read as well or get lower scores on IQ tests—all of which are the case—doesn't mean that TV causes those problems.

TV as baby-sitter

What isn't disputed is that in many homes, TV has become an easy substitute for family interaction. According to the A. C. Nielsen Company, children watch an average of 4 hours a day, and some estimates go as high as 6 hours. "TV saps the energy out of family life," says child expert Marie Winn, author of *The Plug-In Drug*. For America's estimated 7 million latchkey kids under age 14—who are alone until their parents return from work—TV can be the only friendly voice in an otherwise empty home. TV also may be partly responsible for raising false expectations. "The people on TV are so well off economically that many kids don't understand why they can't buy clothes and live like that," says Aletha Huston, codirector of the University of Kansas Center for Research on the Influence of Television on Children. "There's this notion that everybody should be happy all the time [on TV]," adds Jo Ann Allen, professor of social work at the University of Michigan. "People can't be happy all the time."

For some, the strains of growing up in the '80s prove too much. Mimicking adult behavior, more children experiment with alcohol and other drugs at an earlier age than ever before. "It's not uncommon to find kids initiating alcohol or marijuana use at ages 9 or 10," says Catherine Bell, chief of the National Institute of Drug Abuse (NIDA) prevention branch. The blend of peer pressure and curiosity can unleash self-destructive habits. "When children like themselves, they do positive things like eat healthy foods and won't take drugs or alcohol," says Dr. Lee Salk, professor of psychology, psychiatry and pediatrics at Cornell University Medical School. Although there are signs that drug use by older teens has leveled off—except for cocaine—more of their younger siblings are experimenting with drugs, NIDA estimates. "They tell you if you do it [smoke marijuana], it will make you feel good," says Adam Royder, 10, a Houston fifth grader.

Sex has become an issue for more kids at an earlier age, too. "Your parents tell you, 'Don't have sex until you're married.' But people do on television," says Michigan ninth grader La Vere. Nearly a quarter of 15-to-19-year-old girls had sexual intercourse before age 16. In 1984, the rate of illegitimate births among teenage girls was almost double that in 1970.

Along with grown-up problems like drug abuse, suicide and depression are becoming kid stuff. America's overall suicide rate peaked in 1977 and has decreased for all groups except kids under 15. Last year, about 300 children under age 15 committed suicide, double the number in 1980. Dr. Brain of the Psychiatric Institute of Washington says that, in reviewing a two-year period of admissions, 80 percent of children age 5 through 13 were clinically depressed, had attempted suicide or were experiencing suicidal thoughts. Among the cases were an 8-year-old who deliberately ran in front of a car, a 9-year-old who tried to hang himself and a 12-year-old who took a drug overdose.

Three years ago, Judie Smith, director of educational services at the Dallas Suicide and Crisis Center, encountered her youngest suicide attempt: A 5-year-old boy, whose 8-year-old brother dialed the hot line. The parents were contacted before the boy carried out his threat.

More cases of depression among children are being reported. While some experts say that indicates better diagnosis, not a higher incidence of troubles, a number of studies suggest that at least 5 percent of all children suffer significant episodes of depression. The rate of serious depression for children age 6 to 12 is about 2 percent, and for adolescents the rate may go as high as 15 percent, says Andres Pumariega, director of child and adolescent psychiatry at the University of Texas. "Only in the last decade did we acknowledge that kids could be depressed," Smith says.

The stresses felt by children with families go double for children of broken homes. In the last 15 years, 15.6 million marriages ended in divorce, disrupting the lives of 16.3 million children under age 18. The number of preschoolers living with a single divorced parent soared 111 percent in the last 15 years. Half of new marriages this year will fail, predict government statisticians. "I have one friend whose parents are so angry at each other she says she doesn't ever want to get married," says an 11-year-old private-school student in San Francisco. "So many kids in my class have divorced parents."

Divorce's legacy

Not surprisingly, children of divorce show more signs of stress than others. According to a 10-year study by Judith Wallerstein, executive director of the Center for The Family in Transition near San Francisco, who has looked at the impact of divorce on 130 middle-class children, 37 percent of them were more emotionally troubled five years after the divorce than they were initially. Fear of divorce affects their love lives later, Wallerstein says. The subjects in her study "enter relationships with a high level of anxiety and have more trouble with marriage—they want a stable relationship more and are more worried about it [than children of nondivorced parents]."

The remarriage of a divorced parent often adds to, rather than subtracts from, a child's stress, by shattering his or her last hopes for a reconciliation and rearranging family loyalties. An estimated 1,300 stepfamilies are formed every day, points out the Stepfamily Association of America. "The hardest thing is getting along with my new stepmother's daughter," says a 12-year-old Atlanta girl who lives with her father, his new wife, and a 6-year-old stepsister. "I can talk to my dad about it sometimes, but mostly he's busy." The mere thought of parents' remarrying upsets many kids. "I'd come home from dates and find one of my daughters camped out in a sleeping bag beside my bed," says Mary Wudtke, a Chicago insurance executive who divorced seven years ago when her girls were 10 and 12.

In single-parent homes, children are often expected to stand in for the absent adult by doing household chores and caring for younger siblings. Although most middleclass kids under age 14 live in two-parent families, the percentage of these children living in female-headed middleclass families increased from 9 percent to 13 percent in the last decade. "Some of these children lead two lives. They're grown-ups at home and they come to school to be children," says elementary-school counselor Ynolia Bell Trammell of John Codwell Elementary School in Houston. One of her students, for instance, 11-year-old Joseph Jackson, often dozed in class. His teacher finally discovered the reason: Joseph's mother works until midnight, and he baby-sits for his 1-year-old brother. "Sometimes, he wakes up early in the morning, and I have to, too," Joseph says.

143

4. FAMILY, SCHOOL, AND CULTURAL INFLUENCES: Stress and Maltreatment

Dr. Benjamin Spock on bringing up today's children
Don't push your kids too hard

Q Dr. Spock, why are today's children under stress?

Partly because we've given up so many of the comforts and sources of security of the past, such as the extended family and the small, tightly knit community and the comfort and guidance that people used to get from religion.

Q How do working mothers affect kids?

It is stressful to children to have to cope with groups, with strangers, with people outside the family. That has emotional effects, and, if the deprivation of security is at all marked, it will have intellectual effects, too.

We know now that if there's good day care it can substitute pretty well for parental care. But, though we're the richest country the world has ever known, we have nowhere near the amount of subsidized day care we need. We're harming our children emotionally and intellectually to the degree that they're in substandard day care.

Q Are children raised in single-parent homes more stressed than other kids?

It's not that a single parent can't raise a child well but that it's harder to raise a child in most cases with one parent than it is with two parents. The parents can comfort and consult and back up each other.

Q Do parents harm kids by pushing them to achieve?

Our emphasis on fierce competition and getting ahead minimizes the importance of cooperation, helpfulness, kindness, lovingness. These latter qualities are the things that we need much more than competitiveness. I'm bothered, for instance, at the way we coach young children in athletics and, even more ludicrous, the interest we focus on superkids. It hasn't gone very far, but there are parents who, when they hear that other children are learning to read at the age of 2, think, "My God, we should be providing reading instruction, too," without ever asking the most significant question: "Does it make the child a better reader or is there any other advantage to learning to read at 2 rather than waiting until age 6?" It imposes strains on children. It teaches them that winning is the important thing. We've gone much too far in stressing winning.

I was in Japan lecturing a few years ago, and they told me that the rate of suicide among elementary schoolchildren is shockingly high and that Japanese elementary schoolchildren commit suicide because they are afraid that they aren't getting grades high enough to satisfy their parents. We haven't gotten that far in the United States, but we're certainly headed that way.

Q What child-rearing advice is best suited to the 1980s?

We can at least bring up children with a strong feeling that they're in the world not just for their own fulfillment—although I think fulfillment is fine—but also to be useful and to help others. Children should be brought up with a strong feeling that there are lots of problems in the neighborhood, the nation and the world, and that they're growing up to help solve those problems.

That emphasis on helpfulness should begin at a very early age with things as simple as letting them help set the table. Never say, "It's easier for me to do it myself." You should encourage children to be helpful, and not by scolding them or forcing them but by supporting them or complimenting when they're helpful.

Q Are there specific things to avoid?

Absolutely no violence on television. Don't give war toys. These are poisonous to children. This whole Rambo spirit is a distressing thing, especially in the most violent country in the world.

Q Is watching television harmful to kids?

A lot of what they see brutalizes sexuality. In simpler societies, you don't see people smashing each other in the face or killing each other. The average American child on reaching the age of 18 has watched 18,000 murders on TV. Yet we know that every time a child or an adult watches brutality, it desensitizes and brutalizes them to a slight degree. We have by far the highest crime rates in the world in such areas as murders within the family, rape, wife abuse, child abuse. And yet we're turning out more children this way, with this horrible profusion of violence that children watch on TV. It's a terrible thing.

Q Do the new stresses on kids make them better equipped to deal with adult stresses?

No. Human beings do make some adjustment to stresses, but that doesn't mean that they're doing better by being brought up with stresses. It's going to make them more tense, more harsh, more intensely competitive and more greedy. I don't think people can live by that. It is a spiritual malnutrition, just like a lack of vitamins or a lack of calories.

Q What kind of parents will today's children make?

If they're brought up with tension and harshness, then they'll do the same with their children. Everybody acquires his attitude and behavior toward his children by how he was treated in his own childhood. What was done to you in childhood, you are given permission to do. To put it more positively, parental standards are what make for a better society, and poor parental standards are what makes for a deteriorating society.

Q Is it harder to be a parent today?

Yes. When I started pediatric practice in '33, parents worried about polio and pneumonia and ear abscesses. Now they have to worry about drugs and teenage pregnancy and nuclear annihilation.

30. Children Under Stress

Time bombs or strong characters?

The long-term effects of stress on tomorrow's adults are yet to be determined, but there are some disturbing clues. Because studies show that children tend to repeat parental patterns with their own children—child abuse is a prime example—a self-perpetuating cycle of destructive behavior may already be in motion. "We're going to be delivering time bombs of potential destruction into society," warns Dr. Salk. Adds Lois Lane, a clinical social worker in Marin County, Calif. "What worries me a lot is that these stressed, disturbed kids aren't going to be a whole lot better before they have kids of their own. So you've got the problem of sick parents and sicker kids. It will be a generational crisis."

Other research, which emphasizes children's amazing ability to snap back from adversity, offers a more hopeful prognosis. Says Jerome Kagan, a developmental psychologist at Harvard and the leading proponent of the resiliency theory: "Ninety percent of the kids experiencing stress today will learn to cope with it. A child is remarkably plastic." Some studies bolster the view that kids who are victimized can overcome early traumas. After World War II, for instance, orphans from Balkan countries who had roamed homeless and eaten out of garbage cans were later adopted by middle-class families in America and fared well. Similar reports exist for orphans of the Korean War.

The advice most experts offer parents for minimizing stress on children is really just common sense: Communicate. Respect the child's individuality instead of making comparisons with others. Set a good example, and don't push them.

A small number of parents who can afford to are even re-evaluating whether the stresses of two-income families are worth it. A case in point: Carol Orsborn and her husband Dan, who have two children and run a public-relations agency in San Francisco, recently scaled back their work hours and living standards, including moving to a smaller house, to do this. "If we didn't have it all, we felt guilty. If we did, we felt exhausted," says Carol Orsborn, the author of *Enough Is Enough: Exploding the Myth of Having It All.*

There are other signs that a backlash against pushing kids too fast may be building. A few states, such as New Hampshire and Oklahoma, promote "developmental screening" tests designed to place kids in appropriately paced classes or signal parents to hold their children back a year. New Hampshire has two decades of experience in creating "readiness classes" that use an informal, play-oriented approach for 6-year-olds who don't score well on developmental tests. At L. J. Campbell School in Atlanta, educators have broken down the kindergarten curriculum into steps for kids to reach at their own pace. Soon, first, second and third grades will be subdivided into academic steps that allow slower kids more time to catch up without being stigmatized.

Small ripples perhaps, but taken together, they reflect second thoughts about the cult of the superkid. Still, the social changes that have pushed the fast-forward button on childhood are irreversible. The age of innocence is shorter than it has ever been before.

---

by Beth Brophy with Maureen Walsh of the Economic Unit, Art Levine, Andrea Gabor, Betsy Bauer, Lisa Moore and bureau reports

The hidden obstacles to black success.
RUMORS OF INFERIORITY

Jeff Howard and Ray Hammond

Jeff Howard is a social psychologist; Ray Hammond is a physician and ordained minister.

TODAY'S black Americans are the beneficiaries of great historical achievements. Our ancestors managed to survive the brutality of slavery and the long history of oppression that followed emancipation. Early in this century they began dismantling the legal structure of segregation that had kept us out of the institutions of American society. In the 1960s they launched the civil rights movement, one of the most effective mass movements for social justice in history. Not all of the battles have been won, but there is no denying the magnitude of our predecessors' achievement.

Nevertheless, black Americans today face deteriorating conditions in sharp contrast to other American groups. The black poverty rate is triple that of whites, and the unemployment rate is double. Black infant mortality not only is double that of whites, but may be rising for the first time in a decade. We have reached the point where more than half of the black children born in this country are born out of wedlock—most to teenage parents. Blacks account for more than 40 percent of the inmates in federal and state prisons, and in 1982 the probability of being murdered was six times greater for blacks than for whites. The officially acknowledged high school dropout rate in many metropolitan areas is more than 30 percent. Some knowledgeable observers say it is over 50 percent in several major cities. These problems not only reflect the current depressed state of black America, but also impose obstacles to future advancement.

The racism, discrimination, and oppression that black people have suffered and continue to suffer are clearly at the root of many of today's problems. Nevertheless, our analysis takes off from a forward-looking, and we believe optimistic, note: we are convinced that black people today, because of the gains in education, economic status, and political leverage that we have won as a result of the civil rights movement, are in a position to substantially improve the conditions of our communities using the resources already at our disposal. Our thesis is simple: the progress of any group is affected not only by public policy and by the racial attitudes of society as a whole, but by that group's capacity to exploit its own strengths. Our concern is about factors that prevent black Americans from using those strengths.

It's important to distinguish between the specific circumstances a group faces and its capacity to marshal its own resources to change those circumstances. Solving the problems of black communities requires a focus on the factors that hinder black people from more effectively managing their own circumstances. What are some of these factors?

Intellectual Development. Intellectual development is the primary focus of this article because it is the key to success in American society. Black people traditionally have understood this. Previous generations decided that segregation had to go because it relegated blacks to the backwater of American society, effectively denying us the opportunities, exposure, and competition that form the basis of intellectual development. Black intellectual development was one of the major benefits expected from newly won access to American institutions. That development, in turn, was expected to be a foundation for future advancement.

YET NOW, three decades after *Brown v. Board of Education*, there is pervasive evidence of real problems in the intellectual performance of many black people. From astronomical high school dropout rates among the poor to substandard academic and professional performance among those most privileged, there is a disturbing consistency in reports of lagging development. While some black people perform at the highest levels in every field of endeavor, the percentages who do so are small. Deficiencies in the process of intellectual development are one effect of the long-term suppression of a people; they are also, we believe, one of the chief causes of continued social and economic underdevelopment. Intellectual underdevelopment is one of the most pernicious effects of racism, because it limits the people's ability to solve problems over which they are capable of exercising substantial control.

Black Americans are understandably sensitive about discussions of the data on our performance, since this kind of information has been used too often to justify attacks on affirmative action and other government efforts to improve the position of blacks and other minorities. Nevertheless, the importance of this issue demands that black people and all others interested in social justice overcome our sensitivities, analyze the problem, and search for solutions.

The Performance Gap. Measuring intellectual performance requires making a comparison. The comparison may be with the performance of others in the same situation, or with some established standard of excellence, or both. It is typically measured by grades, job performance ratings, and scores on standardized and professional tests. In recent years a flood of articles, scholarly papers, and books have documented an intellectual performance gap between blacks and the population as a whole.

• In 1982 the College Board, for the first time in its history, published data on the performance of various groups on the Scholastic Aptitude Test (SAT). The difference between the combined median scores of blacks and whites on the verbal and math portions of the SAT was slightly more than 200 points. Differences in family income don't explain the gap. Even at incomes over $50,000, there remained a 120-point difference. These differences persisted in the next two years.

• In 1983 the NCAA proposed a requirement that all college athletic recruits have a high school grade-point average of at least 2.0 (out of a maximum of 4.0) and a minimum combined SAT score of 700. This rule, intended to prevent the exploitation of young athletes, was strongly opposed by black college presidents and civil rights leaders. They were painfully aware that in recent years less than half of all black students have achieved a combined score of 700 on the SAT.

• Asian-Americans consistently produce a median SAT score 140 to 150 points higher than blacks with the same family income.

• The pass rate for black police officers on New York City's sergeant's exam is 1.6 percent. For Hispanics, it's 4.4 percent. For whites, it's 10.6 percent. These are the results *after* $500,000 was spent, by court order, to produce a test that was job-related and nondiscriminatory. No one, even those alleging discrimination, could explain how the revised test was biased.

• Florida gives a test to all candidates for teaching positions. The pass rate for whites is more than 80 percent. For blacks, it's 35 percent to 40 percent.

This is just a sampling. All these reports demonstrate a real difference between the performance of blacks and other groups. Many of the results cannot be easily explained by socioeconomic differences or minority status per se.

WHAT IS the explanation? Clear thinking about this is inhibited by the tendency to equate performance with ability. Acknowledging the performance gap is, in many minds, tantamount to inferring that blacks are intellectually inferior. But inferior performance and inferior ability are not the same thing. Rather, the performance gap is largely a behavioral problem. It is the result of a remediable tendency to avoid intellectual engagement and competition. Avoidance is rooted in the fears and self-doubt engendered by a major legacy of American racism: the strong negative stereotypes about black intellectual capabilities. Avoidance of intellectual competition is manifested most obviously in the attitudes of many black youths toward academic work, but it is not limited to children. It affects the intellectual performance of black people of all ages and feeds public doubts about black intellectual ability.

I. INTELLECTUAL DEVELOPMENT

The performance gap damages the self-confidence of many black people. Black students and professional people cannot help but be bothered by poor showings in competitive academic and professional situations. Black leaders too often have tried to explain away these problems by blaming racism or cultural bias in the tests themselves. These factors haven't disappeared. But for many middle-class black Americans who have had access to educational and economic opportunities for nearly 20 years, the traditional protestations of cultural deprivation and educational disadvantage ring hollow. Given the cultural and educational advantages that many black people now enjoy, the claim that all blacks should be exempt from the performance standards applied to others is interpreted as a tacit admission of inferiority. This admission adds further weight to the questions, in our own minds and in the minds of others, about black intelligence.

The traditional explanations—laziness or inferiority on the one hand; racism, discrimination, and biased tests on the other—are inaccurate and unhelpful. What is required is an explanation that accounts for the subtle influences people exert over the behavior and self-confidence of other people.

Developing an explanation that might serve as a basis for corrective action is important. The record of the last 20 years suggests that waiting for grand initiatives from the outside to save the black community is futile. Blacks will have to rely on our own ingenuity and resources. We need local and national political leaders. We need skilled administrators and creative business executives. We need a broad base of well-educated volunteers and successful people in all fields as role models for black youths. In short, we need a large number of sophisticated, intellectually developed people who are confident of their ability to operate on an equal level with anyone. Chronic mediocre intellectual performance is deeply troubling because it suggests that we are not developing enough such people.

The Competitive Process. Intellectual development is not a fixed asset that you either have or don't have. Nor is it based on magic. It is a process of expanding mental strength and reach. The development process is demanding. It requires time, discipline, and intense effort. It almost always involves competition as well. Successful groups place high value on intellectual performance. They encourage the drive to excel and use competition to sharpen skills and stimulate development in each succeeding generation. The developed people that result from this competitive process become the pool from which leadership of all kinds is drawn. Competition, in other words, is an essential spur to development.

Competition is clearly not the whole story. Cooperation and solitary study are valuable, too. But of the various keys to intellectual development, competition seems to fare worst in the estimation of many blacks. Black young people, in particular, seem to place a strong negative value on intellectual competition.

Black people have proved to be very competitive at some activities, particularly sports and entertainment. It is our sense, however, that many blacks consider intellectual competition to be inappropriate. It appears to inspire little interest or respect among many youthful peer groups. Often, in fact, it is labeled "grade grubbing," and gives way to sports and social activity as a basis for peer acceptance. The intellectual performance gap is one result of this retreat from competition.

II. THE PSYCHOLOGY OF PERFORMANCE

Rumors of Inferiority. The need to avoid intellectual competition is a psychological reaction to an image of black intellectual inferiority that has been projected by the larger society, and to a less than conscious process of internalization of that image by black people over the generations.

The rumor of black intellectual inferiority has been around for a long time. It has been based on grounds as diverse as twisted biblical citations, dubious philosophical arguments, and unscientific measurements of skull capacity. The latest emergence of this old theme has been in the controversy over race and IQ. For 15 years newsmagazines and television talk shows have enthusiastically taken up the topic of black intellectual endowment. We have watched authors and critics debate the proposition that blacks are genetically inferior to whites in intellectual capability.

Genetic explanations have a chilling finality. The ignorant can be educated, the lazy can be motivated, but what can be done for the individual thought to have been born without the basic equipment necessary to compete or develop? Of course the allegation of genetic inferiority has been hotly disputed. But the debate has touched the consciousness of most Americans. We are convinced that this spectacle has negatively affected the way both blacks and whites think about the intellectual capabilities of black people. It also has affected the way blacks behave in intellectually competitive situations. The general expectation of black intellectual inferiority, and the fear this expectation generates, cause many black people to avoid intellectual competition.

OUR HYPOTHESIS, in short, is this. (1) Black performance problems are caused in large part by a tendency to avoid intellectual competition. (2) This tendency is a psychological phenomenon that arises when the larger society projects an image of black intellectual inferiority and when that image is internalized by black people. (3) Imputing intellectual inferiority to genetic causes, especially in the face of data confirming poorer performance, intensifies the fears and doubts that surround this issue.

Clearly the image of inferiority continues to be projected. The internalization of this image by black people is harder to prove empirically. But there is abundant evidence in the expressed attitudes of many black youths toward intellectual competition; in the inability of most black communities to inspire the same commitment to intellectual excellence that is routinely accorded athletics and entertainment; and in the fact of the performance gap itself—especially when that gap persists among the children of economically and educationally privileged households.

Expectancies and Performance. The problem of black intellectual performance is rooted in human sensitivity to a particular kind of social interaction known as "expectancy communications." These are expressions of belief—verbal or nonverbal—from one person to another about the kind of performance to be expected. "Mary, you're one of the best workers we have, so I know that you won't have any trouble with this assignment." Or, "Joe, since everyone else is busy with other work, do as much as you can on this. When you run into trouble, call Mary." The first is a positive expectancy; the second, a negative expectancy.

Years of research have clearly demonstrated the powerful impact of expectancies on performance. The expectations of teachers for their students have a large effect on academic achievement. Psychological studies under a variety of circumstances demonstrate that communicated expectations induce people to believe that they will do well or poorly at a task, and that such beliefs very often trigger responses that result in performance consistent with the expectation. There is also evidence that "reference group expectancies"—directed at an entire category of people rather than a particular individual—have a similar impact on the performance of members of the group.

EXPECTANCIES do not always work. If they come from a questionable source or if they predict an outcome that is too inconsistent with previous experience, they won't have much effect. Only credible expectancies—those that come from a source considered reliable and that address a belief or doubt the performer is sensitive to—will have a self-fulfilling impact.

The widespread expectation of black intellectual inferiority—communicated constantly through the projection of stereotyped images, verbal and nonverbal exchanges in daily interaction, and the incessant debate about genetics and intelligence—represents a credible reference-group expectancy. The message of the race/IQ controversy is: "We have scientific evidence that blacks, because of genetic inadequacies, can't be expected to do well at tasks that require great intelligence." As an explanation for past black intellectual performance, the notion of genetic inferiority is absolutely incorrect. As an expectancy communication exerting control over our present intellectual strivings, it has been powerfully effective. These expectancies raise fear and self-doubt in the minds of many blacks, especially when they are young and vulnerable. This has resulted in avoidance of intellectual activity and chronic underperformance by many of our most talented people. Let us explore this process in more detail.

The Expectancy/Performance Model. The powerful effect of expectancies on performance has been proved, but the way the process works is less well understood. Expectancies affect behavior, we think, in two ways. They affect performance behavior: the capacity to marshal the sharpness and intensity required for competitive success. And they influence cognition: the mental processes by which people make sense of everyday life.

Behavior. As anyone who has experienced an "off day" knows, effort is variable; it is subject to biological cycles, emotional states, motivation. Most important for our discussion, it depends on levels of confidence going into a task. Credible expectancies influence performance behavior. They affect the intensity of effort, the level of concentration or distractibility, and the willingness to take reasonable risks—a key factor in the development of self-confidence and new skills.

Cognition. Expectations also influence the way people think about or explain their performance outcomes. These explanations are called "attributions." Research in social psychology has demonstrated that the causes to which people attribute their successes and failures have an important impact on subsequent performance.

All of us encounter failure. But a failure we have been led to expect affects us differently from an unexpected failure. When people who are confident of doing well at a task are confronted with unexpected failure, they tend to attribute the failure to inadequate effort. The likely response to another encounter with the same or a similar task is to work harder. People who come into a task expecting to fail, on the other hand, attribute their failure to lack of ability. Once you admit to yourself, in effect, that "I don't have what it takes," you are not likely to approach that task again with great vigor.

Indeed, those who attribute their failures to inadequate effort are likely to conclude that more effort will produce a better outcome. This triggers an adaptive response to failure. In contrast, those who have been led to expect failure will attribute their failures to lack of ability, and will find it difficult to rationalize the investment of greater effort. They will often hesitate to continue "banging my head against the wall." They often, in fact, feel depressed when they attempt to work, since each attempt represents a confrontation with their own feared inadequacy.

THIS COMBINED EFFECT on behavior and cognition is what makes expectancy so powerful. The negative expectancy first tends to generate failure through its impact on behavior, and then induces the individual to blame the failure on lack of ability, rather than the actual (and correctable) problem of inadequate effort. This misattribution in turn becomes the basis for a new negative expectancy. By this process the individual, in effect, internalizes the low estimation originally held by others. This internalized negative expectancy powerfully affects future competitive behavior and future results.

The process we describe is not limited to black people. It goes on all the time, with individuals from all groups. It helps to explain the superiority of some groups at some areas of endeavor, and the mediocrity of those same groups in other areas. What makes black people unique is that they are singled out for the stigma of genetic intellectual inferiority.

The expectation of intellectual inferiority accompanies a black person into each new intellectual situation. Since each of us enters these tests under the cloud of predicted failure, and since each failure reinforces doubts about our capabilities, all intellectual competition raises the specter of having to admit a lack of intellectual capacity. But this particular expectancy goes beyond simply predicting and inducing failure. The expectancy message explicitly ascribes the expected failure to genes, and amounts to an open suggestion to black people to understand any failure in intellectual activity as confirmation of genetic inferiority. Each engagement in intellectual competition carries the weight of a test of one's own genetic endowment and that of black people as a whole. Facing such a terrible prospect, many black people recoil from any situation where the rumor of inferiority might be proved true.

For many black students this avoidance manifests itself in a concentration on athletics and socializing, at the expense of more challenging (and anxiety-provoking) academic work. For black professionals, it may involve a ten-

4. FAMILY, SCHOOL, AND CULTURAL INFLUENCES: Cultural Influences

dency to shy away from competitive situations or projects, or an inability to muster the intensity—or commit the time—necessary to excel. This sort of thinking and behavior certainly does not characterize all black people in competitive settings. But it is characteristic of enough to be a serious problem. When it happens, it should be understood as a less than conscious reaction to the psychological burden of the terrible rumor.

The Intellectual Inferiority Game. There always have been constraints on the intellectual exposure and development of black people in the United States, from laws prohibiting the education of blacks during slavery to the Jim Crow laws and "separate but equal" educational arrangements that persisted until very recently. In dismantling these legal barriers to development, the civil rights movement fundamentally transformed the possibilities for black people. Now, to realize those possibilities, we must address the mental barriers to competition and performance.

The doctrine of intellectual inferiority acts on many black Americans the way that a "con" or a "hustle" like three-card monte acts on its victim. It is a subtle psychological input that interacts with characteristics of the human cognitive apparatus—in this case, the extreme sensitivity to expectancies—to generate self-defeating behavior and thought processes. It has reduced the intellectual performance of millions of black people.

Intellectual inferiority, like segregation, is a destructive idea whose time has passed. Like segregation, it must be removed as an influence in our lives. Among its other negative effects, fear of the terrible rumor has restricted discussion by all parties, and has limited our capacity to understand and improve our situation. But the intellectual inferiority game withers in the light of discussion and analysis. We must begin now to talk about intellectual performance, work through our expectations and fears of intellectual inferiority, consciously define more adaptive attitudes toward intellectual development, and build our confidence in the capabilities of all people.

THE expectancy/performance process works both ways. Credible positive expectancies can generate self-confidence and result in success. An important part of the solution to black performance problems is converting the negative expectancies that work against black development into positive expectancies that nurture it. We must overcome our fears, encourage competition, and support the kind of performance that will dispel the notion of black intellectual inferiority.

III. THE COMMITMENT TO DEVELOPMENT

In our work with black high school and college students and with black professionals, we have shown that education in the psychology of performance can produce strong performance improvement very quickly. Black America needs a nationwide effort, now, to ensure that all black people—but especially black youths—are free to express their intellectual gifts. That effort should be built on three basic elements:

• Deliberate control of expectancy communications. We must begin with the way we talk to one another: the messages we give and the expectations we set. This includes the verbal and nonverbal messages we communicate in day-to-day social intercourse, as well as the expectancies communicated through the educational process and media images.

• Definition of an "intellectual work ethic." Black communities must develop strong positive attitudes toward intellectual competition. We must teach our people, young and mature, the efficacy of intense, committed effort in the arena of intellectual activity and the techniques to develop discipline in study and work habits.

• Influencing thought processes. Teachers, parents, and other authority figures must encourage young blacks to attribute their intellectual successes to ability (thereby boosting confidence) and their failures to lack of effort. Failures must no longer destroy black children's confidence in their intelligence or in the efficacy of hard work. Failures should be seen instead as feedback indicating the need for more intense effort or for a different approach to the task.

The task that confronts us is no less challenging than the task that faced those Americans who dismantled segregation. To realize the possibilities presented by their achievement, we must silence, once and for all, the rumors of inferiority.

Who's Responsible? Expectations of black inferiority are communicated, consciously or unconsciously, by many whites, including teachers, managers, and those responsible for the often demeaning representations of blacks in the media. These expectations have sad consequences for many blacks, and those whose actions lead to such consequences may be held accountable for them. If the people who shape policy in the United States, from the White House to the local elementary school, do not address the problems of performance and development of blacks and other minorities, all Americans will face the consequences: instability, disharmony, and a national loss of the potential productivity of more than a quarter of the population.

However, when economic necessity and the demands of social justice compel us toward social change, those who have the most to gain from change—or the most to lose from its absence—should be responsible for pointing the way.

It is time that blacks recognize our own responsibility. When we react to the rumor of inferiority by avoiding intellectual engagement, and when we allow our children to do so, black people forfeit the opportunity for intellectual development that could extinguish the debate about our capacities, and set the stage for group progress. Blacks must hold ourselves accountable for the resulting waste of talent—and valuable time. Black people have everything to gain—in stature, self-esteem, and problem-solving capability—from a more aggressive and confident approach to intellectual competition. We must assume responsibility for our own performance and development.

BIOLOGY, DESTINY, AND ALL THAT

Grabbing hold of a tar baby
of research findings, our writer
tries to pull apart truth from myth
in ideas about the differences
between the sexes.

Paul Chance

Paul Chance is a psychologist, writer, and contributing editor of *Psychology Today*.

In the 1880s, scholars warned against the hazards of educating women. Some experts of the day believed that too much schooling could endanger a woman's health, interfere with her reproductive ability, and cause her brain to deteriorate. In the 1980s we laugh at such absurd ideas, but have we (men and women alike) really given up the ancient idea that a woman is fundamentally an inferior sort of man? It seems not.

It's hard to find evidence these days of gross discrimination against women as a company policy. Successful lawsuits have made that sort of prejudice expensive. Yet evidence of more subtle forms of bias abound. The sociologist Beth Ghiloni conducted a study while she was a student at the University of California, Santa Cruz, that shows that some corporations are meeting the demands of affirmative action by putting women into public relations posts. Public relations is important but distant from the activities that generate revenue, so PR assignments effectively keep women out of jobs that include any real corporate power. Thus, women increasingly complain of facing a "glass ceiling" through which they can see, but cannot reach, high level corporate positions.

It seems likely that such discrimination reflects some very old ideas about what men and women are like. Men, the stereotype has it, are aggressive and self-confident. They think analytically, and are cool under fire. They enjoy jobs that offer responsibility and challenge. They are, in other words, ideally suited for important, high-level positions. Put them on the track that may one day lead to corporate vice president.

Women, the thinking goes, are passive and filled with self-doubt. They think intuitively, and are inclined to become emotionally distraught under pressure. They therefore enjoy jobs that involve working with people. Give them the lower-rung jobs in the personnel department, or send them to public affairs.

Figuring out how truth and myth intertwine in these stereotypes is difficult. The research literature on sex differences is a tar baby made of numbers, case studies, and anecdotal impressions. To paraphrase one researcher, "If you like ambiguity, you're gonna love sex-difference research." Nevertheless, let us be brave and take the stereotypes apart piece by piece.

Aggression

On this there is no argument: Everyone agrees that men are more aggressive than women. In a classic review of the literature, the psychologists Eleanor Maccoby of Stanford University and Carol Jacklin of the University of Southern California found that boys are more aggressive than girls both physically and verbally, and the difference begins to show up by the age of 3. Boys are more inclined to rough-and-tumble play; girls have tea parties and play with dolls.

The difference in aggressiveness is most clearly seen in criminal activity. There are far more delinquent boys than girls, and prisons are built primarily to contain men. The most aggressive crimes, such as murder and assault, are especially dominated by men.

It seems hardly likely that male aggressiveness, as it is documented by the research, would win favor among personnel directors and corpo-

4. FAMILY, SCHOOL, AND CULTURAL INFLUENCES: Cultural Influences

rate headhunters. Yet aggressiveness is considered not only a virtue but an essential trait for many jobs. Vice President George Bush learned the value of aggressiveness when he managed to put aside his wimp image by verbally attacking the CBS news anchorman Dan Rather on national television. The assumption seems to be that the same underlying trait that makes for murderers, rapists, and strong-arm bandits also, in more moderate degree or under proper guidance, makes people more competitive and motivated to achieve great things.

But the research suggests that women may be just as aggressive in this more civilized sense as men. For instance, most studies of competitiveness find no differences between the sexes, according to the psychologists Veronica Nieva and Barbara Gutek, co-authors of *Women and Work*. And many studies of achievement motivation suggest that women are just as eager to get things done as men.

Self-confidence

Men are supposed to be self-confident, women full of self-doubt. Again there is some evidence for the stereotype, but it isn't particularly complimentary to males. Various studies show that males overestimate their abilities, while females underestimate theirs. Researchers have found, for instance, that given the option of choosing tasks varying in difficulty, boys erred by choosing those that were too difficult for them, while girls tended to select tasks that were too easy.

Other research shows that women are not only less confident of their ability to do a job, when they succeed at it they don't give themselves credit. Ask them why they did well and they'll tell you it was an easy task or that they got lucky. Ask men the same thing and they'll tell you they did well because of their ability and hard work.

It is perhaps this difference in self-confidence that makes women better risks for auto insurance, and it may have something to do with the fact that almost from the day they can walk, males are more likely to be involved in pedestrian accidents.

Rational Thinking

A great many studies have found that men do better than women on tests of mathematical reasoning. Julian Stanley, a psychologist at Johns Hopkins University, has been using the mathematics portion of the Scholastic Aptitude Test to identify mathematically gifted youths. He and his colleagues have consistently found that a majority of the high scoring students are boys. Stanley reports that "mathematically gifted boys outnumber gifted girls by a ratio of about 13 to 1." Moreover, the very best scores almost inevitably come from boys.

But while some findings show that men are better at mathematical reasoning, there is no evidence that they are more analytical or logical in general. In fact, the superiority of men at mathematical problem solving seems not to reflect superior analytical thinking but a special talent men have for visualizing objects in space. Women are every bit the match of men at other kinds of problems such as drawing logical conclusions from written text. Indeed, girls have an advantage over boys in verbal skills until at least adolescence.

As for the idea that women are more intuitive, forget it. Numerous studies have shown that women are better at reading body language, and it is probably this skill (born, perhaps, of the need to avoid enraging the more aggressive sex) that gives rise to the myth of women's intuition. In reality, women and men think alike.

Emotionality

In Victorian England, to judge by the novels of the day, a woman was no woman at all if she didn't feel faint or burst into tears at least once a week. The idea that women are more emotional than men, that they feel things more deeply and react accordingly, persists. Is it true?

No. Carol Tavris and Carole Wade, psychologists and co-authors of *The Longest War: Sex Differences in Perspective*, write that the sexes are equally likely to feel anxious in new situations, to get angry when insulted, to be hurt when a loved one leaves them, and to feel embarrassed when they make mistakes in public.

The sexes are equally emotional, but there are important differences in how willing men and women are to express emotions, which emotions they choose to express, and the ways in which they express them. If you ask men and women in an emotional situation what they are feeling, women are likely to admit that they are affected, while men are likely to deny it. Yet studies show that when you look at the psychological correlates of emotion—heartbeat, blood pressure, and the like—you find that those strong silent men are churning inside every bit as much as the women.

Men are particularly eager to conceal feelings such as fear, sorrow, and loneliness, according to Tavris and Wade. Men often bottle these "feminine" feelings even with those they hold most dear.

Another difference comes in how emotions are expressed. Women behave differently depending upon what they feel. They may cry if sad, curse if angry, pout if their pride is hurt. Men tend to respond to such situations in a more or less uniform way. Whether they have been jilted, frightened, or snubbed, they become aggressive. (Men, you will recall, are very good at aggression.) As for the notion that keeping one's head is characteristic of men, well, don't get sore fellows, but it ain't necessarily so.

> "It's unlikely that male aggressiveness, as documented by research, would win favor among personnel directors."

32. Biology, Destiny, and All That

Men's math superiority reflects a special talent for visualizing objects in space.

Job Interests

If you ask men and women what they like about their work, you will get different answers. In 1957, Frederick Herzberg published a study of what made work enjoyable to employees. Men, he concluded, enjoyed work that offered responsibility and challenge. For women, on the other hand, the environment was the thing. They wanted an attractive work area, and some pleasant people to talk to while they did whatever needed to be done. Give your secretary an office with some nice wallpaper, put a flower on her desk once in a while, and she'll be happy.

Experts now agree, however, that such differences probably reflect differences in the jobs held by the people studied. You may get such findings, suggest *Women and Work* co-authors Nieva and Gutek, if you compare female file clerks and male engineers—but not if you compare female and male engineers. In a study of workers in various jobs, Daphne Bugental, a psychologist at the University of California, Santa Barbara, and the late Richard Centers found no consistent differences in the way men and women ranked "intrinsic" job characteristics such as responsibility and challenge and "extrinsic" characteristics such as pleasant surroundings and friendly co-workers. In other words, the differences in what men and women find interesting about work reflect differences in the kinds of work men and women characteristically do. People in relatively high level jobs enjoy the responsibility and challenge it offers; people in low level jobs that offer little responsibility and challenge look elsewhere for satisfaction.

Where Do the Differences Come From?

So, what differences separate the boys from the girls? Men are reported to be more aggressive than women in a combative sense, but they are not necessarily more competitive. Men are more confident, perhaps recklessly so, and women may be too cautious. Men are not more likely to be cool under fire, but they are more likely to become aggressive regardless of what upsets them. Men are not more analytic in their thinking, nor women more intuitive, but men are better at solving mathematical problems, probably because they are better at spatial relationships. Finally, women are less interested in the responsibility and challenge of work, but only because they usually have jobs that offer little responsibility and challenge.

While this seems to be the gist of the matter, it leaves open the question of whether the differences are due to biology or to environment. Are

---

THE DEVELOPMENT OF THE "WEAKER SEX"
(AND THE DEMORALIZATION OF THE DUDE).

VASSAR GRADUATE.—"These are the dumb-bells I used last term in our gymnasium; won't one of you gentleman just put them up? It's awfully easy."

4. FAMILY, SCHOOL, AND CULTURAL INFLUENCES: Cultural Influences

"All-male groups may be better at brainstorming, while all-female groups may be better at finding a solution."

men more aggressive because they are born that way or because of lessons that begin in the cradle? Are women less confident than men because different hormones course through their blood, or because for years people have told them that they can't expect much from themselves?

The research on the nature-nurture question is a candy store in which people of varying biases can quickly find something to their liking. Beryl Lieff Benderly, an anthropologist and journalist, critiqued the physiological research for her new book, *The Myth of Two Minds*. The title tells the story. She even challenges the notion that men are stronger than women. "The plain fact is that we have no idea whether men are 'naturally' stronger than women," she writes. Short of conducting an experiment along the lines of *Lord of the Flies*, we are unlikely to unravel the influences of nature and nurture to everyone's satisfaction. Nevertheless, research does offer hints about the ways that biology and environment affect stereotypical behavior. Take the case of aggression. There are any number of studies linking aggression to biological factors. Testosterone, a hormone found in much higher levels in men than in women, has been found in even larger quantities in criminally aggressive males. Castration, which decreases the level of testosterone, has been used for centuries to produce docile men and animals. And girls who have had prenatal exposure to high levels of testosterone are more tomboyish than other girls.

Yet biology is not quite destiny. In her famous *Six Cultures* study, the anthropologist Beatrice Whiting and her colleagues at Harvard University found that in each of the societies studied, boys were more aggressive than girls. But the researchers also found such wide cultural differences that the girls in a highly aggressive society were often more aggressive than the boys in another, less aggressive society.

The same mix of forces applies wherever we look. Biology may bend the twig in one direction, but the environment may bend it in another. But whether biology or environment ultimately wins the hearts and minds of researchers is less important than how the differences, wherever they come from, affect behavior in the workplace. If someone explodes in anger (or breaks down in tears) in the midst of delicate contract negotiations, it matters little to the stockholders whether the lost business can, in the end, be blamed on testosterone or bad toilet training. The more important question is, what are the implications of sex difference research for business?

What Difference Do the Differences Make?

The research on sex differences suggest three points that people in business can usefully consider. First, the differences between the sexes are small. Researchers look for "statistically significant" differences. But statistically significant differences are not necessarily practically significant. There is a great deal of overlap between the sexes on most characteristics and especially on the characteristics we have been considering. Men are, on average, better at solving mathematical problems, but there are many women who are far above the average man in this area. Women are, on average, less confident than men. But there are many men who doubt themselves far more than the average woman. A study of aggressiveness is illustrative. The psychologist D. Anthony Butterfield and the management expert Gary N. Powell had college students rate the ideal U.S. President on various characteristics, including aggressiveness. Then they had them rate people who were running for President and Vice President at the time: Ronald Reagan, Walter Mondale, George Bush, and Geraldine Ferraro. Researchers found that the ideal president was, among other things, aggressive. They also found that Geraldine Ferraro was judged more aggressive than the male candidates. The point is that it is impossible to predict individual qualities from group differences.

Indeed, Carol Jacklin suggests that differences in averages may give a quite distorted view of both sexes. She notes that while, on average, boys play more aggressively than girls, her research shows that the difference is due to a small number of very aggressive boys. Most of the boys are, in fact, very much like the girls. "I'd be willing to bet," Jacklin says, "that much of the difference in aggressiveness between men and women is due to a small number of extremely aggressive men—many of whom are in prison—and that the remaining men are no more aggressive than most women."

Second, different doesn't necessarily mean inferior. It is quite possible that feminine traits (in men or women) are assets in certain situations, while masculine traits may be advantageous in other situations. The psychologist Carol Gilligan, author of *In a Different Voice*, says that women are more comfortable with human relationships than men are, and this may sometimes give them an edge. For instance, Roderick Gilkey and Leonard Greenhalgh, psychologists at Dartmouth University, had business students simulate negotiations over the purchase of a used car and television advertising time. The women appeared better suited to the task than the men. They were more flexible, more willing to compromise, and less deceptive. "Women can usually come to an agreement on friendly terms," says Greenhalgh. "They're better at avoiding impasses."

In another study, the psychologist Wendy Wood of Texas A&M University asked college students to work on problems in groups of three.

32. Biology, Destiny, and All That

"Biology may bend the twig in one direction, but the environment may bend it in another."

Some groups consisted only of men, others of women. The groups tried to solve problems such as identifying the features to consider in buying a house. The men, it turns out, came up with more ideas, while the women zeroed in on one good idea and developed it. Wood concluded that all-male groups might be better for brainstorming, while all-female groups might be better for finding the best solution to a problem.

Third, sometimes people lack the characteristics needed for a job until they are in the job. Jobs that offer responsibility and challenge, for example, tend to create a desire for more responsibility and challenge. While there are research studies to support this statement, an anecdote from the sociologist Rosabeth Moss Kanter is more telling. Linda, a secretary for 17 years in a large corporation, had no interest in being anything but a secretary, and when she was offered a promotion through an affirmative action program, she hesitated. Her boss persuaded her to take the job and she became a successful manager and loved the additional responsibility and challenge. She even set her sights on a vice president position. As a secretary, Linda would no doubt have scored near the female stereotype. But when she became a manager, she became more like the stereotypical male.

Kanter told that story a dozen years ago, but we are still struggling to learn its lesson. A discrimination case against Sears, Roebuck and Company recently made news. The Equal Employment Opportunity Commission (EEOC) argued that Sears discriminated against women because nearly all of the employees in the company who sell on commission are men. Sears presented evidence that women expressed little interest in commission-sales work, preferring the less risky jobs in salaried sales. In the original decision favoring Sears, the U.S. District Court had found that "noncommission saleswomen were generally happier with their present jobs at Sears, and were much less likely than their male counterparts to be interested in other positions, such as commission sales. . . ." But in the U.S. Court of Appeals, Appellate Judge Cudahy, who dissented in part from the majority, wrote that this reasoning is "of a piece with the proposition that women are by nature happier cooking, doing the laundry, and chauffeuring the children to softball games than arguing appeals or selling stocks."

The point is not that employees must be made to accept more responsible positions for their own good, even if it is against their will. The point is that business should abandon the stereotypes that lock men and women into different, and often unequal, kinds of work. If it finally does, it will discover the necessity—and value—of finding ways of enticing men and women into jobs for which, according to the stereotypes, they are not suited. If business doesn't do that, it may discover that differences between men and women really do separate the sexes.

ALIENATION
AND THE FOUR WORLDS OF CHILDHOOD

The forces that produce youthful alienation are growing in strength and scope, says Mr. Bronfenbrenner. And the best way to counteract alienation is through the creation of connections or links throughout our culture. The schools can build such links.

Urie Bronfenbrenner

Urie Bronfenbrenner is Jacob Gould Shurman Professor of Human Development and Family Studies and of Psychology at Cornell University, Ithaca, N.Y.

To be alienated is to lack a sense of belonging, to feel cut off from family, friends, school, or work—the four worlds of childhood.

At some point in the process of growing up, many of us have probably felt cut off from one or another of these worlds, but usually not for long and not from more than one world at a time. If things weren't going well in school, we usually still had family, friends, or some activity to turn to. But if, over an extended period, a young person feels unwanted or insecure in several of these worlds simultaneously or if the worlds are at war with one another, trouble may lie ahead.

What makes a young person feel that he or she doesn't belong? Individual differences in personality can certainly be one cause, but, especially in recent years, scientists who study human behavior and development have identified an equal (if not even more powerful) factor: the circumstances in which a young person lives.

Many readers may feel that they recognize the families depicted in the vignettes that are to follow. This is so because they reflect the way we tend to look at families today: namely, that we see parents as being good or not-so-good without fully taking into account the circumstances in their lives.

Take Charles and Philip, for example. Both are seventh-graders who live in a middle-class suburb of a large U.S. city. In many ways their surroundings seem similar; yet, in terms of the risk of alienation, they live in rather different worlds. See if you can spot the important differences.

CHARLES

The oldest of three children, Charles is amiable, outgoing, and responsible. Both of his parents have full-time jobs outside the home. They've been able to arrange their working hours, however, so that at least one of them is at home when the children return from school. If for some reason they can't be home, they have an arrangement with a neighbor, an elderly woman who lives alone. They can phone her and ask her to look after the children until they arrive. The children have grown so fond of this woman that she is like another grandparent—a nice situation for them, since their real grandparents live far away.

Homework time is one of the most important parts of the day for Charles and his younger brother and sister. Charles's parents help the children with their homework if they need it, but most of the time they just make sure that the children have a period of peace and quiet—without TV—in which to do their work. The children are allowed to watch television one hour each night—but only after they have completed their homework. Since Charles is doing well in school, homework isn't much of an issue, however.

Sometimes Charles helps his mother or father prepare dinner, a job that everyone in the family shares and enjoys. Those family members who don't cook on a given evening are responsible for cleaning up.

Charles also shares his butterfly collection with his family. He started the collection when he first began learning about butterflies during a fourth-grade science project. The whole family enjoys picnicking and hunting butterflies together, and Charles occasionally asks his father to help him mount and catalogue his trophies.

Charles is a bit of a loner. He's not a very good athlete, and this makes him somewhat self-conscious. But he does have one very close friend, a boy in his class who lives just down the block. The two boys have been good friends for years.

Charles is a good-looking, warm, happy young man. Now that he's beginning to be interested in girls, he's gratified to find that the interest is returned.

PHILIP

Philip is 12 and lives with his mother, father, and 6-year-old brother. Both of his parents work in the city, commuting more than an hour each way. Pandemonium strikes every weekday morning as the entire family prepares to leave for school and work.

Philip is on his own from the time school is dismissed until just before dinner, when his parents return after stopping to pick up his little brother at a nearby day-care home. At one time, Philip took care of his little brother after school, but he resented having to do so. That arrangement ended one day when Philip took his brother out to play and the little boy wandered off and got lost. Philip didn't even notice for several hours that his brother was missing. He felt guilty at first about not having done a better job. But not having to mind his brother freed him to hang out with his friends or to watch television, his two major after-school activities.

The pace of their life is so demanding that Philip's parents spend their weekends just trying to relax. Their favorite weekend schedule calls for watching a ball game on television and then having a cookout in the back yard. Philip's mother resigned herself long ago to a messy house; pizza, TV dinners, or fast foods are all she can manage in the way of meals on most nights. Philip's father has made it clear that she can do whatever she wants in managing the house, as long as she doesn't try to involve him in the effort. After a hard day's work, he's too tired to be interested in housekeeping.

Philip knows that getting a good education is important; his parents have stressed that. But he just can't seem to concentrate in school. He'd much rather fool around with his friends. The thing that he and his friends like to do best is to ride the bus downtown and go to a movie, where they can show off, make noise, and make one another laugh.

Sometimes they smoke a little marijuana during the movie. One young man in Philip's social group was arrested once for having marijuana in his jacket pocket. He was trying to sell it on the street so that he could buy food. Philip thinks his friend was stupid to get caught. If you're smart, he believes, you don't let that happen. He's glad that his parents never found out about the incident.

Once, he brought two of his friends home during the weekend. His parents told him later that they didn't like the kind of people he was hanging around with. Now Philip goes out of his way to keep his friends and his parents apart.

THE FAMILY UNDER PRESSURE

In many ways the worlds of both

> Institutions that play important roles in human development are rapidly being eroded, mainly through benign neglect.

teenagers are similar, even typical. Both live in families that have been significantly affected by one of the most important developments in American family life in the postwar years: the employment of both parents outside the home. Their mothers share this status with 64% of all married women in the U.S. who have school-age children. Fifty percent of mothers of preschool children and 46% of mothers with infants under the age of 3 work outside the home. For single-parent families, the rates are even higher: 53% of all mothers in single-parent households who have infants under age 3 work outside the home, as do 69% of all single mothers who have school-age children.[1]

These statistics have profound implications for families — sometimes for better, sometimes for worse. The determining factor is how well a given family can cope with the "havoc in the home" that two jobs can create. For, unlike most other industrialized nations, the U.S. has yet to introduce the kinds of policies and practices that make work life and family life compatible.

It is all too easy for family life in the U.S. to become hectic and stressful, as both parents try to coordinate the disparate demands of family and jobs in a world in which everyone has to be transported at least twice a day in a variety of directions. Under these circumstances, meal preparation, child care, shopping, and cleaning — the most basic tasks in a family — become major challenges. Dealing with these challenges may sometimes take precedence over the family's equally important child-rearing, educational, and nurturing roles.

But that is not the main danger. What threatens the well-being of children and young people the most is that the external havoc can become internal, first for parents and then for their children. And that is exactly the sequence in which the psychological havoc of families under stress usually moves.

Recent studies indicate that conditions at work constitute one of the major sources of stress for American families.[2] Stress at work carries over to the home, where it affects first the relationship of parents to each other. Marital conflict then disturbs the parent/child relationship. Indeed, as long as tensions at work do not impair the relationship between the parents, the children are not likely to be affected. In other words, the influence of parental employment on children is indirect, operating through its effect on the parents.

That this influence is indirect does not make it any less potent, however. Once the parent/child relationship is seriously disturbed, children begin to feel insecure — and a door to the world of alienation has been opened. That door can open to children at any age, from preschool to high school and beyond.

My reference to the world of school is not accidental, for it is in that world that the next step toward alienation is likely to be taken. Children who feel rootless or caught in conflict at home find it difficult to pay attention in school. Once they begin to miss out on learning, they feel lost in the classroom, and they begin to seek acceptance elsewhere. Like Philip, they often find acceptance in a group of peers with similar histories who, having no welcoming place to go and nothing challenging to do, look for excitement on the streets.

OTHER INFLUENCES

In contemporary American society the growth of two-wage-earner families is not the only — or even the most serious — social change requiring accommodation through public policy and practice in order to avoid the risks of alienation. Other social changes include lengthy trips to and from work; the loss of the extended family, the close neighborhood, and other support systems previously available to families; and the omnipresent threat of television and other media to the family's traditional role as the primary transmitter of culture and values. Along with most families today, the families of Charles and Philip are experiencing the unraveling and disintegration of social institutions that in the

4. FAMILY, SCHOOL, AND CULTURAL INFLUENCES: Cultural Influences

past were central to the health and well-being of children and their parents.

Notice that both Charles and Philip come from two-parent, middle-class families. This is still the norm in the U.S. Thus neither family has to contend with two changes now taking place in U.S. society that have profound implications for the future of American families and the well-being of the next generation. The first of these changes is the increasing number of single-parent families. Although the divorce rate in the U.S. has been leveling off of late, this decrease has been more than compensated for by a rise in the number of unwed mothers, especially teenagers. Studies of the children brought up in single-parent families indicate that they are at greater risk of alienation than their counterparts from two-parent families. However, their vulnerability appears to have its roots not in the single-parent family structure as such, but in the treatment of single parents by U.S. society.[3]

In this nation, single parenthood is almost synonymous with poverty. And the growing gap between poor families and the rest of us is today the most powerful and destructive force producing alienation in the lives of millions of young people in America. In recent years, we have witnessed what the U.S. Census Bureau calls "the largest decline in family income in the post-World War II period." According to the latest Census, 25% of all children under age 6 now live in families whose incomes place them below the poverty line.

COUNTERING THE RISKS

Despite the similar stresses on their families, the risks of alienation for Charles and Philip are not the same. Clearly, Charles's parents have made a deliberate effort to create a variety of arrangements and practices that work against alienation. They have probably not done so as part of a deliberate program of "alienation prevention" — parents don't usually think in those terms. They're just being good parents. They spend time with their children and take an active interest in what their children are thinking, doing, and learning. They control their television set instead of letting it control them. They've found support systems to back them up when they're not available.

Without being aware of it, Charles's parents are employing a principle that the great Russian educator Makarenko employed in his extraordinarily successful programs for the reform of wayward adolescents in the 1920s: "The maximum of support with the maximum of challenge."[4] Families that produce effective, competent children often follow this principle, whether they're aware of it or not. They neither maintain strict control nor allow their children total freedom. They're always opening doors — and then giving their children a gentle but firm shove to encourage them to move on and grow. This combination of support and challenge is essential, if children are to avoid alienation and develop into capable young adults.

From a longitudinal study of youthful alienation and delinquency that is now considered a classic, Finnish psychologist Lea Pulkkinen arrived at a conclusion strikingly similar to Makarenko's. She found "guidance" — a combination of love and direction — to be a critical predictor of healthy development in youngsters.[5]

No such pattern is apparent in Philip's family. Unlike Charles's parents, Philip's parents neither recognize nor respond to the challenges they face. They have dispensed with the simple amenities of family self-discipline in favor of whatever is easiest. They may not be indifferent to their children, but the demands of their jobs leave them with little energy to be actively involved in their children's lives. (Note that Charles's parents have work schedules that are flexible enough to allow one of them to be at home most afternoons. In this regard, Philip's family is much more the norm, however. One of the most constructive steps that employers could take to strengthen families would be to enact clear policies making such flexibility possible.)

But perhaps the clearest danger signal in Philip's life is his dependence on his peer group. Pulkkinen found heavy reliance on peers to be one of the strongest predictors of problem behavior in adolescence and young adulthood. From a developmental viewpoint, adolescence is a time of challenge — a period in which young people seek activities that will serve as outlets for their energy, imagination, and longings. If healthy and constructive challenges are not available to them, they will find their challenges in such peer-group-related behaviors as poor school performance, aggressiveness or social withdrawal (sometimes both), school absenteeism or dropping out, smoking, drinking, early and promiscuous sexual activity, teenage parenthood, drugs, and juvenile delinquency.

This pattern has now been identified in a number of modern industrial societies, including the U.S., England, West Germany, Finland, and Australia. The pattern is both predictable from the circumstances of a child's early family life and predictive of life experiences still to come, e.g., difficulties in establishing relationships with the opposite sex, marital discord, divorce, economic failure, criminality.

If the roots of alienation are to be found in disorganized families living in disorganized environments, its bitter fruits are to be seen in these patterns of disrupted development. This is not a harvest that our nation can easily afford. Is it a price that other modern societies are paying, as well?

A CROSS-NATIONAL PERSPECTIVE

The available answers to that question will not make Americans feel better about what is occurring in the U.S. In our society, the forces that produce youthful alienation are growing in strength and scope. Families, schools, and other institutions that play important roles in human development are rapidly being eroded, *mainly through benign neglect*. Unlike the citizens of other modern nations, we Americans have simply not been willing to make the necessary effort to forestall the alienation of our young people.

As part of a new experiment in higher education at Cornell University, I have been teaching a multidisciplinary course for the past few years titled "Human Development in Post-Industrial Societies." One of the things we have done in that course is to gather comparative data from several nations, including France, Canada, Japan, Australia, Germany, England, and the U.S. One student summarized our findings succinctly: "With respect to families, schools, children, and youth, such countries as France, Japan, Canada, and Australia have more in common with each other than the United States has with any of them." For example:

- The U.S. has by far the highest rate of teenage pregnancy of any industrialized nation — twice the rate of its nearest competitor, England.
- The U.S. divorce rate is the highest in the world — nearly double that of its nearest competitor, Sweden.
- The U.S. is the only industrialized society in which nearly one-fourth of all infants and preschool children live in families whose incomes fall below the

33. Alienation and the Four Worlds of Childhood

poverty line. These children lack such basics as adequate health care.

• The U.S. has fewer support systems for individuals in all age groups, including adolescence. The U.S. also has the highest incidence of alcohol and drug abuse among adolescents of any country in the world.[6]

All these problems are part of the unraveling of the social fabric that has been going on since World War II. These problems are not unique to the U.S., but in many cases they are more pronounced here than elsewhere.

WHAT COMMUNITIES CAN DO

The more we learn about alienation and its effects in contemporary post-industrial societies, the stronger are the imperatives to counteract it. If the essence of alienation is disconnectedness, then the best way to counteract alienation is through the creation of connections or links.

For the well-being of children and adolescents, the most important links must be those between the home, the peer group, and the school. A recent study in West Germany effectively demonstrated how important this basic triangle can be. The study examined student achievement and social behavior in 20 schools. For all the schools, the researchers developed measures of the links between the home, the peer group, and the school. Controlling for social class and other variables, the researchers found that they were able to predict children's behavior from the number of such links they found. Students who had no links were alienated. They were not doing well in school, and they exhibited a variety of behavioral problems. By contrast, students who had such links were doing well and were growing up to be responsible citizens.[7]

In addition to creating links within the basic triangle of home, peer group, and school, we need to consider two other structures in today's society that affect the lives of young people: the world of work (for both parents and children) and the community, which provides an overarching context for all the other worlds of childhood.

Philip's family is one example of how the world of work can contribute to alienation. The U.S. lags far behind other industrialized nations in providing child-care services and other benefits designed to promote the well-being of children and their families. Among the most needed benefits are maternity and paternity leaves, flex-time, job-sharing arrangements, and personal leaves for parents when their children are ill. These benefits are a matter of course in many of the nations with which the U.S. is generally compared.

In contemporary American society, however, the parents' world of work is not the only world that both policy and practice ought to be accommodating. There is also the children's world of work. According to the most recent figures available, 50% of all high school students now work part-time — sometimes as much as 40 to 50 hours per week. This fact poses a major problem for the schools. Under such circumstances, how can teachers assign homework with any expectation that it will be completed?

The problem is further complicated by the kind of work that most young people are doing. For many years, a number of social scientists — myself included — advocated more work opportunities for adolescents. We argued that such experiences would provide valuable contact with adult models and thereby further the development of responsibility and general maturity. However, from their studies of U.S. high school students who are employed, Ellen Greenberger and Lawrence Steinberg conclude that most of the jobs held by these youngsters are highly routinized and afford little opportunity for contact with adults. The largest employers of teenagers in the U.S. are fast-food restaurants. Greenberger and Steinberg argue that, instead of providing maturing experiences, such settings give adolescents even greater exposure to the values and lifestyles of their peer group. And the adolescent peer group tends to emphasize immediate gratification and consumerism.[8]

Finally, in order to counteract the

> Caring is surely an essential aspect of education in a free society; yet we have almost completely neglected it.

mounting forces of alienation in U.S. society, we must establish a working alliance between the private sector and the public one (at both the local level and the national level) to forge links between the major institutions in U.S. society and to re-create a sense of community. Examples from other countries abound:

• Switzerland has a law that no institution for the care of the elderly can be established unless it is adjacent to and shares facilities with a day-care center, a school, or some other kind of institution serving children.

• In many public places throughout Australia, the Department of Social Security has displayed a poster that states, in 16 languages: "If you need an interpreter, call this number." The department maintains a network of interpreters who are available 16 hours a day, seven days a week. They can help callers get in touch with a doctor, an ambulance, a fire brigade, or the police; they can also help callers with practical or personal problems.

• In the USSR, factories, offices, and places of business customarily "adopt" groups of children, e.g., a day-care center, a class of schoolchildren, or a children's ward in a hospital. The employees visit the children, take them on outings, and invite them to visit their place of work.

We Americans can offer a few good examples of alliances between the public and private sectors, as well. For example, in Flint, Michigan, some years ago, Mildred Smith developed a community program to improve school performance among low-income minority pupils. About a thousand children were involved. The program required no change in the regular school curriculum; its principal focus was on building links between home and school. This was accomplished in a variety of ways.

• A core group of low-income parents went from door to door, telling their neighbors that the school needed their help.

• Parents were asked to keep younger children out of the way so that the older children could complete their homework.

• Schoolchildren were given tags to wear at home that said, "May I read to you?"

• Students in the high school business program typed and duplicated teaching materials, thus freeing teachers to work directly with the children.

• Working parents visited school classrooms to talk about their jobs and

159

about how their own schooling now helped them in their work.

WHAT SCHOOLS CAN DO

As the program in Flint demonstrates, the school is in the best position of all U.S. institutions to initiate and strengthen links that support children and adolescents. This is so for several reasons. First, one of the major — but often unrecognized — responsibilities of the school is to enable young people to move from the secluded and supportive environment of the home into responsible and productive citizenship. Yet, as the studies we conducted at Cornell revealed, most other modern nations are ahead of the U.S. in this area.

In these other nations, schools are not merely — or even primarily — places where the basics are taught. Both in purpose and in practice, they function instead as settings in which young people learn "citizenship": what it means to be a member of the society, how to behave toward others, what one's responsibilities are to the community and to the nation.

I do not mean to imply that such learnings do not occur in American schools. But when they occur, it is mostly by accident and not because of thoughtful planning and careful effort. What form might such an effort take? I will present here some ideas that are too new to have stood the test of time but that may be worth trying.

Creating an American classroom. This is a simple idea. Teachers could encourage their students to learn about schools (and, especially, about individual classrooms) in such modern industrialized societies as France, Japan, Canada, West Germany, the Soviet Union, and Australia. The children could acquire such information in a variety of ways: from reading, from films, from the firsthand reports of children and adults who have attended school abroad, from exchanging letters and materials with students and their teachers in other countries. Through such exposure, American students would become aware of how attending school in other countries is both similar to and different from attending school in the U.S.

But the main learning experience would come from asking students to consider what kinds of things *should* be happening — or not happening — in American classrooms, given our nation's values and ideals. For example, how should children relate to one another and to their teachers, if they are doing things in an *American* way? If a student's idea seems to make sense, the American tradition of pragmatism makes the next step obvious: try the idea to see if it works.

The curriculum for caring. This effort also has roots in our values as a nation. Its goal is to make caring an essential part of the school curriculum. However, students would not simply learn about caring; they would actually engage in it. Children would be asked to spend time with and to care for younger children, the elderly, the sick, and the lonely. Caring institutions, such as daycare centers, could be located adjacent to or even within the schools. But it would be important for young caregivers to learn about the environment in which their charges live and the other people with whom their charges interact each day. For example, older children who took responsibility for younger ones would become acquainted with the younger children's parents and living arrangements by escorting them home from school.

Just as many schools now train superb drum corps, they could also train "caring corps" — groups of young men and women who would be on call to handle a variety of emergencies. If a parent fell suddenly ill, these students could come into the home to care for the children, prepare meals, run errands, and serve as an effective source of support for their fellow human beings. Caring is surely an essential aspect of education in a free society; yet we have almost completely neglected it.

Mentors for the young. A mentor is someone with a skill that he or she wishes to teach to a younger person. To be a true mentor, the older person must be willing to take the time and to make the commitment that such teaching requires.

We don't make much use of mentors in U.S. society, and we don't give much recognition or encouragement to individuals who play this important role. As a result, many U.S. children have few significant and committed adults in their lives. Most often, their mentors are their own parents, perhaps a teacher or two, a coach, or — more rarely — a relative, a neighbor, or an older classmate. However, in a diverse society such as ours, with its strong tradition of volunteerism, potential mentors abound. The schools need to seek them out and match them with young people who will respond positively to their particular knowledge and skills.

The school is the institution best suited to take the initiative in this task, because the school is the only place in which all children gather every day. It is also the only institution that has the right (and the responsibility) to turn to the community for help in an activity that represents the noblest kind of education: the building of character in the young.

There is yet another reason why schools should take a leading role in rebuilding links among the four worlds of childhood: schools have the most to gain. In the recent reports bemoaning the state of American education, a recurring theme has been the anomie and chaos that pervade many U.S. schools, to the detriment of effective teaching and learning. Clearly, we are in danger of allowing our schools to become academies of alienation.

In taking the initiative to rebuild links among the four worlds of childhood, U.S. schools will be taking necessary action to combat the destructive forces of alienation — first, within their own walls, and thereafter, in the life experience and future development of new generations of Americans.

---

1. Urie Bronfenbrenner, "New Worlds for Families," paper presented at the Boston Children's Museum, 4 May 1984.
2. Urie Bronfenbrenner, "The Ecology of the Family as a Context for Human Development," *Developmental Psychology*, in press.
3. Mavis Heatherington, "Children of Divorce," in R. Henderson, ed., *Parent-Child Interaction* (New York: Academic Press, 1981).
4. A.S. Makarenko, *The Collective Family: A Handbook for Russian Parents* (New York: Doubleday, 1967).
5. Lea Pulkkinen, "Self-Control and Continuity from Childhood to Adolescence," in Paul Baltes and Orville G. Brim, eds., *Life-Span Development and Behavior*, Vol. 4 (New York: Academic Press, 1982), pp. 64-102.
6. S.B. Kamerman, *Parenting in an Unresponsive Society* (New York: Free Press, 1980); S.B. Kamerman and A.J. Kahn, *Social Services in International Perspective* (Washington, D.C.: U.S. Department of Health, Education, and Welfare, n.d.); and Lloyd Johnston, Jerald Bachman, and Patrick O'Malley, *Use of Licit and Illicit Drugs by America's High School Students — 1975-84* (Washington, D.C.: U.S. Government Printing Office, 1985).
7. Kurt Aurin, personal communication, 1985.
8. Ellen Greenberger and Lawrence Steinberg, *The Work of Growing Up* (New York: Basic Books, forthcoming).

Tracked to Fail

In today's schools, children who test poorly may lose the chance for a quality education. Permanently.

Sheila Tobias

Sheila Tobias is the author of Breaking the Science Barrier *(in preparation),* Succeed With Math *(1987) and* Overcoming Math Anxiety *(1978).*

No one who has ever read Aldous Huxley's anti-utopian novel, *Brave New World,* can forget the book's opening scene, a tour of the "Hatchery and Conditioning Centre." There human embryos in their first hours of existence are transformed into Alphas, Betas, Gammas, Deltas and Epsilons—the five social classes that collectively meet the economy's manpower needs. Arrested in their development, the Gamma, Delta and Epsilon embryos are programmed *in vitro* for a lower-class future. After "birth," whatever individuality remains with these preordained proletarians will be conditioned out of each child, until there is no one in this brave new world who does not grow up accepting and even loving his bleak servitude.

Huxley's totalitarian embryology may seem fanciful to us, but his real message was political, not technological. Huxley understood, as he wrote in the foreword to the 1946 edition of *Brave New World,* that any "science of human differences" would enable the authorities to assess the relative capacities of each of us and then assign everybody his or her appropriate place in society. Huxley's vision of the modern state, with its desire for social control, implies that the discovery that ability can be measured will suggest that it *should* be. Similarly, the knowledge that people can be sorted by ability will lead irresistibly to the belief that they ought to be.

Today, many educators contend that a "science of human differences" does exist in the form of standardized tests for intelligence and ability. And, as Huxley foresaw, the pressures have grown to put these discriminating instruments to use. Education in this country is becoming a process of separating the "gifted" from the "average," the "intelligent" from the "slow"—one is tempted to say, the wheat from the chaff. From an early age, children are now ranked and sorted (a process known variably as tracking, ability grouping or screening) as they proceed through school. Those who test well are encouraged and expected to succeed and offered the most challenging work. Those who do not, get a watered-down curriculum that reflects the system's minimal expectations of them.

All this is a far cry from the vision of schooling that America's founding educators had in mind. Horace Mann, the father of American public education and the influential first secretary of the Massachusetts board of education from 1837 to 1848, thought public education would be "the great equalizer" in a nation of immigrants. For over a century now, Mann's egalitarian vision, translated into educational policy, has helped millions of immigrants to assimilate and to prosper here. But this vision is now threatened by a competing view of individual potential—and worth. We are becoming a society where test-taking skills are the prerequisites for a chance at getting a good education, and where hard work, hope and ambition are in danger of becoming nothing more than meaningless concepts.

A poor showing on tests was once a signal to all concerned—child, teacher, parents—that greater effort was needed to learn, or to teach, what was required. It didn't mean that a child *couldn't* learn. But the damaging assumption behind testing and tracking as they are now employed in many schools is that *only* those who test well are capable of learning what is needed to escape an adult life restricted to menial, dead-end jobs. This new message imparted by our schools is profoundly inegalitarian: that test-measured ability, not effort, is what counts. What many students are learning is that they are *not* equal to everybody else. Gammas, Deltas and Epsilons shouldn't even try to compete with Alphas. Alphas are better, *born* better, and it is impossible for others to catch up. What's tragic about this change is not just that it's unjust—but that it's untrue.

A Lifetime of Testing

In a private Los Angeles primary school, a 4-year-old is being taught to play a game-like test he is going to have to pass to show that he is ready for kindergarten. This is the first in an endless series of evaluations that will determine who he is, what he can learn and how far he will go in school. Just before the test begins, the counselor hands him the red plastic cube he will use. But he doesn't need her cube. He has taken this test so often, as his parents drag him around from his preschool admissions screenings, that when the time comes to play, he pulls his *own* bright red cube out of his pocket. Whether or not he is ready for this particular school, he is more than ready for the test.

Each year after this child's admission to kindergarten, he will take "norm-referenced tests" to show his overall achievement against those of his age group and "criterion-referenced tests," which examine the specific skills he is supposed to have learned in each grade. Even if he and his parents are not told his test scores (a practice that varies from school to school), ability-grouping in elementary school will soon let him know where he stands. "By the second or third grade," says Susan Harter, a psychology professor at the University of Denver who studies social development in children, "children know precisely where they stand on the 'smart or dumb' continuum, and since most children at this age want to succeed in school, this knowledge profoundly affects their self-esteem."

The point is that today "smart or dumb" determinations are made very early. "Those who come to school knowing how to read or who learn very quickly are pronounced bright," says Jeannie Oakes, author of *Keeping Track: How Schools Structure Inequality*. "Those for whom reading is still a puzzle at the end of the first grade are judged slow." And these early decisions stick. As children proceed through the elementary grades, more and more of their course work is grouped by ability. By ninth grade, 80% to 90% of students are in separate classes determined by whether they are judged to be "fast," "average" or "slow."

Magnifying Our Differences

Tracking in all its variants is rarely official policy, and the validity and fairness of standardized testing have long been under fire. Nevertheless, both tracking and testing are becoming more common. As a result, argues University of Cincinnati education professor Joel Spring (in unwitting resonance with Huxley), education in America has become a "sorting machine."

Moreover, the stunting effects of this machine may remain with students for a lifetime. "Adults can remember well into middle age whether they were 'sharks' or 'goldfish' in reading," says Bill Kelly, professor of education at Regis College in Denver. Students learn whether they have good verbal skills or mathematical ones. They learn whether or not they are musically or mechanically inclined, and so on. There are millions of adults who carry with them the conviction that they "can't do math" or play an instrument or write well. And it may all be the result of assessments made of them and internalized as children — long before they had any idea of what they wanted from life. Their sense of inadequacy may prevent them from exploring alternative careers or simply narrow their experiences.

Why are testing and tracking on the rise? Oakes, who has studied more than 13,000 junior- and senior-high-school students, their schools and their teachers, suggests that the answer has several components. They range from the focus on educational excellence during the last decade to widespread public confidence that testing is an accurate, appropriate way of gauging educational potential. Oakes also believes that testing and tracking comprise a not-so-subtle effort to resegregate desegregated schools. But they reflect as well a preference among teachers for "homogeneous groupings" of students, which are easier to teach than classes composed of students of varying abilities.

Whatever the motives, Oakes is convinced that the basic premise of the whole system is wrong. There is no way, she says, to determine accurately the potential of young or even older children by standardized tests. One key reason: Such examinations are always fine-tuned to point out differences, not similarities. They eliminate those items that everyone answers the same way — either right or wrong. Thus, small differences that may or may not measure ability in general are amplified to give the test makers what they want, namely ease of sorting. Test results, then, will make any group of individuals appear to be more different than they really are.

Benjamin Bloom, Distinguished Service Professor Emeritus of Education at the University of Chicago, agrees. "I find that many of the individual differences in school learning are man-made and accidental rather than fixed in the individual at the time of conception," he writes in his book *All Our Children Learning*. "When students are provided with unfavorable learning conditions, they become even more dissimilar." Bloom concedes that some longitudinal studies show that between grades 3 and 11, for example, children's rank in class remains virtually the same. But this is not because intelligence is fixed, he argues. It is the result of the unequal, unsupportive education the schools provide. So long as schools think there is little they can do about "learning ability," says Bloom, they will see their task as weeding out the poorer learners while encouraging the better learners to get as much education as they can.

Watered-Down Education

Research generated by Oakes and others supports Bloom, revealing that placement in a low track has a corroding impact on students' self esteem. Worse yet, because there are real differences not just in level but in the *content* of what is being taught, tracking may in fact contribute to academic failure.

Students in low-track courses are almost never exposed to what educators call "high-status knowledge," the kind that will be useful in colleges and universities. They do not read works of great literature in their English classes, Oakes's team found, and instead of critical-thinking skills and expository writing, low-track students are taught standard English usage and "functional literacy," which involves mainly filling out forms, job applications and the like. In mathematics, high-track students were exposed to numeration, computational systems, mathematical models, probability and statistics in high school. "In contrast," writes Oakes, "low-track classes focused grade after grade on basic computational skills and arithmetic facts" and sometimes on simple measurement skills and converting English to metric.

More generally, Oakes's team also found that high-track classes emphasize reasoning ability over simple memorization of disembodied facts. Low-track students, meanwhile, are taught by rote, with an emphasis on conformity. "Average" classes — the middle track — resembled those in the high track, but they are substantially "watered down."

Is this discriminatory system the only way to handle differences in ability among students? One innovative program is challenging that notion. Called "accelerated learning," it is the creation of Henry M. Levin, a professor of education and economics at Stanford University. Levin, an expert on worker-managed companies, decided to apply the principles of organizational psychology to an analysis of the crisis in education. He began with a two-year-study, during which he surveyed the literature on education and looked at hundreds of evaluations of at-risk students at elementary and middle schools. Fully one-third of all students, he estimated, were "educationally disadvantaged" in some way, were consigned to a low track and were falling farther and farther behind in one or more areas. These children needed remedial help, but that help, Levin writes, treated "such students and their educators as educational discards, marginal to mainstream education." For them, the pace of instruction was slowed to a crawl and progressed by endless repetition. The whole system seemed designed to demoralize and fail everyone who was a part of it. As Levin told one reporter, "As soon as you begin to talk about kids needing remediation, you're talking about damaged merchandise. And as soon as you have done that, you have lost the game."

To try to change the game, Levin designed and is helping to implement the Accelerated Schools Program. Now being tested in California, Utah, Missouri and (this fall) Illinois, the project accepts that elementary school children who are having academic problems *do* need special assistance, but it departs radically from traditional tracking in every other respect. First, Accelerated Schools are expected to have all their students learning at grade level by the time they reach the sixth grade. In other words, the remedial track exists only to get students off it. Collectively, the teachers and administrators at each school are allowed to design their own curricula, but they must create a clear set of measurable (and that means testable) goals for students to meet each year they are in the program. Finally, it is expected that the curriculum, whatever its specifics, will be challenging and fast-paced and will emphasize abstract reasoning skills and a sophisticated command of English.

Levin's program reflects the current administration's view that business practice has much to contribute to schooling. Levin wants schools to find a better way to produce what might be called their product—that is, children willing and able to get the quality education they will need in life. To do this, he recognizes that schools must offer better performance incentives to students, teachers and administrators. "Everyone benefits from the esprit de corps," explains Levin, "and the freedom to experiment with curriculum and technique—which we also encourage—is an incentive for teachers." By insisting upon school and teacher autonomy, the regular attainment of measurable goals and the development of innovative, engaging curricula, Accelerated Schools also hope to erase the stigma associated with teaching or needing remediation. The early results of this six-year test program are encouraging: The Hoover Elementary School in Redwood City, CA, one of the first schools to embark on the project, is reporting a 22 percentile increase in sixth-grade reading scores, actually outperforming state criteria. Both Levin and Ken Hill, the district superintendent, caution that these results are preliminary and the improved scores could be due to many factors other than the Accelerated Schools Program. But regardless of the program's measurable impact, Hill sees real changes in the school. "Teachers are now working with the kids on science projects and developing a literature-based reading program. There's a positive climate, and all the kids are learners."

Another alternative to tracking is what Bloom calls "mastery learning." He believes that it is the rate of learning, not the capacity to learn, that differentiates students with "high" or "low" abilities. This is a critical distinction, for we are rapidly approaching the day when all but the most menial jobs will require relatively complex reasoning and technical skills.

In a mastery class, children are given as much time as they need to become competent at a certain skill or knowledge level. Teachers must take 10% to 15% more time with their classes and break the class down into small groups in which the fast learners help their peers along. In time, the slower students catch up both in the amount of knowledge acquired and in the rate at which they learn. Though slow students may start out as much as five times slower than their classmates, Bloom says, "in mastery classes, fast and slow students become equal in achievement and increasingly similar in their learning rates."

At present, fewer than 5% of the nation's schools are following either of these promising strategies, estimates Gary Fenstermacher, dean of the University of Arizona's College of Education. He is a firm believer that de-tracking in some form must be the educational wave of the future. "There are ethical and moral imperatives for us to do whatever we can to increase the equality of access to human knowledge and understanding," he says.

Second Class and Dropping Out

Until society responds to those ethical and moral imperatives, however, the educational system, with its testing, tracking and discriminatory labeling, will continue on its questionable course. Today, around 25% of America's teenagers—40% to 60% in inner-city schools—do not graduate from high school, according to Jacqueline P. Danzberger of the Institute for Educational Leadership in Washington, DC. Most of the attrition occurs by the third year of high school, and many educators believe increased testing is a contributing factor.

Norman Gold, former director of research for the District of Columbia's public school system, says school dropouts are linked to the raising of standards (with no compensatory programs) in the late 1970s and the end of "social promotions"—the habit of routinely allowing failing students to move to a higher grade. "Studies show," he says, "that the risk of dropping out goes up 50% if a child fails one school year." Neil Shorthouse, executive director of Atlanta's Cities in Schools, which enrolls 750 teenagers on the point of dropping out, agrees. "Most of these kids quit school," he says of his students, "because they repeatedly get the message that they are bad students, 'unteachables.'"

Ending social promotions was long overdue. What purpose is served by graduating high-school students who can't read, write or do simple arithmetic? But schools have done little to help these failing students catch up. The present system is continuing to produce a whole class of people, particularly inner-city blacks and Hispanics, who have little economic role in our society. High school, Gold observes, has become an obstacle course that a significant number of young people are unable to negotiate. "We expect them to fail. We have to have greater expectations, and equally great support."

These failing students are missing what John Ogbu, an educational anthropologist at the University of California, Berkeley, calls "effort optimism," the faith that hard work will bring real rewards in life. Ogbu's ethnographic studies of black and Hispanic schoolchildren in Stockton, CA, suggest that one reason today's inner-city children do poorly in tests is that "they do not bring to the test situation serious attitudes and do not persevere to maximize their scores." The fault lies neither with their intelligence, Ogbu argues, nor with the absence of the "quasi-academic training" that middle-class children experience at home. Rather, it is **their lower caste status and the limited job prospects of their parents that lower their sights. Tracking formalizes this caste humiliation and leads to disillusionment about school and what school can do for their lives.**

4. FAMILY, SCHOOL, AND CULTURAL INFLUENCES: Education

What Parents Can Do

If you are worried that your own child is losing his or her enthusiasm for schoolwork as a result of being put in a lower, "dumber" track, Susan Harter of the University of Denver advises you to watch for the following signs of trouble:

Decline in intrinsic motivation, the kind of curiosity and involvement in school work that promises long-term academic success, and its replacement with *extrinsic* motivation, doing just enough to get by while depending too much on the teacher for direction and help.

Indifference to school and schoolwork; losing homework on the way to school, or homework assignments on the way back; delivering homework that is crumpled, dirty or incomplete.

Constant self-deprecation: "I'm no good." "I can't do long division."

Signs of helplessness: unwillingness to try a task, especially new ones; starting but not finishing work; difficulty in dealing with frustration.

Avoiding homework, or school, altogether. (The most frequent cause of truancy, says Olle Jane Sahler of the pediatrics department at the University of Rochester, is low self-esteem with regard to school subjects.)

Should parents whose kids have problems undertake compensatory home instruction? Sherry Ferguson and Lawrence E. Mazin, authors of *Parent Power: A Program to Help Your Child Succeed in School,* think so, not because parents can make their children "smart," but because they have the power to make their kids persistent, competitive and eager. Here are some specific steps parents can take at home to achieve this end, according to Abigail Lipson, Ph.D., a clinical psychologist at the Harvard University Bureau of Study Counsel:

Praise your child for effort, not just for achievement. Children learn about persistence from many contexts, not just academic ones, so praise your child for hard work at any task: developing a good hook shot, painting a picture, etc.

Ask your child to explain her homework, or the subjects she is studying at school, to you. Try to learn *from* your child, don't just instruct her.

Find a regular time when you and your child can work in the same space. When children are banished to their rooms to do homework, they are cut off from social interaction. It can be very lonely. Setting up a special study time together can help both (or all) of you focus on accomplishing difficult or onerous tasks. While your child is doing homework, you can balance your checkbook, pay bills, whatever.

Help your child find a learning activity he feels good about. If the subject is animals, go to a zoo. If it's cars, select some car books together from the library. Encourage him to pursue his natural interests.

Games of all kinds are good for teaching children about persistence and achievement. Competitive games emphasize strategies for competing effectively and fairly, while noncompetitive games provide children with a sense of accomplishment through perseverance.

—S.T.

Who Is "Smart"?
Who Will "Succeed"?

The consequences of increased testing and tracking are only now beginning to be felt. First there is personal trauma, both for students who do reasonably well but not as well as they would like, and for those who fail. "When a child is given to understand that his or her worth resides in what he or she achieves rather than in what he or she is, academic failure becomes a severe emotional trauma," David Elkind writes in *The Child and Society.*

But the most severe consequence may be what only dropouts are so far demonstrating—an overall decline in Ogbu's effort optimism. Its potential social effects extend well beyond the schoolroom. Intelligence and ability, says writer James Fallows, have become legally and socially acceptable grounds for discrimination, and both are measured by the testing and tracking system in our schools. Doing well in school has thus come to be the measure of who is intelligent and who has ability. Beyond that, Fallows writes, our culture increasingly accepts that "he who goes further in school will go further in life." Many of the best jobs and most prestigious professions are restricted to those with imposing academic and professional degrees, thus creating a monopoly on "positions of privilege."

At a time when our economy requires better-educated workers than ever before, can we afford to let abstract measures of ability curtail the educational aspirations and potential accomplishments of our children? Quite aside from questions of national prosperity, do we really want to become a culture whose fruits are not available to most of its citizens? Despite income disparities and more classism than many observers are willing to admit, there has always been the *belief* in America that success, the good life, is available to all who are willing to work for it. But with our current fixation on testing and tracking, and what Fallows calls credentialism, we may be abandoning that belief and, with it, the majority of our young people.

PROFILE BENJAMIN S. BLOOM

Master of Mastery

THIS 73-YEAR-OLD SCHOLAR IN A BUSINESS SUIT WOULD GLADLY RUIN AMERICAN EDUCATION.

Paul Chance

Paul Chance is a contributing editor of Psychology Today.

"What are you working on now?" a friend asked. "An article on Benjamin Bloom," I replied. "Ah," she said, "the man who ruined American education."

The man who *ruined* American education? Could we be talking about the same person? Could she mean the Benjamin S. Bloom who is Charles H. Swift Distinguished Service Professor Emeritus in Education at the University of Chicago and professor of education at Northwestern University; one of the founders of the International Association for the Evaluation of Educational Achievement; the educator whose name is linked to some of the most popular educational buzzwords, including time-on-task, educational objectives and mastery learning? Could she be talking about *that* Benjamin Bloom?

The reason for my friend's comment is that much of the current back-to-basics movement in education is a revolt against the kinds of changes that Bloom and like-minded people have tried to bring about. Bloom thinks, for example, that there is too much drill, too much rote learning, too little active participation by students, too much emphasis on lower-level "basic" skills, too much attention to "minimum" standards, too much competition and, most of all, too much failure in today's schools. He believes that the current educational system is structurally flawed and should be thoroughly rehabilitated, like an old house that is in danger of collapsing and killing its occupants. In this sense, Bloom would gladly plead guilty to having tried to "ruin" American education.

Bloom does not look the part of an educational Karl Marx. If he were to show up at central casting, he would be pegged as Mr. Anyman, the butcher, the baker, the undertaker. But a man who could be accused of trying to tear down our nation's school system? Never.

But rebels rarely look the part; that's their disguise. If they looked like Jack Palance, we would be on guard. They look like Tevye, the harmless milkman in *Fiddler on the Roof*. So when you meet Benjamin Bloom and you see this 73-year-old scholar in a business suit sitting across a table talking about education in soft, loving tones, it is easy to miss the fact that he is a kind of quiet rebel. His antiestablishment views probably had their origin in his early days at the University of Chicago, where he has spent nearly all his professional life. After taking bachelor's and master's degrees from Pennsylvania State University, he went on to earn a Ph.D. in education at Chicago in 1942. He stayed on to become assistant to the University Examiner and later filled the position of Examiner.

It was the responsibility of the University Examiner to make up and administer the comprehensive examinations taken by all of the university's undergraduates. In that post, Bloom became part of a movement to shift the school's emphasis from teaching facts to teaching students how to use knowledge in solving problems.

Bloom soon discovered that some students were very poor problem-solvers and, with research assistant Lois Broder, undertook a study to determine why. They gave students problems like those on the comprehensive exams and had them work on them aloud. What the researchers learned was that the successful problem-solvers attacked problems in a systematic and analytical way, while the poor problem-solvers simply tried to recall a memorized answer. Bloom characterized the two approaches as active and passive. For instance, Bloom and Broder asked college students to answer this problem aloud: "Give the reasons which would have influenced a typical Virginia tobacco farmer to support the ratification of the Constitution in 1788, and the reasons which would have influenced him to oppose the ratification."

A good problem-solver, Ralph, gave an answer that went, in part, like this: "Well, what rights did the Constitution give him? Well . . . from the standpoint of money, which one would be more to his advantage? Well, prior to the Revolutionary War, he would have to pay taxes to England. . . . I think he would approve of it for patriotic reasons, and from the standpoint of money he wouldn't have to ship his tobacco to England. . . . he wouldn't have to pay the taxes."

A weaker problem-solver, George, gave an answer that went like this: ". . . Well, uh, to tell the truth I never had anything on [that] and at present I couldn't think of any."

From Psychology Today, April 1987, pp. 43-46. Copyright © 1987 by Sussex Publishers, Inc. Reprinted by permission.

Bloom and Broder concluded, "George probably has almost as much real knowledge about the Virginia tobacco farmer as Ralph. However, Ralph keeps working with what is given until he is able to give some semblance of a solution."

Besides identifying differences between successful and unsuccessful problem-solvers, Bloom and Broder found that they were able to teach weak problem-solvers the skills of their more successful classmates.

The Bloom and Broder study, published in 1950, became a classic of educational research. It demonstrated that "higher-level skills," skills that many psychologists and educators took to be largely inherited, could be taught. And it suggested that the wide variability in student achievement commonly seen in classrooms might not be inevitable, that the gap between the top and bottom of a given class might be substantially narrowed.

Our present educational system rests on the assumption that wide variability in achievement is largely the result of wide variability in innate learning ability. Most of us probably find little to quarrel with in this commonsense assumption. A recent survey revealed, for instance, that most mothers in the United States believe that the principal ingredient in school success is the inborn talent of the youngster. Wide variability in student achievement is therefore natural and inevitable. Who can argue with that?

Bloom can. The work he and his colleagues have done over the past 40 years has convinced him that much of the variability seen in student performance is neither natural nor inevitable but the product of our educational system. Bloom admits that there are innate differences in learning ability, but he believes that these differences are much smaller than most of us imagine and do not account for the wide differences in student achievement. What might be called Bloom's dictum states: What one student can learn, nearly all students can learn.

Bloom points to studies of tutoring to support his view. He and his doctoral students have conducted studies in which tutored students are compared with those taught under conventional group instruction. They have found that the average tutored student learns more than do 98 percent of students taught in regular classes. They also found that 90 percent of the tutored students attained levels reached by only the top 20 percent of those in regular classes.

Tutoring shows that the vast majority of students are capable of doing outstanding work. But Bloom doesn't believe that tutoring is a practical approach to instruction: "We simply can't afford a student-to-teacher ratio of 1 to 1 or even 3 to 1." Over the past 25 years, Bloom and his colleagues have worked to develop a system of group instruction that would approximate the effects of tutoring. The system is called mastery learning.

In mastery learning the teacher instructs the class in more or less the usual way, although Bloom likes teachers to involve the students more actively and reinforce their contributions more frequently. At the conclusion of an instructional unit (about every two weeks), the teacher gives a "formative test" to determine the need for "corrective instruction." The test is not used for grading but lets the teacher know what the students haven't yet learned.

The teacher studies the test results to identify common errors—points most students didn't get from the lesson. This material is then retaught, perhaps in a different way, to try to get the ideas across.

After this, the students work in groups of two or three for 20 to 30 minutes. The purpose of this group work is for the students to help one another on points they had missed on the formative test. The student who doesn't understand the procedure for dividing fractions asks classmates for help. If no one in the group is able to provide the answer, they call on the teacher. But usually, Bloom says, the groups are able to work on their own.

Some students need help beyond group work and may be assigned workbook exercises, text reading, the viewing of a videotape or some other activity. It usually takes these students no more than an hour or two a week to complete the extra work necessary to catch up.

The final step in the mastery approach is the "evaluative test." This test is similar but not identical to the formative test; it "counts" toward the student's grade. Grading is not, however, on the curve. Students' grades reflect the extent to which they mastered the unit, not their class rank. This means that every student can earn an A.

Not every student does earn an A, but studies have consistently shown that mastery students learn more than those taught in the conventional manner. In fact, the average mastery learning student does better than about 85 percent of students taught in the traditional way. And 70 percent of mastery students attain levels reached by only the top 20 percent of students in regular classrooms.

Many people would be content with these results, but Bloom is determined to push the limits further. For instance, one reason students differ in achievement is past learning. What a student knows at the beginning of a lesson affects what the student gets out of that lesson. Someone with a good grasp of short division is likely to follow a lesson on long division; a student who is confused about short division will simply become more confused. Those who differ at the outset in what Bloom calls the "prerequisites for learning" will be even further apart at the end of the lesson.

But, Bloom asks, what would happen if all students started the lesson on an equal footing? One of Bloom's students, Fernando Leyton, performed an experiment to answer this question. Students in a second-year algebra class took a test at the beginning of the school year to determine what they recalled from the first-year course. Then, using the corrective instruction method of mastery learning, the teacher taught the students the specific skills they lacked. After this, the teacher taught the first unit using the mastery learning approach. The students in this class did far better on a test of that unit than did students in a comparable class who had merely had a general review of first-year algebra and were taught in the ordinary way. Leyton obtained similar results in a study of second-year French students.

The benefits of prerequisite training, when combined with mastery learning, multiply over a period of weeks. One of Leyton's experimental classes continued using mastery learning for about three months. The average student in this class scored higher than did 95 percent of those in a regular class on the same material.

Impressive as these results are (they nearly match the effects of tutoring), Bloom is not satisfied. He notes that variability in student achievement is not due solely to what takes place in school. Parents who encourage their children to do well in school, who let them know that school learning is important, who show by their own behavior that they value learning, who help with homework, who provide tutors when their own efforts are inadequate—such parents have children who enjoy more success in school. It is not, Bloom emphasizes, demographic characteristics such as parental income, occupation and educational level that need to be changed but parental behavior. There is little we can do to change the demographic features of the parents of students who do not do well, but we can help parents to change their behavior in ways that will help their children in school. We can and, Bloom insists, we should.

Although Bloom is careful about how he deals with this touchy issue, he does advocate having educators work with parents. And he believes that parental training should begin before children start school. "The most rapid period of learning is the period that ends at about the time the student begins first grade. Some parents make good use of this time, but others don't. The result is that some students are far ahead of others before the school bell rings." Programs such as Head Start that try to bring the disadvantaged student up to par are fine, Bloom believes, but training the parents to do the job right in the first place would be even better.

Bloom believes that with parental training, prerequisite instruction and mastery learning, almost all students can master the content of their courses. But, true to his rebel nature, Bloom admits that even this would leave him dissatisfied.

"Having an effective method of instruction is only half the battle," he says. "Once you know how to teach well you have to ask the question, 'What is worth learning well?'" Many years ago Bloom and his colleagues produced two volumes aimed at classifying the kinds of things students are asked to learn. The first volume of *Taxonomy of Educational Objectives* identifies academic objectives that range from lower-level skills, such as the ability to recall and understand facts, to higher-level skills, such as the ability to synthesize and evaluate facts.

Traditional education, Bloom complains, has always devoted itself almost exclusively to lower-level goals. "Studies have shown," he notes, "that over 95 percent of the items on teacher-made tests require nothing more than the recollection of facts." It is no wonder, Bloom observes, that so many of the bright college students he and Broder studied years ago could not solve problems. "If we do not teach higher-level skills such as problem solving," says Bloom, "we cannot reasonably expect students to master them."

The second volume of the *Taxonomy* recognizes that students learn much more in school than academic subjects. Much of what they learn involves interests, attitudes, values and social skills. Bloom believes that the most important learning that takes place in school may have to do with feelings. Students who do well on a task feel good about the task, the school, the teacher and themselves. In discovering that they are good students, they learn that they have value in the eyes of teachers, parents and even other students. Students who do poorly on a task feel unhappy about the task and everything associated with it, including themselves. Bad students learn that they are bad people.

Teachers inevitably convey these judgments; they cannot do otherwise in a system that focuses on class rank instead of mastery of course content. "It's hard to think of any place in our society that is as preoccupied as the schools with comparing people with one another," Bloom observes. He notes, for example, that employees are rarely ranked from highest to lowest, from best to worst. "Yet that is exactly what is done every day in our schools."

Bloom believes that traditional education not only undermines the self-esteem of students who do poorly, it "infects" these students with emotional problems. Bloom emphasizes that this outcome is not a rare event attributable to an occasional insensitive teacher. Rather, it is the inevitable result of a system of education that assumes that large numbers of students must fail or just get by.

Why do our schools persist in walking the same old path? The answer is rooted in our history. Our educational system had its origin in the agricultural age, a time when society had little need of large numbers of well-educated citizens. With industrialization came the need for widespread literacy, but only those who would govern, run businesses or follow professions needed more than a smattering of education. What did a farmer or textile worker need to know of Shakespeare, Newton or Locke? So, to a large extent the purpose of public schooling was to identify those students who would go on to become leaders. Because education was as much concerned with selection as with instruction, a system that encouraged competition and left many students behind made some sense.

But our society has changed. We now face a world in which farming and manufacturing, long our major employers, play minor roles. As Alvin Toffler has shown in *The Third Wave,* increasing numbers of people are employed at tasks that involve the manipulation of information more than physical labor. And many of these workers can expect to return to school for retraining repeatedly during their careers. "We need large numbers of people with high-level skills who like to learn," says Bloom, "and we're not going to get them with an educational system designed to ensure that most students fail." The solution, he suggests, is to replace our antiquated educational system with one that produces very few failures.

His efforts to do just that have led some people to suggest that he is a woolly-headed professor whose rebellious ideas would ruin our schools, if they haven't already. But if Bloom is right, his brand of education would produce results far superior to those of the traditional system. The majority of students would leave our schools knowing much more than students do today. And feeling better about themselves.

Not Just for Nerds

Science is now too important to be left to the technicians, the experts insist. But what can we expect from the crusade to make it user-friendly?

You may not recall the difference between DNA and the PTA, but you've no doubt heard about the science-literacy crisis. Every year brings a fresh set of revelations about America's seemingly boundless superstition and ignorance. Fully 40 percent of the nation's adults think alien creatures have visited Earth, according to studies by Jon Miller, director of Northern Illinois University's Public Opinion Laboratory. Only 45 percent of us know the planet revolves annually around the Sun, and just 46 percent have accepted that humans evolved from earlier species. The prescience era even has its acolytes in the White House. Ronald Reagan admits to being interested in astrology and permitted his wife's obsession with pseudoscience to influence the presidential schedule.

While hocus-pocus thrives in high places, one international study after another places U.S. school kids near the bottom of the heap in mathematical achievement. The Educational Testing Service reported last year that Korean 13-year-olds succeeded twice as often as Americans at solving a two-step mathematical problem, such as determining an average. Three times as many could design a simple scientific experiment. University of Michigan researchers have documented similar achievement gaps between Chinese and American children of various ages. Other investigators have shown that *average* Japanese 12th graders have a better command of mathematics than the top 5 percent of their American counterparts. Indeed, a 1989 report by the National Research Council estimates that three quarters of the nation's graduating high-school seniors leave school without the skills to survive a college-level math or engineering course.

These unrelenting poor results have managed to restore science education to the nation's list of unmet crises. Science—with its drab lab coats, pocket penholders and flaming Bunsen burners—suddenly seems more important. Brains are beautiful! The nation needs its nerds; maybe we all better become nerds. The alternative, say the mighty coalition of teachers, business leaders and public officials, is trouble. Without more and better technical education, the argument runs, future workers will be ill prepared for even the most routine jobs. The United States will lack the engineering talent to compete effectively in the new global economy. Worse yet, voters won't be able to make sensible decisions about waste management or global warming or genetic experimentation or new missile systems. None of those claims is beyond dispute—indeed, some scientists take issue with each of them—but no one denies that something is seriously wrong.

Open eyes: Science is simply a way of looking at the world. At root, it consists of asking questions, proposing answers and testing them rigorously against the available evidence. As the popular astronomer Carl Sagan wrote recently, "Science invites us to let the facts in, even when they don't conform to our preconceptions. It counsels us to carry alternative hypotheses in our heads and see which best match the facts." For all the natural forces it explains, science is a course in analytical thought.

Unfortunately, few American students ever get to taste real science, for few of the nation's schools teach it. All parties now seem to agree that American science education serves not to nurture children's natural curiosity but to extinguish it with catalogs of dreary facts and terms. "The questions we're all interested in concern the universe we live in, the way our bodies work, what the mind is, how all these things are integrated," says Stephen Toulmin, a physicist turned philosopher at Northwestern University. As most of us know firsthand, those aren't the kinds of questions that kids are encouraged to ponder. Instead, says Toulmin, "we teach them to solve differential equations, to handle test tubes without breaking them."

Whether out of boredom, laziness or the allure of other pursuits, American students are fleeing math and science in droves. By the third grade, half of all students don't want to take science anymore, says Edward Pizzini, associate professor of science education at the University of Iowa; by the eighth grade, only 1 in 5 wants to keep going. Fewer than half ever take a math or science course after the 10th grade, and only 1 percent study calculus, a subject pursued by 12 percent of Japan's high-school students. The proportion of college students who major in engineering is six times as high in Japan (4 percent) as in the United States (.7 percent). And the U.S.-born engineering doctorate is almost a thing of the past. Foreign nationals now receive a quarter of the natural-science Ph.D.s and more than half of the engineering Ph.D.s awarded in this country.

Sense of crisis: All of which raises economic concerns. Thanks largely to the influx of foreign graduate students, the nation's annual Ph.D. crop has so far held steady as a percentage of the population. But many analysts predict actual shortages of Ph.D. scientists by the mid-1990s unless more students are drawn into the pipeline—particularly more minority students. Historically, science and engineering have been overwhelmingly white, male enclaves; in a 1986 count, only 2.5 percent of the nation's engineers were black and even fewer Hispanic. But both of those groups are now growing fast as a percentage of the population, and the proportion of white males is declining. To keep up, science will have to attract more minority and female students.

Stirred by the sense of crisis, various groups—school boards, corporations, foundations, the federal government—have launched ambitious initiatives to help create a more science-savvy populace. Project 2061, a massive program sponsored by the American Association for the Advancement of Science, now has six teams of elementary educators working up a plan to revise the nation's primary schools. The National Science Teachers Association has its own nationwide "scope sequence and coordination project"; 10,000 members of the NSTA will meet this week in Atlanta to chart the crisis, among other things. The National Science Foundation is offering grants to states that want to revamp their science- and math-education programs. The National Research Council and the Smithsonian Institution have joined forces to set up a National Sciences Resources Center to train teachers and help bring real science into the schools. The list goes on.

These may all be worthy efforts. The current system is repelling vast numbers of Americans from a fascinating human endeavor, and it may be depriving the nation of a new generation of technical innovators. To the extent that better teaching can alleviate our ignorance and stoke the economy, no one is likely to oppose it. Nonetheless, a few apostates question whether the situation is as dire, or better teaching as sure a cure, as the science boosters proclaim. Will the demand for scientists and engineers really go unmet for any duration, they ask. Is a working knowledge of science in fact essential to productive employment? To good citizenship?

Morris Shamos, an emeritus professor of physics at New York University, thinks not. The impending manpower shortage, he argues, has never been unambiguously documented. "We have seen spot shortages of scientists and engineers in the past and will continue to see them in the future, simply because the supply cannot be turned on and off as quickly as the demand," he argues. "But the same is true of surpluses in the field, and we have seen times when engineers were pumping gas or driving cabs."

Professional scientists are not the education advocates' only concern. They warn that numerically controlled machines are fast replacing old drill presses, and that future autoworkers will have to manipulate robots rather than wrenches. Shamos counters that Americans have mastered all manner of gizmos—from personal computers to electronic machine tools—by reading the instructions or attending on-the-job training courses. The fact that few of us can explain what's going on inside these contraptions has yet to cripple us.

There would seem to be even less evidence that science illiteracy is locking people out of the white-collar work force. As Harper's editor Lewis Lapham has observed, "The society bestows its rewards on the talent for figuring a market, not on the proofs of learning or the subtlety of mind." Ronald Reagan's superstitions didn't slow him down. Nor have Vice President Dan Quayle's scientific failings kept him from becoming chairman of the National Space Council. Why send astronauts to Mars? "We have seen pictures," he explained last fall, "where there are canals, we believe, and water. If there's water, that means there's oxygen. If oxygen, that means we can breathe. And therefore, from the information we have right now, Mars clearly offers the best opportunity to see if a man or a woman can be able to survive on that planet."

Atomic poetry: As for the argument that good citizenship requires a working knowledge of science, it's true that public-policy issues have become increasingly technical. It's also true that scientists are often as bitterly divided on them as the rest of us. People who are familiar with science can argue more effectively about the pros and cons of pesticides or nuclear power. But if *science* is needed for more effective debate, asks Harvey Brooks, emeritus professor of technology and public policy at Harvard, what about other disciplines? Central American history, say, or economics? What about geography? Just last year, young Americans placed last among 10 nations on a National Geographic quiz about the world's political boundaries. When adults were included, Americans placed sixth. But even then, only 32 percent could find Vietnam on a map. Fourteen percent could not find the United States.

None of this suggests that science doesn't matter. To say a president needn't understand biology is not to say he *shouldn't*. And trying to raise public awareness of history or geography hardly precludes doing the same with science. The reasons aren't all utilitarian. The insights of science rival those of poetry or music; the contemplation of atomic structure or speculation on the cosmic reaches offer mystery and elegance that would please any supple mind.

Unfortunately, none of us can begin to take in the expanding web of human knowledge; no amount of schooling will make us expert in every new drug, weapon, device, food additive or environmental issue. Since we can't know what's in every black box science sends our way, some critical perspective seems especially important. Whether we gain that perspective via physics or philosophy is of little

We're Not No. 1

■ On a chemistry achievement test, high-school students in Hong Kong ranked first among 13 countries, followed by England and Singapore. **Americans ranked 11th.**

■ In physics, Hong Kong was first again, followed by England and Hungary. **American students who took two years of physics ranked ninth.**

■ Singapore scored first in biology, followed by England, Hungary and Poland. **In biology, the most popular science course, U.S. kids ranked last.**

SOURCE: THE INTERNATIONAL ASSOCIATION FOR THE EVALUATION OF EDUCATIONAL ACHIEVEMENT

consequence. But gain it we must. The alternative is live by the judgment of experts who are no less fallible than we are.

GEOFFREY COWLEY *with* KAREN SPRINGEN *in Chicago,* TODD BARRETT *in Boston and* MARY HAGER *in Washington*

Development During Adolescence and Early Adulthood

The onset of adolescence is marked by the emergence of secondary sex characteristics and the achievement of reproductive maturity. However, adolescence also brings substantive shifts in memory and problem-solving skills, in preferred activities, and in emotional behavior. "Those Gangly Years" presents the results of a study following a group of adolescents over a three-year period, highlighting early vs. late maturation differences in boys and girls. The timing of puberty affects many events in the adolescent's life, including school achievement, moods, interaction with the opposite sex, and family relationships.

Teenagers today are faced with temptations such as drugs, sex, and mobility at an earlier age than their predecessors. At the same time, they are less likely to get the guidance they need from adults, due to such factors as the high rate of single-parent homes and working mothers. As a consequence, teen-related problems such as drug and alcohol use, teen pregnancy, high school dropout rate, and suicide occur at distressingly high rates. These issues are examined in "A Much Riskier Passage."

In some cultures, a ritualistic ceremony marks the transition to adulthood—a transition that occurs quickly, smoothly, and with relatively few problems. The onset of adulthood is more difficult to distinguish in modern indus-

# Unit 5

trial societies. In American culture, the transition is vague. Does someone become an adult when he or she achieves the right to vote, the privilege of obtaining a driver's license, the ability to legally order an alcoholic drink in a bar, or the right to volunteer for the armed forces?

Adolescence has its ups and downs, but some researchers argue that much of the storm and stress attributed to adolescence is exaggerated. "The Myth About Teenagers" reveals that while adolescents are subject to mood-swings, most are basically well-adapted and happy. Focusing on the negative aspects of adolescent behavior may create a set of expectations that the adolescent strives to fulfill. However, for some adolescents the transition to adulthood is fraught with despair, loneliness, and interpersonal conflict. The pressures of peer group, school, and family may produce conformity or may lead to rebellion against or withdrawal from friends, parents, or society at large. These pressures may peak as the adolescent prepares to separate from the family and assume the independence and responsibilities of adulthood. Bruce A. Baldwin's analysis in "Puberty and Parents" suggests that adolescence spans a 20-year period, roughly between the ages of 10 and 30. Three separate periods can be identified: early adolescence (adhering to tribal loyalties), middle adolescence (testing adult realities), and late adolescence (joining up). Understanding adolescent attitudes and confronting emotional reactions to adolescents can promote parental growth and development as much as it promotes the development of the adolescent. The idea that psychological aspects of adolescence continue into the late 20s is examined further in "Therapists Find Last Outpost of Adolescence in Adulthood."

"Proceeding With Caution" describes the "twentysomething generation," (today's young adults who fall between 18 and 29 years of age). They are a "back-to-basics" group living in the shadow of the baby boomers. They are staying single longer and living at home longer. Many are from families of divorce or were latchkey kids and they do not want the same for their children. This generation likes reinforcement—reviews, grades, performance evaluations, and so on. They are the best-educated group in U.S. history—the means to a middle-class life-style. However, they are reluctant to make vocational commitments.

Although much attention has been given to the problems of adolescence and the transition to adulthood, developmentalists have shown far less interest in the early years of adulthood. Yet, during early adulthood, many individuals experience significant changes in their lives. Marriage, parenthood, divorce, single parenting, employment, and the effects of sexism may be powerful influences on ego development, self-concept, and personality. Negative emotions such as jealousy and envy subvert efforts to establish effective interpersonal relationships, and often lead to hostility or isolation. In "Jealousy and Envy," Jon Queijo describes three strategies—self-reliance, selective ignoring, and self-bolstering—that individuals can use in their attempts to cope with negative emotions. Some individuals seem to grow stronger when confronted by the stresses of daily living, whereas others have great difficulty coping. The challenge for developmentalists is to discover the factors that contribute to one's ability to cope with stress and the natural crises of life with minimum disruption to the integrity of one's personality.

Looking Ahead: Challenge Questions

Why does adolescence in current times seem to be fraught with so many difficulties, even for the well-adjusted? Do you think adolescence really continues into the late 20s?

What techniques do you use to deal with your emotional ups and downs? Which do you think are effective and which are ineffective? Do you see any signs of growth or change in yourself?

Why do you think so many adolescents and young adults find it easier to lose themselves in drugs or cults than to confront their problems and take steps to develop self-control and self-reliance? What kinds of parenting techniques might have given such individuals sufficient self-esteem and coping skills to combat their self-doubts and loneliness?

Do generations really differ as much as suggested in "Proceeding With Caution"? In your experience, do the characteristics of the "twentysomething" generation described in this article ring true? Or are people, regardless of similarities in age, too variable to categorize?

# The Myth About Teen-Agers

Richard Flaste

Richard Flaste is Science and Health Editor of The New York Times.

A father I know tells of one unsettling moment when he was sure he would never understand the teen-age mind. The mind in question was that of his own teen-ager, a 15-year-old blonde possessed of considerable charm and an aggressive reticence. The pivotal moment was this effort at conversation:

"So, how was your day at school?"

"Good."

"Was it more than just good? I mean, did anything actually happen that was interesting?"

"No, it was just good."

"What about the bio test? Didn't you have a test?"

(The answer this time came with a gleam of irritation in her eyes.) "I told you, everything was *good*."

He backed away, feeling foolish, sorry he had tried, sorry he'd stirred her pique. What could possibly be going on in her mind? And he wondered about himself, too, as he slipped away into a friendlier room. Was he actually afraid of her?

Later, having thought about that scene many times, he concluded that he had in fact been frightened, but not so much of her — after all, they had their good times, and she did seem to love and respect him at least every now and then. Rather it was the condition of adolescence that scared him. On occasions like that abortive effort at conversation he wasn't just confronting a teen-ager who might or might not be in the mood to talk to him but also everything he had ever heard about adolescence, the Sturm, the Drang and the plain old orneriness of it. That flash of annoyance in her eyes was all that was necessary to evoke an image of those infamous raging hormones boiling inside a pubescent caldron.

The burden was a heavy one to take into a small inquiry about a person's day. It was an unnecessary burden, too. For, although most of us aren't aware of it, the concept of adolescence as a period of angry, dark turmoil has largely been overthrown, and along with it the idea that kids need to be tormented and perverse to make the storm-tossed transition to adulthood. That concept is being replaced by a new psychology of adolescence, a growing body of work that reflects a vigorous attempt to find out what life is genuinely like for normal teen-agers, and which reveals that delight plays as large a role among most adolescents as misery does.

A substantial number of psychotherapists and personality theorists still believe that adolescence is a trial by fire, as do many writers of juvenile fiction. But what is now the mainstream of psychological research dismisses this emphasis on turmoil as balderdash. It is a misguided emphasis, many researchers say, which grew out of the overwrought imaginations of romanticists ranging from the Freuds (mostly Anna) to Goethe (particularly, "The Sorrows of Young Werther").

In the 1970's, I wrote regularly about child-development issues for this newspaper, and I accepted the idea that any teen-ager was necessarily a little mad. To a large extent this notion was promoted by the psychoanalytic literature, but it was embraced by many of us as a handy way to explain hard times with our teen-agers. We could say that noxious behavior such as naked aggression was normal, and so we didn't have to worry. We reassured ourselves constantly, usually with a knowing laugh about the craziness of teen-agers. But I did not believe it completely. Why should only this age group carry the label of madness? Coming back to the question now, I am struck by how different the mood among many psychologists and psychiatrists is. By and large, they are more determined to draw an empirical and detailed picture of the varied and complex adolescent experience.

Some researchers have surveyed thousands of teen-agers to learn what they believe about themselves and their families. Others have given youngsters beepers to carry around, so that when they are signaled they will report their moods at that moment. And still others have worked with families, asking parents, teen-agers and their siblings to take batteries of tests.

Among the most influential of this cadre of researchers is Dr. Daniel Offer of the University of Chicago, a psychiatrist who believes that the widespread idea that the teen years are unavoidably insane has mischaracterized the lives of millions of people. Moreover, he contends that the emphasis on turmoil has created the expectation of Sturm and Drang for every teen-ager, thereby masking the serious emotional difficulties of a significant minority of adolescents who need professional help. Dr. Offer was moved to declare a "Defense of Adolescents" in The Journal of the American Medical Association, and he made a presentation along the same lines to the convention of the American Psychiatric Association last May.

Dr. Offer's team was among the earliest to plumb the day-to-day feelings of large numbers of teen-agers. Their first explorations, in the 1960's, began to reveal that, incredible as it might have seemed, most teen-agers were happy most of the time. After years of confirmatory work, he confidently told the psychiatrists' convention that "the vast majority of adolescents are well-adjusted, get along well with their peers and their parents, adjust well to the mores and values of their social environment and cope well with their internal and external worlds." Dr. Offer recalled in a recent interview that this message disturbed some of his colleagues, because they didn't believe that teen-agers' responses to questions could be trusted.

The work of Dr. Offer and his team has provided a starting point for many of the nation's researchers. By no means do they make adolescence out to be an easy time, any more than life as a whole is easy. Adolescence is a period of rapid and profound change in the body and mind. It is a time to find out who you are and to begin to move toward what you will become. Family bickering is bound to escalate during this period, but it usually centers on what one psychologist calls the "good-citizen topics," such as chores, dress and schoolwork. Most researchers feel that this conflict is useful, because it allows a teen-ager to assert his or her individuality over relatively minor issues.

There are several explanations for this rise in family quarreling. A widely cited cognitive explanation comes from the work of the Swiss psychologist Jean Piaget, who showed that children do not have the capacity for the abstract, analytical thinking he called "formal operations" until the teen years. The arrival of that tool is what enables them, in the view of some experts, to question their parents' thinking.

RESEARCH BY JUDITH G. SMETANA, AN ASSOCIATE PROfessor of education, psychology and pediatrics at the University of Rochester, has recently aroused much interest among psychologists. She contends that concentrating on the development of a child's ability to think logically isn't useful in elucidating family relationships. Instead, she focuses on the way parents and children conceptualize their experiences. According to her research, there are two fundamentally different world views at the core of family conflict: adolescents tend to see much of their behavior as a "personal" matter, affecting no one but themselves and therefore up to them entirely, while their parents tend to hold to what she calls "conventional thinking," which sees society's rules and expectations as primary. The dichotomy provides for commonplace clashes:

"Clean up your room. This family does not live in a hovel."

"It's my room and I like hovels."

In a recent paper Smetana charts the typical evolution from personal to conventional thinking in a teenager. A child of 12 or 13 generally has no use for conventions when it comes to family issues. Between 14 and 16 the teen-ager comes to recognize conventions as the way society regulates itself. Then comes a brief period in which conventions are rejected again, but more thoughtfully. Between 18 and 25 conventions are seen as playing an important and admirable role in facilitating the business of society.

As the teen-ager moves in fits and starts in the direction of his or her parents, the parents generally stick to their guns. Nevertheless, the adolescent has begun to reason like them, and so, by the age of 15 or 16, the quarreling usually subsides, at least for a while. It is replaced by a period in which members of the family are able to negotiate more successfully than in the past. The teen-ager learns how to "work on mom and dad" to achieve goals like staying out late at night or using the family car.

Indeed, "negotiation" has emerged as one of the key words in the new psychology of adolescence. Instead of talking about rebellion and a painful separation from the family, many psychologists now see adolescence as a time in which parents and children negotiate new relationships with one another. The teen-ager must gain more authority over his or her own life; the parents must come to see their child as more nearly an equal, with a right to differing opinions.

ANOTHER WAY OF LOOKING AT THIS PERIOD OF NEGOTIAtion has been formulated by psychologists Harold D. Grotevant at the University of Texas at Austin and Catherine R. Cooper at the University of California, Santa Cruz. They believe that typically there is an elaborate interplay between a teen-ager's striving to be an individual who is separate from the family and his attempts to maintain a close, caring relationship with his parents. In a recent study they found that teen-agers who had the strongest sense of themselves as individuals were raised in families where the parents offered guidance and comfort but also permitted their children to develop their own points of view.

For some families, this can be a terrible period. Parents who try to exert too much control over their children and find it impossible to yield in a conflict can be driven into a frenzy by the efforts of their teen-agers to establish their own identities. But most parents are more pliable, and find ways to compromise.

A neighbor of mine, unlucky enough to have a son who became a teen-ager when punk was hot, remembers how uncontrollably rattled she would get when she saw him dressed for school in the morning. "He would throw on any rag, this way and that," she recalls. She couldn't contain her exasperation and her fear that this monstrous style of dressing would somehow reflect on her. She imagined that others might pity her for her misfortune. After repeated failed confrontations, she decided not to come downstairs until her son was gone. It seems to her now that he dressed a little more sensibly if he knew she wouldn't be there to see it.

The influence of peers in everything from dress to sexual mores is undeniably strong during the teen-age years, although some experts downplay it because, like turmoil, it's been given more press than they think it's worth, and because they believe the emphasis on peers underestimates the importance of continuing attachments to the family. Parents may find it comforting to

5. ADOLESCENCE AND EARLY ADULTHOOD

realize that this growing influence of friends is not the first assault on their authority. Throughout most of a child's life—not just in adolescence—parents share control with others: teachers, friends, siblings. After the earliest years parents are no longer in a position to know about whole segments of their children's lives, because so much takes place outside the home. Gerald R. Adams, a psychologist at Utah State University, making this point in a recent conversation, said, "If I interviewed your family, you'd be shocked at how little you know about the life of your child—school life, social life. Most parents don't really know the world of their children."

There are many things that children don't tell us because they know we won't approve, even if we are trying very hard to give them a measure of greater freedom. Who among us really wants to know everything about a child's sexual experimentation or moments of embarrassment?

But that isn't the only reason for reticence. Sometimes, teen-agers, like the rest of us, just don't feel like talking. The notorious moodiness of teen-agers is one of the most interesting areas of the latest psychological investigations. Dr. Offer says that he has found little tendency among normal teen-agers to plummet into deep and dark despair. (He points out that, although teen-age suicide is a deeply troubling phenomenon, the suicide rate for adolescents is lower than it is for people in their 20's and far lower than it is for people in their 70's.)

Reed Larson, a psychologist at the University of Illinois, has found a middle ground between coloring all of adolescence with dark moods or dismissing moodiness altogether. In studies carried out by giving kids beepers that signaled them when to report their feelings, he found that mood swings are a fact of teen-age life. On average, adolescents feel more delight and more sadness at any given moment than adults and move from one mood to another more rapidly. For teen-agers, emotional states generally last no more than 15 minutes. Even the strongest feelings tend not to last more than a half-hour, while the same kinds of feelings may last for two hours or longer among adults. But Larson and his colleagues concluded that these mood swings are healthy and natural, a reasonable response to a time of life filled with fast-paced events. "The typical adolescent may be moody," they wrote in a recent paper, "but not in turmoil."

As word of the new insights into teen-age life gets out, parents are bound to benefit. They will learn to expect a certain amount of bickering in early adolescence, and they'll realize that it has a normal course to run. Parents might find it helpful to abandon the old vocabulary of rebellion and defiance. The words have been so widely misapplied that they are even used to characterize things like the piercing of ears against mother's wishes or staying out late. When self-assertion and self-indulgence are overdramatized by parents and viewed as rebellion or defiance, they take on the aura of criminal acts instead of being part of the fascinating interplay between parent and child in which the child eventually becomes an adult and the parent eventually accepts that.

In the case of the teen-ager who resists parental efforts at conversation, parents should realize that kids are sometimes out of sorts, or they may be too bewildered by the hectic events of their lives to find the words to describe them. Some teen-agers are more apt to engage in friendly conversation at dinner time, or just before they go to bed, than at times when the stresses of the day are still fresh. (Of course, when a teen-ager senses that his or her parents' attempts at conversation are prompted by their need to be in control, conversation will often be resisted on that ground alone.)

While a momentary rebuff or a bit of surliness should not be reason for deep concern, parents ought to worry about those moods that don't change. Teen-agers who are depressed for long periods of time, relentlessly combative, friendless, reclusive or miserable in other ways are not going through a normal adolescence; they and their families need help, perhaps professional.

Implicit in much of the new work on adolescence is the belief that parents must take a strong grip on their own sense of themselves, their own worth, so that they are not so easily shaken by every normal challenge to their control, and so that they can hold on to their children while confidently letting out some line.

Parents should find life a little easier, in any event, if the bogeyman of necessary insanity in adolescence is finally vanquished. For many parents that will mean they no longer need to fear their children's adolescence but can relax and maybe even enjoy it.

Article 38

Those Gangly Years

NEW BODIES, NEW SCHOOLS AND NEW EXPECTATIONS NOTWITHSTANDING, MOST EARLY ADOLESCENTS WEATHER THE EXPERIENCE SURPRISINGLY WELL.

Anne C. Petersen

Anne C. Petersen, Ph.D., a developmental psychologist, is a professor of human development at Pennsylvania State University, where she heads the interdisciplinary department of individual and family studies.

"How can you stand studying adolescents? My daughter has just become one and she's impossible to live with. Her hormones may be raging, but so am I!" A colleague at a cocktail party was echoing the widespread view that the biological events of puberty necessarily change nice kids into moody, rebellious adolescents. The view has gained such a foothold that some parents with well-behaved teenagers worry that their kids aren't developing properly.

They needn't worry. My research, and that of many others, suggests that although the early teen years can be quite a challenge for normal youngsters and their families, they're usually not half as bad as they are reputed to be. And even though the biological changes of puberty do affect adolescents' behavior, attitudes and feelings in many important ways, other, often controllable, social and environmental forces are equally important.

One 14-year-old, for example, who tried to excuse his latest under-par report card by saying, "My problem is testosterone, not tests," only looked at part of the picture. He ignored, as many do, the fact that, because of a move and the shift to junior high school, he had been in three schools in as many years.

My colleagues and I at Pennsylvania State University looked at a three-year span in the lives of young adolescents to find out how a variety of biological and social factors affected their behavior and their feelings about themselves. A total of 335 young adolescents were randomly selected from two suburban school districts, primarily white and middle- to upper-middle-class. Two successive waves of these kids were monitored as they moved from the sixth through the eighth grade. Twice a year we interviewed them individually and gave them psychological tests in groups. When the youngsters were in the sixth and eighth grades, we also interviewed and assessed their parents. Just recently we again interviewed and assessed these young people and their

175

5. ADOLESCENCE AND EARLY ADULTHOOD

I DIDN'T LIKE BEING EARLY. BUT BY EIGHTH GRADE, EVERYONE WORE A BRA AND HAD THEIR PERIOD. I WAS NORMAL.

parents during the adolescents' last year of high school.

We followed the children's pubertal development by asking them to judge themselves every six months on such indicators as height, pubic hair and acne in both boys and girls; breast development and menstruation in girls; and voice change and facial-hair growth in boys. We also estimated the timing of puberty by finding out when each youngster's adolescent growth spurt in height peaked, so we could study the effects of early, on-time or late maturing.

Although we have not yet analyzed all the data, it's clear that puberty alone does not have the overwhelming psychological impact that earlier clinicians and researchers assumed it did (see "The Puzzle of Adolescence," this article). But it does have many effects on body image, moods and relationships with parents and members of the opposite sex.

Being an early or late maturer (one year earlier or later than average), for example, affected adolescents' satisfaction with their appearance and their body image—but only among seventh- and eighth-graders, not sixth-graders. We found that among students in the higher two grades, girls who were physically more mature were generally less satisfied with their weight and appearance than their less mature classmates.

A seventh-grade girl, pleased with being still childlike, said, "You can do more things—you don't have as much weight to carry around." A girl in the eighth grade, also glad to be a late maturer, commented, "If girls get fat, they have to worry about it." In contrast, an early-maturing girl subsequently commented, "I didn't like being early. A lot of my friends didn't understand." Another girl, as a high school senior, described the pain of maturing extremely early: "I tried to hide it. I was embarrassed and ashamed." However, her discomfort ended in the eighth grade, she said, because "by then everyone wore a bra and had their period. I was normal."

We found the reverse pattern among boys: Those who were physically more mature tended to be more satisfied with their weight and their overall appearance than their less mature peers. One already gangling seventh-grade boy, for example, said he liked being "a little taller and having more muscle development than other kids so you can beat them in races." He conceded that developing more slowly might help "if you're a jockey" but added, "Really, I can't think of why [developing] later would be an advantage." In reflecting back from the 12th grade, a boy who had matured early noted that at the time the experience "made me feel superior."

For seventh- and eighth-grade boys, physical maturity was related to mood. Boys who had reached puberty reported positive moods more often than their prepubertal male classmates did. Pubertal status was less clearly and consistently related to mood among girls, but puberty did affect how girls got along with their parents. As physical development advanced among sixth-grade girls, their relationships with their parents declined; girls who were developmentally advanced talked less to their parents and had less positive feelings about family relationships than did less developed girls. We found a similar pattern among eighth-grade girls, but it was less clear in the seventh grade, perhaps because of the many other changes occurring at that time, such as the change from elementary to secondary school format and its related effects on friendship and school achievement.

The timing of puberty affected both school achievement and moods. Early maturers tended to get higher grades than later maturers in the same class. We suspect that this may stem from the often documented tendency of teachers to give more positive ratings to larger pupils. Although early maturers had an edge academically, those who matured later were more likely to report positive moods.

As we have noted, among relatively physically mature adolescents, boys and girls had opposite feelings about their appearance: The boys were pleased, but the girls were not. We believe that, more generally, pubertal change is usually a positive experience for boys but a negative one for girls. While advancing maturity has some advantages for girls, including gaining some of the rights and privileges granted to maturing boys, it also brings increased limitations and restrictions related to their emerging womanhood. One sixth-grade girl stated emphatically, "I don't like the idea of getting older or any of that. If I had my choice, I'd rather stay 10." Or, as one seventh-grade boy graphically explained the gender differences, "Parents let them [boys] go out later than girls because they don't have to worry about getting raped or anything like that."

Differences in the timing of puberty also affect interactions with members of the opposite sex. But it takes two to tango, and in the sixth grade, although many girls have reached puberty and are ready to socialize with boys, most boys have not yet made that transition. Thus, as one girl plaintively summed up the sixth-grade social scene, "Girls think about boys more than boys think about girls."

In the seventh and eighth grades, the physically more mature boys and girls are likely to be pioneers in exploring social relations with members of the opposite sex, including talking with them on the phone, dating, having a boyfriend or girlfriend and "making out." We had the sense that once these young people began looking like teenagers, they wanted to act like them as well.

SHIFTING SCHOOLS EXPOSES TEENS TO NEW EXTRACURRICULAR ACTIVITIES—LICIT AND ILLICIT

176

38. Those Gangly Years

But puberty affects the social and sexual activity of individual young adolescents both directly and indirectly; the pubertal status of some students can have consequences for the entire peer group of boys and girls. Although dating and other boy-girl interactions are linked to pubertal status, and girls usually reach puberty before boys do, we found no sex differences in the rates of dating throughout the early-adolescent period. When the early-maturing kids began socializing with members of the opposite sex, the pattern quickly spread throughout the entire peer group. Even prepubertal girls were susceptible to thinking and talking about boys if all their girlfriends were "boy crazy."

The physical changes brought on by puberty have far-reaching effects, but so do many other changes in the lives of adolescents. One we found to be particularly influential is the change in school structure between the sixth and eighth grades. Most young adolescents in our country shift from a relatively small neighborhood elementary school, in which most classes are taught by one teacher, to a much larger, more impersonal middle school or junior high school (usually farther from the child's home), in which students move from class to class and teacher to teacher for every subject. This shift in schools has many ramifications, including disrupting the old peer-group structure, exposing adolescents to different achievement expectations by teachers and providing opportunities for new extracurricular activities—licit and illicit.

Both the timing and number of school transitions are very important. In our study, for example, students who changed schools earlier than most of their peers, as well as those who changed schools twice (both experiences due to modifications of the school system), suffered an academic slump that continued through eighth grade. Therefore, early or double school transition seemed stressful, beyond the usual effects of moving to a junior high school.

Puberty and school change, which appear to be the primary and most pervasive changes occurring during early adolescence, are often linked to other important changes, such as altered family relations. Psychologist Laurence Steinberg of the University of Wisconsin has found that family re-

THE PUZZLE OF ADOLESCENCE

At the turn of the century, psychologist G. Stanley Hall dignified adolescence with his "storm and stress" theory, and Anna Freud subsequently argued influentially that such storm and stress is a normal part of adolescence. Ever since, clinicians and researchers have been trying—with only limited success—to develop a coherent theory of what makes adolescents tick.

Psychoanalytic theorist Peter Blos added in the late 1960s and 1970s that adolescents' uncontrolled sexual and aggressive impulses affect relationships with their parents. He suggested that both adolescents and their parents may need more distant relationships because of the unacceptable feelings stimulated by the adolescents' sexuality.

Research conducted in the 1960s showed that not all adolescents experience the storm and stress psychoanalytic theory predicts they should. Many studies, including those of Roy Grinker; Joseph Adelson and Elizabeth Douvan; Daniel Offer; and Albert Bandura, demonstrated that a significant proportion of adolescents make it through this period without appreciable turmoil. These findings suggest that pubertal change per se cannot account for the rocky time some adolescents experience.

Other theories of adolescent development have also been linked to pubertal change. For example, in his theory of how children's cognitive capacities develop, Swiss psychologist Jean Piaget attributed the emergence of "formal operational thought," that is, the capacity to think abstractly, to the interaction of pubertal and environmental changes that occur during the same developmental period.

Some researchers have linked the biological events of puberty to possible changes in brain growth or functioning. Deborah Waber, a psychologist at Boston Children's Hospital, has shown that the timing of pubertal change is related to performance differences between the right- and left-brain hemispheres on certain tasks and to the typical adult pattern of gender-related cognitive abilities: Later maturers, including most men, have relatively better spatial abilities, and earlier maturers, including most women, have relatively better verbal abilities.

It has also been suggested that pubertal change affects adolescent behavior through the social consequences of altered appearance. Once young adolescents look like adults, they are more likely to be treated as adults and to see themselves that way, too.

Coming also from a social psychological perspective, psychologist John Hill of Virginia Commonwealth University, together with former Cornell University doctoral student Mary Ellen Lynch, have proposed that pubertal change leads parents and peers to expect more traditional gender-role behavior from adolescents than from younger children; they suggest that both boys and girls become more aware of these gender stereotypes in early adolescence and exaggerate their gender-related behavior at this age.

Despite all these theories, most studies that look at how puberty affects adolescent development are finding that puberty per se is not as important as we once thought. Puberty does specifically affect such things as body image and social and sexual behavior, but it does not affect all adolescent behavior, and it affects some adolescents more strongly than others. In fact, many studies, like ours, are revealing that other changes in early adolescence, particularly social and environmental ones, are at least as important as biological ones.

5. ADOLESCENCE AND EARLY ADULTHOOD

lationships shift as boys and girls move through puberty. During mid puberty, he says, conflict in family discussions increases; when the conflict is resolved, boys usually become more dominant in conversations with their mothers. (Psychologist John Hill of Virginia Commonwealth University has found that family conflict increases only for boys.) Other research, however, suggests that adolescents wind up playing a more equal role relative to both parents.

In our study, the parents of early-maturing girls and late-maturing boys reported less positive feelings about their children in the sixth and eighth grades than did parents of boys and girls with other patterns of pubertal timing. (These effects were always stronger for fathers than for mothers.) The adolescents, however, reported that their feelings about their parents were unrelated to pubertal timing.

The feelings of affection and support that adolescents and their parents reported about one another usually declined from the sixth to the eighth grades, with the biggest decline in feelings between girls and their mothers. But importantly, the decline was from very positive to less positive—but still not negative—feelings.

Early adolescence is clearly an unusual transition in development because of the number of changes young people experience. But the impact of those changes is quite varied; changes that may challenge and stimulate some young people can become overwhelming and stressful to others. The outcome seems to depend on prior strengths and vulnerabilities—both of the individual adolescents and their families—as well as on the pattern, timing and intensity of changes.

Youngsters in our study who changed schools within six months of peak pubertal change reported more depression and anxiety than those whose school and biological transitions were more separated in time. Students who experienced an unusual and negative change at home—such as the death of a parent or divorce of parents—reported even greater difficulties, a finding that supports other research. Sociologists Roberta Simmons and Dale Blyth have found that the negative effects of junior high school transitions, especially in combination with other life changes, continue on into high school, particularly for girls.

Many of the negative effects of transitions and changes seen in our study were tempered when adolescents had particularly positive and supportive relationships with their peers and family. The effects of all these early-adolescent changes were even stronger by the 12th grade than in 8th grade.

Overall, we found that the usual pattern of development in early adolescence is quite positive. More than half of those in the study seemed to be almost trouble-free, and approximately 30 percent of the total group had only intermittent problems during their early teen years. Fifteen percent of the kids, however, did appear to be caught in a downward spiral of trouble and turmoil.

Gender played an important role in how young adolescents expressed and dealt with this turmoil. Boys generally showed their poor adjustment through external behavior, such as being rebellious and disobedient, whereas girls were more likely to show internal behavior, such as having depressed moods. But since many poorly adjusted boys also showed many signs of depression, the rates of such symptoms did not differ between the sexes in early adolescence.

By the 12th grade, however, the girls were significantly more likely than the boys to have depressive symptoms, a sex difference also found among adults. Boys who had such symptoms in the 12th grade usually had had them in the sixth grade as well; girls who had depressive symptoms as high school seniors usually had developed them by the eighth grade.

For youngsters who fell in the troubled group, the stage was already set—and the pathways distinguishable—at the very beginning of adolescence. There is an overall tendency for academic decline in the seventh and eighth grades (apparently because seventh- and eighth-grade teachers adopt tougher grading standards than elementary school teachers do). But the grades of boys with school behavior problems or depressive symptoms in early adolescence subsequently declined far more than those of boys who did not report such problems. Thus, for youngsters whose lives are already troubled, the changes that come with early adolescence add further burdens—and their problems are likely to persist through the senior year of high school.

One 12th-grade boy who followed this pathway described the experience: "My worst time was seventh to ninth grade. I had a lot of growing up to do and I still have a lot more to do. High school was not the 'sweet 16' time everyone said it would be. What would have helped me is more emotional support in grades seven through nine." In explaining that particularly difficult early-adolescent period he said, "Different teachers, colder environment, changing classes and detention all caused chaos in the seventh to ninth grades."

We did not find the same relationship between academic failure and signs of emotional turmoil in girls as in boys. For example, those seventh-grade girls particularly likely to report poor self-image or depressive symptoms were those who were academically successful. Furthermore, when these girls lowered their academic achievement by eighth grade, their depression and their self-image tended to improve. These effects occurred in many areas of girls' coursework but were particularly strong in stereotypically "masculine" courses such as

---

THE VAST MAJORITY OF EARLY TEENS WE STUDIED WERE TROUBLE-FREE OR HAD ONLY INTERMITTENT PROBLEMS. ONLY 15 PERCENT WERE PLAGUED BY TROUBLE AND TURMOIL.

mathematics and science. Like the pattern of problems for boys, the girls' pattern of trading grades to be popular and feel good about themselves persisted into the 12th grade. (Some girls, of course, performed well academically and felt good about themselves both in junior high school and high school.)

We think that for certain girls, high achievement, especially in "masculine" subjects, comes with social costs—speculation supported by the higher priority these particular girls give to popularity. They seem to sacrifice the longer-term benefits of high achievement for the more immediate social benefits of "fitting in." Other studies have revealed a peak in social conformity at this age, especially among girls, and have shown that many adolescents reap immediate, but short-term, social benefits from many types of behavior that adults find irrational or risky.

Our most recent research is focused

PARENTS LET BOYS GO OUT LATER THAN GIRLS BECAUSE THEY DON'T HAVE TO WORRY ABOUT THEM GETTING RAPED.

on exploring further whether the developmental patterns established during early adolescence continue to the end of high school. We are also trying to integrate our observations into a coherent theory of adolescent development and testing that theory by seeing whether we can predict the psychological status of these students at the end of high school based on their characteristics in early adolescence. Other key concerns include discovering early warning signs of trouble and identifying ways to intervene to improve the course of development.

The biological events of puberty are a necessary—and largely uncontrollable—part of growing up. But we may be able to understand and control the social and environmental forces that make adolescence so difficult for a small but troubled group of youngsters. The adolescent's journey toward adulthood is inherently marked by change and upheaval but need not be fraught with chaos or deep pain.

A Much RISKIER PASSAGE

DAVID GELMAN

There was a time when teenagers believed themselves to be part of a conquering army. Through much of the 1960s and 1970s, the legions of adolescence appeared to command the center of American culture like a victorious occupying force, imposing their singular tastes in clothing, music and recreational drugs on a good many of the rest of us. It was a hegemony buttressed by advertisers, fashion setters, record producers suddenly zeroing in on the teen multitudes as if they controlled the best part of the country's wealth, which in some sense they did. But even more than market power, what made the young insurgents invincible was the conviction that they were right: from the crusade of the children, grown-ups believed, they must learn to trust their feelings, to shun materialism, to make love, not money.

In 1990 the emblems of rebellion that once set teenagers apart have grown frayed. Their music now seems more derivative than subversive. The provocative teenage styles of dress that adults assiduously copied no longer automatically inspire emulation. And underneath the plumage, teens seem to be more interested in getting ahead in the world than in clearing up its injustices. According to a 1989 survey of high-school seniors in 40 Wisconsin communities, global concerns, including hunger, poverty and pollution, emerged last on a list of teenage worries. First were personal goals: getting good grades and good jobs. Anything but radical, the majority of teens say they're happy and eager to get on with their lives.

One reason today's teens aren't shaking the earth is that they can no longer marshal the demographic might they once could. Although their sheer numbers are still growing, they are not the illimitably expanding force that teens appeared to be 20 years ago. In 1990 they constitute a smaller percentage of the total population (7 percent, compared with nearly 10 percent in 1970). For another thing, almost as suddenly as they became a highly visible, if unlikely, power in the world, teenagers have reverted to anonymity and the old search for identity. Author Todd Gitlin, a chronicler of the '60s, believes they have become "Balkanized," united less by a common culture than by the commodities they own. He says "it's impossible to point to an overarching teen sensibility."

But as a generation, today's teenagers face more adult-strength stresses than their predecessors did—at a time when adults are much less available to help them. With the divorce rate hovering near 50 percent, and 40 to 50 percent of teenagers living in single-parent homes headed mainly by working mothers, teens are more on their own than ever. "My parents let me do anything I want as long as I don't get into trouble," writes a 15-year-old high-schooler from Ohio in an essay submitted for this special issue of NEWSWEEK. Sociologists have begun to realize, in fact, that teens are more dependent on grown-ups than was once believed. Studies indicate that they are shaped more by their parents than by their peers, that they adopt their parents' values and opinions to a greater extent than anyone realized. Adolescent specialists now see real hazards in lumping all teens together; 13-year-olds, for instance, need much more parental guidance than 19-year-olds.

These realizations are emerging just when the world has become a more dangerous place for the young. They have more access than ever to fast cars, fast drugs, easy sex—"a bewildering array of options, many with devastating out-

From *Newsweek*, Special Issue, Summer/Fall 1990, pp. 10-16. Copyright © 1990 by Newsweek, Inc. All rights reserved. Reprinted by permission.

39. Much Riskier Passage

comes," observes Beatrix Hamburg, director of Child and Adolescent Psychiatry at New York's Mount Sinai School of Medicine. Studies indicate that while overall drug abuse is down, the use of lethal drugs like crack is up in low-income neighborhoods, and a dangerous new kick called ice is making inroads in white high schools. Drinking and smoking rates remain ominously high. "The use of alcohol appears to be normative," says Stephen Small, a developmental psychologist at the University of Wisconsin. "By the upper grades, everybody's doing it."

Sexual activity is also on the rise. A poll conducted by Small suggests that most teens are regularly having sexual intercourse by the 11th grade. Parents are generally surprised by the data, Small says. "A lot of parents are saying, 'Not my kids . . .' They just don't think it's happening." Yet clearly it is: around half a million teenage girls give birth every year, and sexually transmitted diseases continue to be a major problem. Perhaps the only comforting note is that teens who are given AIDS education in schools and clinics are more apt to use condoms—a practice that could scarcely be mentioned a few years ago, let alone surveyed.

One reliable assessment of how stressful life has become for young people in this country is the Index of Social Health for Children and Youth. Authored by social-policy analyst Marc Miringoff, of Fordham University at Tarrytown, N.Y., it charts such factors as poverty, drug-abuse and high-school dropout rates. In 1987, the latest year for which statistics are available, the index fell to its lowest point in two decades. Most devastating, according to Miringoff, were the numbers of teenagers living at poverty levels—about 55 percent for single-parent households—and taking their own lives. The record rate of nearly 18 suicides per 100,000 in 1987—a total of 1,901—was double that of 1970. "If you take teens in the '50s—the 'Ozzie and Harriet' generation—those kids lived on a less complex planet," says Miringoff. "They could be kids longer."

The social index is only one of the yardsticks used on kids these days. In fact, this generation of young people is surely one of the most closely watched ever. Social scientists are tracking nearly everything they do or think about, from dating habits (they prefer going out in groups) to extracurricular activities (cheerleading has made a comeback) to general outlook (45 percent think the world is getting worse and 62 percent believe life will be harder for them than it was for their parents). One diligent prober, Reed Larson of the University of Illinois, even equipped his 500 teen subjects with beepers so he could remind them to fill out questionnaires about how they are feeling, what they are doing and who they are with at random moments during the day. Larson, a professor of human development, and psychologist Maryse Richards of Loyola University, have followed this group since grade school. Although the results of the high-school study have not been tabulated yet, the assumption is that young people are experiencing more stress by the time they reach adolescence but develop strategies to cope with it.

Without doubt, any overview of teenage problems is skewed by the experience of the inner cities, where most indicators tilt sharply toward the negative. Especially among the minority poor, teen pregnancies continue to rise, while the institution of marriage has virtually disappeared. According to the National Center for Vital Statistics, 90 percent of black teenage mothers are unmarried at the time of their child's birth, although about a third eventually marry. Teenage mothers, in turn, add to the annual school-dropout rate, which in some cities reaches as high as 60 percent. Nationwide, the unemployment rate for black teenagers is 40 to 50 percent; in some cities, it has risen to 70 percent. Crack has become a medium of commerce and violence. "The impact of crack is worse in the inner city than anywhere else," says psychiatrist Robert King, of the Yale Child Study Center. "If you look at the homicide rate among young, black males, it's frighteningly high. We also see large numbers of young mothers taking crack."

Those are realities unknown to the majority of white middle-class teenagers. Most of them are managing to get through the adolescent years with relatively few major problems. Parents may describe them as sullen and self-absorbed. They can also be secretive and rude. They hang "Do Not Disturb" signs on their doors, make phone calls from closets and behave churlishly at the dinner table if they can bring themselves to sit there at all. An earlier beeper study by Illinois's Larson found that in the period between ages 10 and 15, the amount of time young people spend with their families decreases by half. "This is when the bedroom door becomes a significant marker," he says.

Yet their rebelliousness is usually overstated. "Arguments are generally about whether to take out the garbage or whether to wear a certain hairstyle," says Bradford Brown, an associate professor of human development at the University of Wisconsin. "These are not earth-shattering issues, though they are quite irritating to parents." One researcher on a mission to destigmatize teenagers is Northwestern University professor Ken Howard, author of a book, "The Teenage World," who has just completed a study in Chicago's Cook County on where kids go for help. The perception, says Howard, is that teenagers are far worse off than they really are. He believes their emotional disturbances are no different from those of adults, and that it is only 20 percent who have most of the serious problems, in any case.

The findings of broad-based studies of teenagers often obscure the differences in their experience. They are, after all, the product of varied ethical and cultural influences. Observing adolescents in 10 communities over the past 10 years, a team of researchers headed by Frances Ianni, of Columbia University's Teachers College, encountered "considerable diversity." A key finding, reported Ianni in a 1989 article in Phi Delta Kappan magazine, was that the people

5. ADOLESCENCE AND EARLY ADULTHOOD

in all the localities reflected the ethnic and social-class lifestyles of their parents much more than that of a universal teen culture. The researchers found "far more congruence than conflict" between the views of parents and their teenage children. "We much more frequently hear teenagers preface comments to their peers with 'my mom says' than with any attributions to heroes of the youth culture," wrote Ianni.

For years, psychologists also tended to overlook the differences between younger and older adolescents, instead grouping them together as if they all had the same needs and desires. Until a decade ago, ideas of teen behavior were heavily influenced by the work of psychologist Erik Erikson, whose own model was based on older adolescents. Erikson, for example, emphasized their need for autonomy—appropriate, perhaps, for an 18-year-old preparing to leave home for college or a job, but hardly for a 13-year-old just beginning to experience the confusions of puberty. The Erikson model nevertheless was taken as an across-the-board prescription to give teenagers independence, something that families, torn by the domestic upheavals of the '60s and '70s, granted them almost by forfeit.

In those turbulent years, adolescents turned readily enough to their peers. "When there's turmoil and social change, teenagers have a tendency to break loose and follow each other more," says Dr. John Schowalter, president of the American Academy of Child and Adolescent Psychiatry. "The leadership of adults is somewhat splintered and they're more on their own—sort of like 'Lord of the Flies'."

That period helped plant the belief that adolescents were natural rebels, who sought above all to break free of adult influence. The idea persists to this day. Says Ruby Takanishi, director of the Carnegie Council on Adolescent Development: "The society is still permeated by the notion that adolescents are different, that their hormones are raging around and they don't want to have anything to do with their parents or other adults." Yet research by Ianni and others suggests the contrary. Ianni points also to studies of so-called invulnerable adolescents—those who develop into stable young adults in spite of coming from troubled homes, or other adversity. "A lot of people have attributed this to some inner resilience," he says. "But what we've seen in practically all cases is some caring adult figure who was a constant in that kid's life."

Not that teenagers were always so dependent on adults. Until the mid-19th century, children labored in the fields alongside their parents. But by the time they were 15, they might marry and go out into the world. Industrialization and compulsory education ultimately deprived them of a role in the family work unit, leaving them in a state of suspension between childhood and adulthood.

To teenagers, it has always seemed a useless period of waiting. Approaching physical and sexual maturity, they feel capable of doing many of the things adults do. But they are not treated like adults. Instead they must endure a prolonged childhood that is stretched out even more nowadays by the need to attend college—and then possibly graduate school—in order to make one's way in the world. In the family table of organization, they are mainly in charge of menial chores. Millions of teenagers now have part-time or full-time jobs, but those tend to be in the service industries, where the pay and the work are often equally unrewarding.

If teenagers are to stop feeling irrelevant, they need to feel needed, both by the family and by the larger world. In the '60s they gained some sense of empowerment from their visibility, their music, their sheer collective noise. They also joined and swelled the ranks of Vietnam War protesters, giving them a feeling of importance that evidently they have not had since. In the foreword to "Student Service," a book based on a 1985 Carnegie Foundation survey of teenagers' attitudes toward work and community service, foundation director Ernest Boyer wrote: "Time and time again, students complained that they felt isolated, unconnected to the larger world . . . And this detachment occurs at the very time students are deciding who they are and where they fit." Fordham's Miringoff goes so far as to link the rising suicide rate among teens to their feelings of disconnection. He recalls going to the 1963 March on Washington as a teenager, and gaining "a sense of being part of something larger. That idealism, that energy, was a very stabilizing thing."

Surely there is still room for idealism in the '90s, even if the causes are considered less glamorous. But despite growing instances of teenagers involving themselves in good works, such as recycling campaigns, tutorial programs or serving meals at shelters for the homeless, no study has yet detected anything like a national groundswell of volunteerism. Instead, according to University of Michigan social psychologist Lloyd Johnston, teens seem to be taking their cues from a culture that, up until quite recently at least, has glorified self-interest and opportunism. "It's fair to say that young people are more career oriented than before, more concerned about making money and prestige," says Johnston. "These changes are consistent with the Me Generation and looking for the good life they see on television."

Some researchers say that, indeed, the only thing uniting teenagers these days are the things they buy and plug into. Rich or poor, all have their Walkmans, their own VCRs and TVs. Yet in some ways, those marvels of communication isolate them even more. Teenagers, says Beatrix Hamburg, are spending "a lot of time alone in their rooms."

Other forces may be working to isolate them as well. According to Dr. Elena O. Nightingale, author of a Carnegie Council paper on teen rolelessness, a pattern of "age segregation" is shrinking the amount of time adolescents spend with grown-ups. In place of family outings and vacations, for example, entertainment is now more geared toward specific age groups. (The teen-terrorizing "Freddy" flicks and their ilk would be one example.) Even in the sorts of jobs typically available to teenagers, such as fast-food chains, they are usually supervised by people close to their age, rather than by adults, notes Nightingale. "There's a real need for places for teenagers to go where there's a modicum of adult involvement," she says.

Despite the riskier world they face, it would be a mistake to suggest that all adolescents of this generation are feeling more angst than their predecessors. Middle-class teenagers, at least, seem content with their lot on the whole: According to recent studies, 80 percent—the same proportion as 20 years ago—profess satisfaction with their own lives, if not with the state of the world. Many teenagers, nevertheless, evince wistfulness for what they think of as the more heroic times of the '60s and '70s—an era, they believe, when teenagers had more say in the world. Playwright Wendy Wasserstein, whose Pulitzer Prize-winning "The Heidi Chronicles" was about coming of age in those years, says she has noticed at least a "stylistic" nostalgia in the appearance of peace-sign earrings and other '60s artifacts. "I guess that comes from the sense of there having been a unity, a togetherness," she says. "Today most teens are wondering about what they're going to do when they grow up. We had more of a sense of liberation, of youth—we weren't thinking about getting that job at Drexel." Pop-culture critic Greil Marcus, however, believes it was merely the "self-importance" of the '60s generation—his own contemporaries—"that has oppressed today's kids into believing they've missed something. There's something sick about my 18-year-old wanting to see Paul McCartney or the Who. We would never have emulated our parents' culture."

But perhaps that's the point: the teens of the '90s do emulate the culture of their parents, many of whom are the very teens who once made such an impact on their own parents. These parents no doubt have something very useful to pass on to their children—maybe their lost sense of idealism rather than the preoccupation with going and getting that seems, so far, their main legacy to the young. Mom and Dad have to earn a living and fulfill their own needs—they are not likely to be coming home early. But there must be a time and place for them to give their children the advice, the comfort and, most of all, the feelings of possibility that any new generation needs in order to believe in itself.

With MARY TALBOT *and* PAMELA G. KRIPKE

Puberty and Parents

Understanding Your Early Adolescent

Dr. Bruce A. Baldwin

Dr. Baldwin is a practicing psychologist who heads Direction Dynamics, a consulting service specializing in promoting professional development and quality of life in achieving men and women. He responds to many requests each year for seminars on topics of interest to professional organizations and businesses.

For busy achievers and involved parents, Dr. Baldwin has authored a popular, positive parenting cassette series and the book, It's All In Your Head: Lifestyle Management Strategies for Busy People. *Both are available in bookstores or from Direction Dynamics in Wilmington, N.C.*

In the large auditorium, concerned parents wait for the program to begin. The speaker appears to talk about the problems of parenting in the eighties. The program begins with a question to the audience: "How many of you would choose to live your adolescent years over if you had the chance?" Relatively few hands are raised and some of them waver indecisively. For just a few, the adolescent years are some of the best. The majority, however, are happy to have reached adulthood and put those tumultuous years behind them.

Then a second question: "How many of you would choose to live your adolescent years over if you had to do it *right now*?" This time, practically no hands are raised. The fact is that in any era, early adolescence is a most difficult time of life. On the other hand, there is ample evidence that this critical period of growth and change for young people is steadily becoming more difficult to negotiate emotionally. Caring parents seem to sense this and they are afraid for their children. Sadly, their intuitive awareness is quite accurate: what they remember as the simpler world of their own youth has changed irrevocably.

Still, beyond the social environment and value system characteristic of this decade resides the basic adolescent. Understanding the changes that occur and the behaviors that are typical of a young man or woman growing up, regardless of time or place, provides parents with a backdrop of awareness that is most reassuring. It also provides the basis for the necessarily changed relationship with a child who is rapidly growing physically and emotionally. Armed with such understanding, parents can better cope with the many issues that are presented by the changes in their adolescent. At times, they can even manage a knowing smile at the many typical reactions they observe.

Parents who have survived the perils of puberty know, though, that dealing with one or more adolescents is not fun and games. Looking after the kids is relatively easy when they are small and dependent and the immediate neighborhood is their whole world. Three parental apprehensions, however, are forced into the forefront of consciousness by the onset of puberty and fueled daily by powerful adolescent strivings for independence.

Parental apprehension #1: "My adolescent will do the same things I did when I was young." With the wisdom of the years, parents look back at their adolescent antics with a bit of amusement tempered by a fair share of "only by the grace of God . . ." feelings. These parents, now mature individuals, simply don't want their children to take the same chances.

Parental apprehension #2: "The world my teen must live in is much more dangerous than it was years ago." This absolutely valid fear is constantly reinforced by public awareness of high suicide rates in adolescents, life-threatening sexually transmitted diseases and the easy availability of drugs. Mistakes and missteps can be much more serious than they were in the past.

Parental apprehension #3: "My child now has a private life that I can't directly control anymore." A reality is that teens force parents to trust them. Adolescence brings increased mobility and an expansion of time spent outside the sphere of direct family influence. Parents are forced to let go and hope that their teen will handle unknown and possibly dangerous situations well.

With impending puberty, the drama of early adolescence begins to unfold relentlessly. Responsible parents struggle to safeguard their teen's present and future. At the same time, their adolescent precociously lays claim to all adult prerogatives and privileges. In the background, a chosen peer group powerfully influences a child to do its immature bidding. Peers, parents and puberty all interact to produce the conflict-laden "adolescent triangle." It's normal but not easy.

The complex relationships of the adolescent triangle have been a perplexing part of the family life for centuries. It is incumbent upon parents to try to understand the developmental processes being experienced by their growing teen. Only then can they effectively modify their parenting relationship to their child-cum-adult in ways that will promote healthy growth toward maturity. And they must persevere without thanks in the face of active resistance by their teen. To set the stage for effective parental coping, here's an overview of the normal changes that occur during early adolescence.

THE STAGES OF ADOLESCENCE

If the typical individual is asked where adolescence begins and ends, the immediate response is "the teen years." Implicit in this response is the assumption that when the early twenties are reached, adolescence has ended and the individual has become an adult. Nothing could be further from the truth. True, in the past there has been an easy biological marker for the beginning of the adolescent years: puberty. And, in generations past, young men and women became financially and emotionally self-sufficient shortly after leaving home in their late teens.

However, the beginning and end of adolescence have become increasingly diffuse and difficult to define clearly. On one hand, we sometimes see a precocious beginning to adolescence that may predate overt signs of puberty. Children frequently begin to act like adolescents before physical changes begin. At the other end of this growth period, the difficulty is obvious. How do you define the moment when a child has become a true adult? Of course the best way is to use emotional maturity as a gauge rather than more obvious but often misleading criteria such as completing an education, earning a living, marrying or becoming a parent.

In short, adolescence in this society at present spans approximately 20 years. For almost two decades young people struggle to become emotionally mature adults. There are three basis stages of the adolescent experience as it exists today. However, for parents and children the most critical and dangerous is the first.

Stage I: Early adolescence (the rise of tribal loyalties). Age span: 10 or 11 through 17 years. In other words, this most tumultuous stage of growth begins in late fifth or sixth grade and typically ends at about the senior year of high school. During this time, your child joins a "tribe" of peers that is highly separate from the adult world. The peer group (tribe) clearly defines itself as a distinct subculture struggling for identity with its own dress codes, language codes, defined meeting places and powerfully enforced inclusion criteria.

During these most difficult years for both parent and child, the most pronounced changes of puberty occur. The core struggle of the child is to become independent—and that means emotionally separating from parents and forging a new adult identity. Initial attempts are awkward and emotionally naive. In three key areas, here's what the early adolescent is like.

A. Relationship to parents. Suspicious and distrustful, the adolescent begins actively to push parents away and resists their attempts to give advice. Life is conducted in a secretive world dominated by peers. Rebelling, pushing limits and constantly testing parental resolve are characteristic.

B. Relationship to peers. The youth experiences emotionally intense "puppy love" relationships with members of the opposite sex and "best friends" relationships with peers of the same sex. These relationships are often superficial, with undue emphasis placed on status considerations: participation in sports, attractiveness, belonging to an *in* group.

C. Relationship to career/future. Largely unrealistic in expectations of the adult world, the adolescent sees making a good living—and getting the training required—as easy and "no problem." Money made by working is often spent on status items such as cars or clothes or just on having a good time. The future is far away.

Stage II: Middle adolescence (testing adult realities). Age span: about 18 through 23 or 24. Beginning late in the high school years, a new awareness, with a subtle accompanying fear, begins to grow within the adolescent: "It's almost over. Soon I'll have to face the world on my own." A personal future and the hard realities it entails can no longer be completely denied. Shortly after high school, this young adult typically leaves home to attend college or technical school, join the service or enter the work force.

5. ADOLESCENCE AND EARLY ADULTHOOD

While on their own, but still basically protected by parents, middle adolescents are actively engaged in testing the self against the real world in ways not possible while living at home. More personal accountability is required and some hard lessons are learned. These sometimes painful experiences help the middle adolescent learn the ways of the world, but many signs of immaturity remain. In more specific terms, here is what's happening.

A. Relationship to parents. This dynamic is improved but still problematic at times. Middle adolescents still aren't really ready to be completely open with parents, but they are less defensive. During visits home, intense conflicts with parents will still erupt about lifestyle, career decisions and responsibility.

B. Relationships with peers. Frequent visits home may be made more with the intention to see the old gang from high school than to see parents. Good buddies remain at home, but new friends are being made in a work or school setting. A deeper capacity for caring is manifested in increasingly mature relationships with both sexes.

C. Relationship to career/future. The economics of self-support are steadily becoming more important. Sights may be lowered and changes in career direction are common. Meeting new challenges successfully brings a growing sense of confidence and self-sufficiency.

Stage III: Late adolescence (joining up). Age span: 23 or 24 to about 30 years. By the mid-twenties, early career experimentation has ended, as has protective parental involvement. The late adolescent is usually financially self-sufficient and remains quite social. Life is relatively simple because there is minimal community involvement, little property needing upkeep and usually an income at least adequate to meet basic needs. The late adolescent years tend to be remembered fondly as having been filled with hard work and good times.

At first glance, the late adolescent may appear to be fully adult, but this perception is deceptive. Significant adjustments to the adult world are still being made but are less obvious than they were. Many insecurities in relationships and at work continue to be faced and resolved. Spurred by a growing commitment to creating a personal niche in the adult world, the individual continues to change in the direction of true adult maturity. Here's how.

A. Relationship to parents. Over 20 years, the late adolescent has come full circle. Now that he or she is emotionally self-sufficient, a closer relationship with parents becomes possible. Mutual respect and acceptance grow. The late adolescent begins to understand parenting behaviors that were resisted earlier.

B. Relationship to peers. Most high school chums have been left behind and are seen only occasionally. A new group of work-related peers has been solidly established. Love relationships are more mature and show increased capacity for give and take. Commitment to a shared future and to a family grows.

C. Relationship to career/future. Active striving toward the good life and personal goals intensifies. At work, there is a continuing need to prove competency and get ahead. At home, a more settled lifestyle, one that is characteristic of the middle-class mainstream, slowly evolves. Limited community involvement is seen.

EARLY ADOLESCENT ATTITUDES

While adolescents struggle for nearly two decades to attain emotional maturity, the period of early adolescence is clearly the most striking. It is during this critical six or seven years that the growing young adult is most vulnerable to major mistakes. It is an emotional, intense, painful and confusing phase. It is also the time remembered by parents as most trying of their ability to cope.

Because the vulnerability of parents and their children is never so high as it is during early adolescence, it is well to define some of the characteristics of the normal teen during these years. Here are listed 15 of the most common adolescent attitudes that make life difficult for parents and children, but which are entirely normal for this age group. (NOTE: "Teen" in this discussion refers specifically to an early adolescent.)

Adolescent attitude #1: Conformity within nonconformity. The early adolescent attempts to separate from parents by rejecting their standards. At the same time there is an absolute need to conform to peer group standards. It is very important to be like peers and unlike parents.

Adolescent attitude #2: Open communication with adults diminishes. The early adolescent doesn't like to be questioned by parents and reveals little about what is really going on. Key items may be conveniently forgotten as a personal life outside the family is protected.

Adolescent attitude #3: Withdrawal from family altogether. With the advent of puberty, there is increasing resistance to the family. The teen would much rather stay home merely in order to be available or spend time doing nothing with friends than participate in anything with the family.

Adolescent attitude #4: Acceptability is linked to externals. Personal acceptability is excessively linked to having the right clothes, friends and fad items in teen culture. Parents are badgered constantly to finance status needs deemed necessary for acceptance by a chosen peer group.

Adolescent attitude #5: Spending more time alone. Ironically, although early adolescents are quite social most of the time, they also like to spend time by themselves. Often, teens will retreat for hours to a bedroom and tell parents in no uncertain terms to respect their privacy and let them alone.

Adolescent attitude #6: A know-it-all pseudo-sophistication. Attempts by parents to give helpful advice are usually met with a weary "I already know that!" More often than not, however, a teen's information about topics important to health and well-being is incomplete, full of distortions or patently false.

Adolescent attitude #7: Rapid emotional changes. One of the most difficult aspects of early adolescence for parents to cope with is rapid mood changes. A teen is on top of the world one minute and sullen or depressed the next. The emotional triggers for such changes are frequent but unclear and unpredictable.

Adolescent attitude #8: Instability in peer relationships. Early adolescent relationships are marked by intensity and change. Overnight, a best friend may become a mortal enemy because of a real or imagined betrayal. Changing loyalties are often triggered by the incessant gossiping characteristic of teen culture.

Adolescent attitude #9: Somatic sensitivity. In other words, a teen's rapidly changing body is cause for great concern. Frequently, an early adolescent will become obsessed with and distraught over a perceived major physical deformity (an asymmetrical nose, not-quite-right ears, two pimples).

Adolescent attitude #10: Personal grooming takes a spectacular upturn. To parents' astonishment, a lackadaisical preadolescent turns practically overnight into a prima donna who spends hours grooming and checking the mirror to make sure that every feature of personal appearance is letter-perfect.

Adolescent attitude #11: Emotional cruelty to one another. Early teens can be incredibly insensitive to one another. Malicious gossip, hurtful teasing, and descriptive nicknames, outright rejection by the peer group—all are reasons why early adolescence is a time of great pain for so many.

Adolescent attitude #12: A highly present-oriented existence. Parents often learn the hard way that seriously discussing the future with an early adolescent is an exercise in futility. Conflict results when an unconcerned teen insists on continuing a day-to-day, pleasure-oriented way of life.

Adolescent attitude #13: A rich fantasy life develops. The adolescent's world is filled with hopes and dreams: knights in shining armor, great achievements, plenty of money, a life of freedom and fun—all without much personal effort. Such fantasies often help deny true realities.

Adolescent attitude #14: There is a strong need for independence. Translation: "I can make my own decisions by myself." Teens take it as an insult to their maturity to have to ask permission for anything. This leads to circumventing established rules or making decisions without parental knowledge.

Adolescent attitude #15: A proclivity for experimentation. With a new body and new feelings, the early adolescent develops an unwarranted sense of personal maturity. This leads to covert experimentation wth adult behaviors (smoking, drug use, sexual activity) aimed at the achievement of status and the satisfaction of curiosity. The knowledge that this is taking place leads to another legitimate parental fear.

THE EMOTIONAL AROUSAL OF PARENTS

It is a given that early adolescence is difficult for parents. To a degree this is attributable to the erratic and challenging behavior of their teens. It is also true that as long as the child is clearly a child, the parents remain weak or lie dormant. However, once puberty begins, a myriad of powerful feelings wells up in the parents.

In many respects, a child's puberty forces parents to deal actively with emotional issues that promote *their* growth and development if handled well. It is as important for parents to understand their suddenly aroused feelings as it is for them to understand what is happening emotionally within their teen.

Aroused emotion #1: Unadulterated fear. I would not be going too far to say that the parents of teens live with fear and constant worry. "What's happened now?" "What am I going to find out about next?" A child's world is quite small. At puberty, it suddenly expands and the teen is gone much of the time. This occurs at about the same time that a teen becomes evasive about what is going on in his or her world. Fears grow.

Aroused emotion #2: A deep sense of helplessness. Parents grow very uncomfortable as they watch their teen experience all the pain and turmoil that early adolescence usually brings. Because adolescents perceive adults as unable *really* to understand anything of importance to themselves, parents may be pushed away when problems occur. The kids don't realize how helpless parents feel when they see their child suffering emotionally but are relegated to the sidelines.

Aroused emotion #3: High levels of frustration. It is a given that many of the behaviors of an early adolescent trigger parental anger. One of a teen's strongest emotional needs is to be separate emotionally. This need is expressed by constantly confronting parents verbally, violating rules and pushing limits right to the brink. This entirely normal adolescent response pattern takes its toll on parents who become highly stressed, frustrated and tired.

Aroused emotion #4: A growing awareness of loss. With the onset of puberty, parents are forced to recognize that in just a few years, their teen will be going into the world and lost forever to the nuclear family. The undeniable fact that "our little girl/boy is growing up" triggers this deepening sense of loss on the part of parents: the sadness is compounded by the withdrawal of the teen from family life. Often this particular feeling is overwhelmed by fleeting wishes that the child would hurry and grow up so parents can have some peace of mind.

Aroused emotion #5: Personal hurt. Parents of teens struggle to do their very best to guide and protect their children. However, no thanks are forthcoming. In fact, parents' efforts are often resented and they are labeled as old-fashioned or Victorian or old fogies who are obviously completely out of touch with reality. Angry confrontations are the norm. Sullen withdrawal is an everyday occurrence. Continued rejection and hurt feelings make it difficult for parents to continue giving their personal best to an unappreciative teen.

IN THE EYE OF THE HURRICANE

At the center of every hurricane is the eye. That's where there is calm despite the intensity of the storm that swirls around it. This is an excellent way to conceptualize the relationship of effective parents to their children during the tumultuous early adolescent years. At puberty, a teen becomes inexorably swept up in the swift winds of change. To help themselves and their child, parents must remain calm and aware in the eye of the hurricane.

In recent years, much has been written about the changing nature of growing up in America. Some authorities emphasize the group's premature sophistication consequent on the fact that teens these days are ex-

5. ADOLESCENCE AND EARLY ADULTHOOD

posed to much more at an earlier age than their parents were. Others who study this special group find that beyond the surface precocity of teens, attaining emotional maturity is steadily becoming a more prolonged and difficult process than ever before. The reality that parents must understand is that these seemingly divergent points of view are both absolutely valid and in no way contradict one another.

To be effective, parents must not be fooled by the misleading sophistication of teens and instead respond to the more complex developmental problems that lie beneath this surface veneer. To be of maximum aid in promoting healthy growth toward maturity, parents must make sure that their responses reflect three important teenage needs.

Teen need #1: "Depth perception" by parents. Basically, parents must be able to see accurately beyond the often erratic surface behaviors of an adolescent to the real issues that simply can't be articulated by a teen. Then parents must respond in caring ways to those emotional needs despite protests, confrontations and denials.

Teen need #2: Consistency of parental responses. Teens are notorious for their inconsistency. One of their deepest needs during these years of turmoil is to have parents who are steady and consistent. Such parents become a stabilizing influence and a center of strength—this helps a teen cope effectively with rapid change in every part of life.

Teen need #3: Strength of parental conviction. At no other time during the entire child-rearing process must parents be surer of their values. Teens focus tremendous pressure on parents to convince them they are wrong or that their values are irrelevant. Far too often the kids succeed in compromising solid parental values, to the detriment of the family and themselves.

One of the most emotionally rigorous tasks that parents face during the adolescent years is to keep doing what is right with very little encouragement and without becoming too insecure. And, after all is said and done and those difficult years are over, most teens do mature to join the ranks of respectable adults. Didn't you? And if you parented well, you will be rewarded eventually when your adult son or daughter thanks you directly for all the sacrifices you made in the face of all the obstacles.

But the progress toward the goal is a nightmare. One frustrated parent put up a sign in the kitchen: "NOTICE TO ALL TEENS! If you are tired of being hassled by unreasonable parents, NOW IS THE TIME FOR ACTION. Leave home and pay your own way WHILE YOU STILL KNOW EVERYTHING!"

These days puberty has perils for parents and for the children in grown-up bodies who are in their charge. And adolescence is no time to cut corners and take the easy road. Perhaps it was a wise parent who remarked that "a shortcut is often the quickest way to get somewhere you weren't going." With adolescents, the best road is always difficult but eventually rewarding. Shortcuts too often lead to dead ends. Or dangerous precipices. Sometimes to places you never expected to visit.

Therapists Find Last Outpost of Adolescence in Adulthood

Daniel Goleman

While adolescence by common reckoning ends with the teen years, it continues psychologically until the end of the 20's, when young people are finally able to establish a fully mature relationship with their parents, psychologists are finding.

Although clinical lore has long suggested that adolescence lingered into the 20's, only now are scientific studies showing just how true that is. One new study has discovered a dramatic shift in psychological maturity that seems to occur in most young adults between 24 and 28.

Before the shift, the men and women studied said they usually relied on their parents to make important choices in life and felt unable to cope with life's difficulties without some help from them. After the change, they felt comfortable making choices based on their own values and confident in their abilities to live on their own.

Such findings are leading developmental psychologists to revise their views of emotional growth in adolescence and early adulthood and to set back the timetable for the end of adolescence. This phenomenon may be a product of this century when, for the first time, most young people remained dependent on their parents for their support through the teen years and beyond.

"Adolescence doesn't really end until the late 20's," said Susan J. Frank, a psychologist at Michigan State University. "The emotional ties that bind children to their parents continue well after they leave home and enter the adult world. There's an astonishing difference between those in their early and late 20's in doing things without leaning on their parents."

Kathleen White, a psychologist at Boston University who has conducted much of the research on maturation in young adults, said this shift also entails a changing view of parents. "By their mid-20's," she said, "most young people are not ready to appreciate their parents as separate individuals, as having needs and strengths and weaknesses in their own right, apart from being parents.

"They still see their parents in egocentric fashion: were they good or bad parents, did they love me or not, were they too restrictive or demanding, and so on."

In a study of 84 adults between the ages of 22 and 29, Dr. White assessed several aspects of maturity, particularly their ability to form intimate relationships and to see their parents independent of their role as parents.

Those in their later 20's had "far more perspective on their parents" than those in their early 20's, Dr. White said.

A similar finding was reported by Dr. Frank and her colleagues in the current issue of Developmental Psychology.

Rage or Dependency

During their 20's, people go through gradual shifts in their sense of independence from parents and in their relationships with them, according to the work of Dr. Frank and other researchers. Emotional maturation in the third decade of life can be charted by these shifts, following general theories of adult development.

For instance, in terms of autonomy, less mature young adults not only tend to rely on their parents to help with decisions, but they also are often overwhelmed by intense feelings of rage or dependency. At its worst, these feelings can lead them to lash out at their parents, even when the parents are trying to be helpful.

Often the less mature adults are emotionally estranged from their parents or have only superficial exchanges with them. They also tend to have little interest in their parents' welfare and little understanding of the complexities of their parents' lives and personalities.

In Control of Emotions

By contrast, more mature young adults tend to have strong confidence in their abilities to make decisions on their own and feel in control of their emotions toward their parents. They also see themselves, rather than their parents, as the best judges of their own worth and so can risk parental disapproval by expressing values that may clash with those of their parents.

They tend to have strong emotional ties to their parents and are able to talk with their parents about feelings and concerns of importance to them, while feeling free to disagree. They also are concerned about their parents' well-being and are able to understand the complexities of their parents' lives, rather than painting them in black-and-white terms. The more mature tend to acknowledge or feel proud of their parents as role models.

"In the 20's, it's not just that you achieve separateness from your parents, but also that you feel connected to them as an adult," Dr. White said. "That includes empathy for them, and seeing things from their points of view. Most young adults don't have this perspective until their late 20's."

Using scales that measure these changes, Dr. Frank and her colleagues assessed the development of 150 men and women in their 20's on their relationships with their parents. All those studied were from middle-class suburbs and lived within a two-hour drive from their parents.

Patterns of Maturation

While Dr. Frank's results bore out the general theories, she found the reality to be more complex, the real-life patterns not fitting a continuum of maturity. There were, she said, several major patterns of maturation among the young adults she studied, with significant differences between men and women.

Women most often fell into a pattern of "competent and connected" relationships with their parents. These people—the pattern fit 40 percent of women and 6 percent of men—had a strong sense of independence and often held views that differed radically from their parents. But even so, they felt more empathy for them, particularly for their mothers, for whom they were often confidants. Among the women, mothers were often seen as demanding and

5. ADOLESCENCE AND EARLY ADULTHOOD

critical, but since they understood their mothers' shortcomings, they were able to keep conflicts from getting out of hand.

Another pattern more common among the women than the men was to be dependent or emotionally enmeshed, most often with their mothers. Although troubled by their inability to handle life without their parents' help, these young people felt trapped by the relationship. Some saw their parents as overbearing and judgmental, others as emotionally detached. Childish power struggles with parents were common in this group.

The largest number of men, however, had "individuated" relationships in which they felt respected by their parents and prepared to meet the challenges of life on their own. Thirty-six percent of the men and 6 percent of the women fell into this group. While they felt a clear boundary between their own lives and those of their parents, they also felt free to seek advice and assistance. Although they enjoyed their parents' company, there was an emotional distance. Their relationships generally were lacking both in discussions of very personal matters and conflicts.

Another pattern more common in men than women was a false autonomy, in which the young adults feigned an indifference to clear conflicts with their parents, which they handled by avoiding confrontations. With their fathers, the main complaint was of mutual disinterest; with their mothers it was the need to hold an intrusive parent at bay. These people resented their parents' offers of help, and often held them in contempt. They also harbored resentment at their parents' inability to accept them as they were.

Reconnecting With Parents

Dr. Frank's findings are consistent with the work of Bertram Cohler, a psychologist at the University of Chicago. Dr. Cohler has proposed that when youths make the transition into adulthood, they become more interdependent with rather than more independent of their parents.

Dr. Frank found that over the course of their 20's, those in her study tended to move in the more mature direction. Although most had made the shift by 29, some still remained in an emotional adolescence.

Dr. Gould said, "There are some eternal adolescents, who never achieve a sense of their own maturity in these ways."

The Question of Marriage

While many influential theories of adult development hold that marriage is a key turning point in emotional maturation, Dr. Frank did not find this to be the case. She found that apart from age, it made no difference in people's psychological growth.

"The clinical literature says that the degree to which you've worked through your emotional relationship with your parents determines your ability to develop a close relationship with your spouse," Dr. White said.

"But that's not what I've found," she added. "People tend to be more mature in their relationships with their spouses than with their parents, if they are mature in either. People are more compelled to work out their marital conflicts than they are to work out their relationships with their parents. Your parents are your parents forever."

Proceeding With Caution

The twentysomething generation is balking at work, marriage and baby-boomer values. Why are today's young adults so skeptical?

DAVID M. GROSS and SOPHFRONIA SCOTT

They have trouble making decisions. They would rather hike in the Himalayas than climb a corporate ladder. They have few heroes, no anthems, no style to call their own. They crave entertainment, but their attention span is as short as one zap of a TV dial. They hate yuppies, hippies and druggies. They postpone marriage because they dread divorce. They sneer at Range Rovers, Rolexes and red suspenders. What they hold dear are family life, local activism, national parks, penny loafers and mountain bikes. They possess only a hazy sense of their own identity but a monumental preoccupation with all the problems the preceding generation will leave for them to fix.

This is the twentysomething generation, those 48 million young Americans ages 18 through 29 who fall between the famous baby boomers and the boomlet of children the baby boomers are producing. Since today's young adults were born during a period when the U.S. birthrate decreased to half the level of its postwar peak, in the wake of the great baby boom, they are sometimes called the baby busters. By whatever name, so far they are an unsung generation, hardly recognized as a social force or even noticed much at all. "I envision ourselves as a lurking generation, waiting in the shadows, quietly figuring out our plan," says Rebecca Winke, 19, of Madison, Wis. "Maybe that's why nobody notices us."

But here they come: freshly minted grownups. And anyone who expected they would echo the boomers who came before, bringing more of the same attitude, should brace for a surprise. This crowd is profoundly different from—even contrary to—the group that came of age in the 1960s and that celebrates itself each week on *The Wonder Years* and *thirtysomething*. By and large, the 18-to-29 group scornfully rejects the habits and values of the baby boomers, viewing that group as self-centered, fickle and impractical.

While the baby boomers had a placid childhood in the 1950s, which helped inspire them to start their revolution, today's twentysomething generation grew up in a time of drugs, divorce and economic strain. They virtually reared themselves. TV provided the surrogate parenting, and Ronald Reagan starred as the real-life Mister Rogers, dispensing reassurance during their troubled adolescence. Reagan's message: problems can be shelved until later. A prime characteristic of today's young adults is their desire to avoid risk, pain and rapid change. They feel paralyzed by the social problems they see as their inheritance: racial strife, homelessness, AIDS, fractured families and federal deficits. "It is almost our role to be passive," says Peter Smith, 23, a newspaper reporter in Ventura, Calif. "College was a time of mass apathy, with pockets of change. Many global events seem out of our control."

The twentysomething generation has been neglected because it exists in the shadow of the baby boomers, usually defined as the 72 million Americans born between 1946 and 1964. Members of the tail end of the boom generation, now ages 26 through 29, often feel alienated from the larger group, like kid brothers and sisters who disdain the paths their siblings chose. The boomer group is so huge that it tends to define every era it passes through, forcing society to accommodate its moods and dimensions. Even relatively small bunches of boomers made waves, most notably the 4 million or so young urban professionals of the mid-1980s. By contrast, when today's 18-to-29-year-old group was born, the baby boom was fading into the so-called baby bust, with its precipitous decline in the U.S. birthrate. The relatively small baby-bust group is poorly understood by everyone from scholars to marketers. But as the twentysomething adults begin their prime working years, they have suddenly become far more intriguing. Reason: America needs them. Today's young adults are so scarce that their numbers could result in severe labor shortages in the coming decade.

Twentysomething adults feel the opposing tugs of making money and doing good works, but they refuse to get caught up in the passion of either one. They reject 70-hour workweeks as yuppie lunacy, just as they shirk from starting another social revolution. Today's young adults want to stay in their own backyard and do their

5. ADOLESCENCE AND EARLY ADULTHOOD

work in modest ways. "We're not trying to change things. We're trying to fix things," says Anne McCord, 21, of Portland, Ore. "We are the generation that is going to renovate America. We are going to be its carpenters and janitors."

This is a back-to-basics bunch that wishes life could be simpler. "We expect less, we want less, but we want less to be better," says Devin Schaumburg, 20, of Knoxville. "If we're just trying to pick up the pieces, put it all back together, is there a label for that?" That's a laudable notion, but don't hold your breath till they find their answer. "They are finally out there, saying 'Pay attention to us,' but I've never heard them think of a single thing that defines them," says Martha Farnsworth Riche, national editor of *American Demographics* magazine.

What worries parents, teachers and employers is that the latest crop of adults wants to postpone growing up. At a time when they should be graduating, entering the work force and starting families of their own, the twentysomething crowd is balking at those rites of passage. A prime reason is their recognition that the American Dream is much tougher to achieve after years of housing-price inflation and stagnant wages. Householders under the age of 25 were the only group during the 1980s to suffer a drop in income, a decline of 10%. One result: fully 75% of young males 18 to 24 years old are still living at home, the largest proportion since the Great Depression.

In a TIME/CNN poll of 18- to 29-year-olds, 65% of those surveyed agreed it will be harder for their group to live as comfortably as previous generations. While the majority of today's young adults think they have a strong chance of finding a well-paying and interesting job, 69% believe they will have more difficulty buying a house, and 52% say they will have less leisure time than their predecessors. Asked to describe their generation, 53% said the group is worried about the future.

Until they come out of their shells, the twentysomething/baby-bust generation will be a frustrating enigma. Riche calls them the New Petulants because "they can often end up sounding like whiners." Their anxious indecision creates a kind of ominous fog around them. Yet those who take a more sanguine view see in today's young adults a sophistication, tolerance and candor that could help repair the excesses of rampant individualism. Here is a guide for understanding the puzzling twentysomething crowd:

FAMILY: THE TIES DIDN'T BIND
"Ronald Reagan was around longer than some of my friends' fathers," says Rachel Stevens, 21, a graduate of the University of Michigan. An estimated 40% of people in their 20s are children of divorce. Even more were latchkey kids, the first to experience the downside of the two-income family. This may explain why the only solid commitment they are willing to make is to their own children—someday. The group wants to spend more time with their kids, not because they think they can handle the balance of work and child rearing any better than their parents but because they see themselves as having been neglected. "My generation will be the family generation," says Mara Brock, 20, of Kansas City. "I don't want my kids to go through what my parents put me through."

That ordeal was loneliness. "This generation came from a culture that really didn't prize having kids anyway," says Chicago Sociologist Paul Hirsch. "Their parents just wanted to go and play out their roles—they assumed the kids were going to grow up all right." Absent parents forced a dependence on secondary relationships with teachers and friends. Flashy toys and new clothes were supposed to make up for this lack but instead sowed the seeds for a later abhorrence of the yuppie brand of materialism. "Quality time" didn't cut it for them either. In a survey to gauge the baby busters' mood and tastes, Chicago's Leo Burnett ad agency discovered that the group had a surprising amount of anger and resentment about their absentee parents. "The flashback was instantaneous and so hot you could feel it," recalls Josh McQueen, Burnett's research director. "They were telling us passionately that quality time was exactly what was not in their lives."

At this point, members of the twentysomething generation just want to avoid perpetuating the mistakes of their own upbringing. Today's potential parents look beyond their own mothers and fathers when searching for child-rearing role models. Says Kip Banks, 24, a graduate student in public policy at the University of Michigan: "When I raise my children, my approach will be my grandparents', much more serious and conservative. I would never give my children the freedoms I had."

MARRIAGE: WHAT'S THE RUSH? The generation is afraid of relationships in general, and they are the ultimate skeptics when it comes to marriage. Some young adults maintain they will wait to get married, in the hope that time will bring a more compatible mate and the maturity to avoid a divorce. But few of them have any real blueprint for how a successful relationship should function. "We never saw commitment at work," says Robert Higgins, 26, a graduate student in music at Ohio's University of Akron.

As a result, twentysomething people are staying single longer and often living together before marrying. Studying the 20-to-24 age group in 1988, the U.S. Census Bureau found that 77% of men and 61% of women had never married, up sharply from 55% and 35%, respectively, in 1970. Among those 25 to 29, the unmarried included 43% of men and 29% of women in 1988, vs. 19% and 10% in 1970. The sheer disposability of marriage breeds skepticism. Kasey Geoghegan, 20, a student at the University of Denver and a child of divorced parents, believes nuptial vows have lost their credibility. Says she: "When people get married, ideally it's permanent, but once problems set in, they don't bother to work things out."

DATING: DON'T STAND SO CLOSE
Finding a date on a Saturday night, let alone a mate, is a challenge for a generation that has elevated casual commitment to an art form. Despite their nostalgia for family values, few in their 20s are eager to revive a 1950s mentality about pairing off. Rick Bruno, 22, who will enter Yale Medical School in the fall, would rather think of himself as a free agent. Says he: "Not getting hurt is a big priority with me." Others are concerned that the generation is too detached to form caring relationships. "People are afraid to like each other," says Leslie Boorstein, 21, a photographer from Great Neck, N.Y.

For those who try to make meaningful connections—often through video dating services, party lines and personals ads—risks of modern love are greater than ever. AIDS casts a pall over a generation that fully expected to reap the benefits of the sexual revolution. Responsibility is the watchword. Only on college campuses do remnants of libertinism linger. That worries public-health officials, who are witnessing an explosion of sexually transmitted diseases, particularly genital warts. "There is a high degree of students who believe oral contraception protects them from the AIDS virus. It doesn't," says Wally Brewer, coordinator of a study of HIV infection on U.S. campuses. "Obviously it's a big educational challenge."

CAREERS: NOT JUST YET, THANKS
Because they are fewer in number, today's young adults have the power to wreak havoc in the workplace. Companies are discovering that to win the best talent, they must cater to a young work force that is considered overly sensitive at best and lazy at worst. During the next several years, employers will have to double their recruiting efforts. According to *American Demographics*, the pool of entry-level workers 16 to 24 will shrink about 500,000 a year through 1995, to 21 million. These youngsters are starting to use their bargaining power to get more of what they feel is coming to them. They want flexibility, access to decision making and a return to the sacredness of work-free weekends. "I want a work environment concerned about my personal growth," says Jennifer Peters, 22, one of the youngest candidates ever to be admitted to the State Bar of California. "I don't want to go to work and feel I'll be burned out two or three years down the road."

42. Proceeding With Caution

Most of all, young people want constant feedback from supervisors. In contrast with the baby boomers, who disdained evaluations as somehow undemocratic, people in their 20s crave grades, performance evaluations and reviews. They want a quantification of their achievement. After all, these were the children who prepped diligently for college-aptitude exams and learned how to master Rubik's Cube and Space Invaders. They are consummate game players and grade grubbers. "Unlike yuppies, younger people are not driven from within, they need reinforcement," says Penny Erikson, 40, a senior vice president at the Young & Rubicam ad agency, which has hired many recent college graduates. "They prefer short-term tasks with observable results."

Money is still important as an indicator of career performance, but crass materialism is on the wane. Marian Salzman, 31, an editor at large for the collegiate magazine *CV*, believes the shift away from the big-salary, big-city role model of the early '80s is an accommodation to the reality of a depressed Wall Street and slack economy. Many boomers expected to have made millions by the time they reached 30. "But for today's graduates, the easy roads to fast money have dried up," says Salzman.

Climbing the corporate ladder is trickier than ever at a time of widespread corporate restructuring. When recruiters talk about long-term job security, young adults know better. Says Victoria Ball, 41, director of Career Planning Services at Brown University: "Even IBM, which always said it would never lay off—well, now they're doing it too." Between 1987 and the end of this year, Big Blue will have shed about 23,000 workers through voluntary incentive programs.

Most of all, young workers want job gratification. Teaching, long disdained as an underpaid and underappreciated profession, is a hot prospect. Enrollment in U.S. teaching programs increased 61% from 1985 to 1989. And more graduates are expressing interest in public-service careers. "The glory days of Wall Street represented an extreme," says Janet Abrams, 29, a Senate aide who regularly interviews young people looking for jobs on Capitol Hill. "Now I'm hearing about kids going to the National Park Service."

Welcome to the era of hedged bets and lowered expectations. Young people increasingly claim they are willing to leave careers in middle gear, without making that final climb to the top. The leitmotiv of the new age: second place seems just fine. But young adults are flighty if they find their workplace harsh or inflexible. "The difference between now and then was that we had a higher threshold of unhappiness," says editor Salzman. "I always expected that a job would be 80% misery and 20% glory, but this generation refuses to pay its dues."

EDUCATION: NO DEGREE, NO DOLLARS Smart and savvy, the twentysomething group is the best-educated generation in U.S. history. A record 59% of 1988 high school graduates enrolled in college, compared with 49% in the previous decade. The lesson they have taken to heart: education is a means to an end, the ticket to a cherished middle-class lifestyle. "The saddest thing of all is that they don't have the quest to understand things, to understand themselves," says Alexander Astin, whose UCLA-based Higher Education Research Institute has been measuring changing attitudes among college freshman for 24 years.

Yet, a fact of life in the 1990s economy is that a college degree is mostly about survival. A person under 30 with a college degree will earn four times as much money as someone without it. In 1973 the difference was only twice as great. With the loss of well-paying factory jobs, there are fewer chances for less-educated young people to reach the middle class. Many dropouts quickly learn this and decide to return to school. But that decision costs money and sends many twentysomethings back to the nest. Others are flocking to the armed services. Private First Class Dorin Vanderjack, 20, of Redding, Calif., left his catering job at a Holiday Inn to join the Army. After two years of racking up credits at the local community college, he was ready for a four-year school and found the Army's offer of $22,800 in tuition assistance too tempting to turn down. "There's no possible way I could save that," he says. "This forced me to grow up."

WANDERLUST: LET'S GET LOST While the recruiters are trying to woo young workers, a generation is out planning its escape from the 9-to-5 routine. Travel is always an easy way out, one that comes cloaked in a mantle of respectability: cultural enrichment. In the TIME/CNN poll, 60% of the people surveyed said they plan to travel a lot while they are young. And it's not just rich students who are doing it. "Travel is an obsession for everyone," says Cheryl Wilson, 21, a University of Pennsylvania graduate who has visited Denmark and Hungary. "The idea of going away, being mobile, is very romantic. It fulfills our sense of adventure."

Unlike previous generations of uppercrust Americans who savored a postgraduate European tour as the ultimate finishing school, today's adventurers are picking places far more exotic. They are seeking an escape from Western culture, rather than further refinement to smooth their entry into society. Katmandu, Dar es Salaam, Bangkok: these are the trendy destinations of many young daydreamers. Susan Costello, 23, a recent Harvard graduate, voyaged to Dharmsala, India, to spend time at the headquarters of the Tibetan government-in-exile, headed by the Dalai Lama. Costello decided to explore Tibetan culture "to see if they really had something in their way of life that we seem to be missing in the West."

ACTIVISM: ART OF THE POSSIBLE People in their 20s want to give something back to society, but they don't know how to begin. The really important problems, ranging from the national debt to homelessness, are too large and complex to comprehend. And always the great, intimidating shadow of 1960s-style activism hovers in the background. Twentysomething youths suspect that today's attempts at political and social action pale in comparison with the excitement of draft dodging or freedom riding.

The new generation pines for a romanticized past when the issues were clear and the troops were committed. "The kids of the 1960s had it easy," claims Gavin Orzame, 18, of Berrien Springs, Mich. "Back then they had a war and the civil rights movement. Now there are so many issues that it's hard to get one big rallying point." But because the '60s utopia never came, today's young adults view the era with a combination of reverie and revulsion. "What was so great about growing up then anyway?" says future physician Bruno. "The generation that had Vietnam and Watergate is going to be known for leaving us all their problems. They came out of Camelot and blew it."

Such views are revisionist, since the '60s were not easy, and the revolution did not end in utter failure. The twentysomething generation takes for granted many of the real goals of the '60s: civil rights, the antiwar movement, feminism and gay liberation. But those movements never coalesced into a unified crusade, which is something the twentysomethings hope will come along, break their lethargy and goad them into action. One major cause is the planet; 43% of the young adults in the TIME/CNN poll said they are "environmentally conscious." At the same time, some young people are joining the ranks of radical-action groups, including ACT UP, the AIDS Coalition to Unleash Power, and Trans-Species Unlimited, the animal-rights group. These organizations have appeal because they focus their message, choose specific targets and use high-stakes pressure tactics like civil disobedience to get things accomplished quickly.

For a generation that has witnessed so much failure in the political system, such results-oriented activism seems much more valid and practical. Says Sean McNally, 20, who headed the Earth Day activities at Northwestern University: "A lot of us are afraid to take an intense stance and then leave it all behind like our parents did. We have to protect ourselves from burning out, from losing faith." Like McNally, the rest of the generation is doing what it can. Its members prefer activities that are small in scope: cleaning up a park over a weekend or teaching literacy to underprivileged children.

LEADERS: HEROES ARE HARD TO FIND Young adults need role models and leaders, but the twentysomething generation has almost no one to look up to.

193

5. ADOLESCENCE AND EARLY ADULTHOOD

Today's new generation of adults has an inate skepticism about social values that evolved over the past 20–25 years. There is a new assessment of what roles work, marriage, and having children play in their lives.

While 58% of those in the TIME/CNN survey said their group has heroes, they failed to agree on any. Ronald Reagan was most often named, with only 8% of the vote, followed by Mikhail Gorbachev (7%), Jesse Jackson (6%) and George Bush (5%). Today's young generation finds no figures in the present who compare with such '60s-era heroes as John F. Kennedy and Martin Luther King. "It seems there were all these great people in the '60s," says Kasi Davidson, 18, of Cody, Wyo. "Now there is nobody."

Today's potential leaders seem unable to maintain their stature. They have a way of either self-destructing or being decimated in the press, which trumpets their faults and foibles. "The media don't really give young people role models anymore," says Christina Chinn, 21, of Denver. "Now you get role models like Donald Trump and all of the moneymakers—no one with real ideals."

SHOPPING: LESS PASSION FOR PRESTIGE Marketers are confounded as they try to reach a generation so rootless and noncommittal. But ad agencies that have explored the values of the twentysomething generation have found that status symbols, from Cuisinarts to BMWs, actually carry a social stigma among many young adults. Their emphasis, according to Dan Fox, marketing planner at Foote, Cone & Belding, will be on affordable quality. Unlike baby boomers, who buy 50% of their cars from Japanese makers, the twentysomething generation is too young to remember Detroit's clunkers of the 1970s. Today's young adult is likely to aspire to a Jeep Cherokee or Chevy Lumina with lots of cup holders. "Don't knock the cup holders," warns Fox. "There's something about them that says, 'It's all right in my world,' That's not a small notion. And Mercedes doesn't have them."

The twentysomething attitude toward consumption in general: get more for less. While yuppies spent money to acquire the best and the rarest toys, young adults believe they can live just as well, and maybe even better, without breaking the bank. They disdain designer anything. "Just point me to the generic aisle," says Jill Mackie, 21, a journalism major at the University of Illinois. Such a no-nonsense outlook has made hay for stores like the Gap, which thrives on young people's desire for casual clothing at a casual price. Similarly, a twentysomething adult picks a Hershey's bar over Godiva chocolates, and Bass Weejuns (price: $75) instead of Lucchese cowboy boots ($500).

CULTURE: FEW FLAVORS OF THEIR OWN Down deep, what frustrates today's young people—and those who observe them—is their failure to create an original youth culture. The 1920s had jazz and the Lost Generation, the 1950s created the Beats, the 1960s brought everything embodied in the Summer of Love. But the twentysomething generation has yet to make a substantial cultural statement. People in their 20s have been handed down everyone else's music, clothes and styles, leaving little room for their own imaginations. Mini-revivals in platform shoes, ripped jeans and urban-cowboy chic all coincide with J. Crew prep, Gumby haircuts and teased-out suburban perms. What young adults have managed to come up with is either *nuevo* hipster or ultra-nerd, but almost always a bland imitation of the past. "They don't even seem to know how to dress," says sociologist Hirsch, "and they're almost unschooled in how to look in different settings."

JEALOUSY & ENVY
The Demons Within Us

JON QUEIJO

Jon Queijo is a free-lance writer who resides in West Roxbury, Massachusetts.

Rick and Liz seemed to have a wonderful marriage; they did everything together. This changed suddenly, however, when Liz's ex-boyfriend began working at her law firm. Besieged by insecurity, Rick began calling Liz's office at odd hours and at night questioned her suspiciously. In a coup de grace, he burst in on her during a business luncheon and falsely accused her of having an affair.

Ann worked extremely hard to achieve success as a real estate agent. Her satisfaction turned sour, however, when a new agent was hired who managed to work less, yet made more sales. Ann hid her dislike of the new agent by offering to take her phone messages while she was out. When Ann began making more sales than her rival, no one made the connection between this turn of events and Ann's tendency to "accidentally" forget to deliver certain phone messages.

The jealous rage of a lover. The shameful actions of envy. Despite our better intentions, most of us feel these emotions dozens of times in our lifetimes. Pulling us apart from lovers, friends, family members, co-workers and even perfect strangers, jealousy and envy can devastate our lives and cause effects ranging from sadness, anger and depression, to estrangement, abuse and even violence.

Beyond our own lives, the power of these emotions has spawned countless works of poetry and prose and triggered numerous historical events. Perhaps for this reason society proclaimed judgment on jealousy and envy thousands of years ago, with the verdict coming down harder on envy. For example, while the pain of jealousy has been forever immortalized in poetry and song, the shame of envy emerges as early as the Ten Commandments: "Thou shalt not covet thy neighbor's house, field, wife or anything that is thy neighbor's." In fact, envy is despicable enough to be considered one of the "Seven Deadly Sins," taking its place alongside pride, gluttony, lust, sloth, anger and greed.

Although we see jealousy and envy arise in numerous situations, their basic definitions are fairly simple: jealousy is "the fear of losing a relationship" (romantic, parental, sibling, friendship); and envy is "the longing for something someone else has" (wealth, possessions, beauty, talent, position).

Despite these definitions and the numerous philosophers, poets and scientists who have pondered these emotions, some remarkably fundamental questions remain: What causes jealousy and envy? Are the emotions actually different? What do they feel like? What are the best ways to cope with these feelings? Why do we often use the terms interchangeably? And what are their implications for society?

Researchers have taken various approaches to answer these questions. The biological view, for example, says that jealousy and envy serve a basic purpose — the emotions lead to biochemical changes that spur the individual to take action and improve the situation. The evolutionary view holds that jealousy may enhance survival by keeping parents together, thus increasing protection of the offspring.

From *Bostonia*, May/June 1988, pp. 31-36. Reprinted with permission of *Bostonia Magazine*.

5. ADOLESCENCE AND EARLY ADULTHOOD

> "People who are dissatisfied with themselves are primed for having other people's talents impinge on them. If, on the other hand, you're satisfied with yourself, then what other people have or do won't unduly raise your expectations, and you should be less likely to feel envy."
>
> RICHARD SMITH

Other explanations range from the reasonable — envy stems from parental attitudes that make a child feel inferior; to the bizarre — the emotions begin in infants when the mother withholds breast-feeding.

Probably the most practical understanding of jealousy and envy, however, emerges from the work of social psychologists — researchers who look at the way people react to each other and society. To them, jealousy and envy arise when the right mix of internal *and* external ingredients are present in society.

"I tend to look at jealousy and envy in terms of motivation and self-esteem," explains Peter Salovey, a social psychologist at Yale University. "It's the interaction between what's important to you and what's happening in the environment. The common denominator is this threat to something that's very important to the person — something that defines self-worth."

Richard Smith, a social psychologist at Boston University who has conducted several studies on jealousy and envy, emphasizes external factors, such as how society affects our view of ourselves. "My perspective is from social comparisons," he explains, "which says we have no objective opinion for evaluating our abilities, so we look at others."

Smith, like Salovey, also stresses internal factors — the role of self-esteem, for example — in determining whether we will feel jealous or envious in any given situation. "People who are dissatisfied with themselves are primed for having other people's talents and possessions impinge on them," Smith points out. "If, on the other hand, you're satisfied with yourself, then what other people have or do won't unduly raise your expectations for yourself, and you should be less likely to feel envy."

Embarrassed by his display of jealousy, Rick apologizes to Liz and they discuss the problem. Soon they realize that while Rick loves Liz and fears losing her, something else is at work here. Because Liz's ex-boyfriend is a lawyer, he possesses skills Rick does not. While Rick is proud of his ability as a store manager, he fears Liz's ex-boyfriend could lure her away with other skills.

Feeling guilty about her actions, Ann calls a friend for support. Ann knows she feels inadequate because the new agent is succeeding in a career that is very important to her, but that doesn't explain everything; others have done better whom Ann has not envied. Then it occurs to her: What bothers her is the way the woman was bettering Ann. She was more outgoing and self-confident — two skills about which Ann has always felt insecure.

As Rick and Ann's situations illustrate, if someone is unsure about an ability — such as Rick and his law knowledge or Ann and her communication skills — then a social situation can bring out that insecurity. "In envy," notes Salovey, "the threat may come from someone else's possessions or attributes. In jealousy the situation is the same, except that the other person's possessions or attributes cause you to fear losing the relationship. Either way, somebody else threatens your self-esteem."

Yet Salovey emphasizes that it is not as simple as saying someone is at risk for these emotions if they have a low opinion of themselves. "It's low self-esteem in a specific *area*," he explains. "If you have a low opinion about your physical looks or occupation, then that's the area in which you're more likely to be vulnerable. You feel it when you confront somebody else who is superior to you in that respect."

From Smith's point of view, the key is how that person compares him or herself to others. In one study, for example, he found that envy was strongest among people who performed below their expectation in an area that was important to them and then confronted someone who functioned better. In a related study, Smith also found that a person's "risk" for feeling jealous or envious increases with the increased importance they put on the quality.

While much of this may sound like common sense, in fact little research has been done to establish even the most basic ground rules of jealousy and envy. For example, are the two emotions actually different? What do people feel when they are jealous or envious? Despite centuries of long-held assumptions, only recently have researchers begun to answer these questions scientifically.

Smith and his colleagues, for example, recently conducted a study to see if the classic distinctions between the two emotions are actually true. Their findings — presented last August at the annual convention of the American Psychological Association — validated what we have always suspected. Jealous people tend to feel a fear of loss, betrayal, loneliness, suspicion and uncertainty. Envious people, on the other hand, tend to feel inferior, longing for what another has, guilt over feeling ill will towards someone, shame and a tendency to deny the emotion.

The study was not an idle exercise in stating the obvious. It was designed to help clear an ongoing debate about jealousy and envy and our curious tendency not only to mix up the terms, but to experience an overlap of both emotions.

Consider, for example, the following uses of the word jealousy: Bruce was jealous when his girlfriend began talking to another man at the party; the boy cried in a fit of jealousy when his parents paid attention to the new baby; Ellen became jealous when her friend began spending more time at the health club. Nothing unusual with any of these uses of jealousy—they all refer to someone's "fear of losing a relationship."

Now, however, consider these uses: The professor, jealous of his colleague's success, broke into his lab and ruined his experiment; Mary is always complaining of being jealous of her sister's beautiful blonde hair; Mark admits that he is jealous of John's athletic ability. All of these situations actually refer to envy, "the longing for something someone else has." Researchers have noticed the mix-up and it has led them to question how different the feelings really are.

"In everyday language it's clear that people use the terms interchangeably," notes Smith, adding, "For that reason, there's naturally some confusion about whether they're different." In a recent study, however, Smith and his colleagues found the mix-up only works one way, with jealousy being the broader term. That is, jealousy is used sometimes in place of envy, but envy is rarely used when referring to jealousy. So while you might say, "I'm jealous of Paul's new Mercedes," you would never say, "When she left her husband, he flew into an *envious* rage."

Is there an underlying reason for why the terms are used interchangeably? One reason people may use jealousy in place of envy, and not vice versa, is because of the social stigma attached to envy. But Salovey and Smith both point to another reason for the overlap.

Dave and Marcia had been dating for a year when Marcia decided she wanted to see other men. Dave was devastated—not only because he cared for her, but because he was older and feared his age was working against him. One evening he bumped into Marcia, arm-in-arm with another man. As Dave talked to the couple, sarcasm led to verbal abuse, until finally Dave took a swing at Marcia's date—a man at least 10 years his junior.

In case you haven't guessed, Dave wanted back more than his relationship with Marcia; he also wanted the return of his youth. He was feeling a painful mixture of jealousy *and* envy. Explains Salovey, "The same feelings emerge when your relationship is threatened by someone else as when you'd like something that person has. I think one reason is that in most romantic situations, envy plays a role. You're jealous because you're going to lose the relationship, but you're also envious because there is something the other person has that allows him to be attractive to the person you care about."

Because this overlap occurs so frequently, Salovey has found the best way to understand jealousy and envy is to examine each *situation*. In addition, because jealousy is the more encompassing term, he views envy as a form of jealousy and distinguishes the two by the terms "romantic jealousy," the fear of losing a relationship; and "social-comparison jealousy," the envy that arises when people compare traits like age, intelligence, possession and talent.

Smith agrees with Salovey that one reason people confuse the terms may be that envy is present in most cases of jealousy. Nevertheless, he takes issue with Salovey's use of the term "social-comparison jealousy." "It may be true that there's almost invariably envy in every case of jealousy," he notes, "but it doesn't mean there's no value in distinguishing the two feelings. The overlap in usage only goes one way, so there's no reason to throw out the term 'envy.'"

Salovey counters, "I'm not saying we need to stop using 'envy.' The reason we use 'social-comparison jealousy' is to emphasize that the situation that creates the feeling is important." And one reason

---

Comparing Ourselves to Others — In Sickness and In Health

Envy, according to Richard Smith, arises when we compare ourselves to others and can't cope with what we see. Indeed, he believes that the way in which we cope with "social comparisons" plays an important role in our physical as well as mental health.

Smith theorizes that we use one of four "comparison styles" to cope with social differences. Two of these styles are "constructive" to well-being, while two are "destructive." "It's a difficult problem to tackle," says Smith, "but we're trying to measure these styles and see if they predict a person's general satisfaction with life or ability to cope with illness." For example, he notes, "There's considerable evidence" that one way people cope with serious illness is by focusing on others who are not doing as well.

Smith has arranged the four comparison styles in a matrix, with the descriptions in each box referring to the characteristics of that style. "*Upward*-Constructive," for example, represents those who compare themselves to others who are better off, and use it as a healthy stimulus. In this category, "You don't feel hostile to others who are better," says Smith, "because you hope to be like them. It suggests that upward comparisons are not necessarily bad."

"*Upward*-Destructive," however, shows how comparing yourself to those doing better can be unhealthy. Envy, resentment, Type A behavior and poor health all fit into this category. Smith points to a study that looked at personalities of people who had heart attacks, "and the only dimension that predicted heart disease was this jealousy-suspicion trait."

What Smith calls "*Downward*-Constructive," refers to people who compare themselves to those worse off, and use it to feel better about themselves. Such people, says Smith, "realize that others aren't doing so well and how lucky they are. There's some solid evidence in the health literature showing the value of that kind of comparison."

Finally, "*Downward*-Destructive," describes those who get pleasure out of comparing themselves to others who are worse off. "I'd call the effect 'schadenfreude,' or joy at the suffering of others," says Smith. "It's akin to sadism and it's probably not conducive to health."

"What's interesting about all four comparison styles," notes Smith, "is that they don't necessarily have any relation to reality. They reflect what people construe and focus on." While Smith stresses, "This is all speculative," he adds that "my feeling is that 'Upward-Destructive' explains people's hostility in terms of their social comparison context. It shows why their relation to people doing better leads to envy and why they'd feel hostile to begin with." J.Q.

Salovey stresses the situation—rather than other mood differences—is that romantic jealousy, since it includes envy, is usually very intense, making it difficult to separate distinct feelings.

Nevertheless, Smith believes the distinction should be made, especially because "In its traditional definition, envy has a hostile component to it." Not everyone would agree with that. After all, in envying others, we can also admire *them* and even use *them* as role models to spur ourselves to greater abilities. There is no hostility in that, yet these cases, Smith contends, are not precisely envy. Indeed, Smith believes envy differs from jealousy not only because of its hostility, but because of another distinct ingredient: privacy.

By the time Bill was 40, he was vice president at his firm and owned a luxurious house in an affluent neighborhood. Nevertheless, Bill had never married and was lonely. He envied his brother Jim, who lived a modest but happy life with his wife and children. One day Jim asked Bill to write a reference letter to help him get a bank loan for a new home. Bill said he'd be delighted, but soon realized he could send the letter to the bank without Jim ever seeing it. Bill wrote the letter and his brother, never knowing why, was refused the loan.

Although this anecdote is fictional, Smith has shown in his research that the principles illustrated are probably true. In a recent study, Smith had subjects identify with an envious person. He then gave them the option of dividing a "resource" between themselves and the envied person. Among the many options were: dividing the resource equally; dividing it so the subject kept the most and gave the least to the envied person; and dividing it so they sacrificed the amount they could otherwise keep for themselves if it meant giving the least to the envied person. Most subjects chose the last option, but only when they could do so in private, rather than public, circumstances.

Although Smith admits the findings need to be verified, he believes the results are strongly "suggestive." The envious person's choice, he says, "was unambiguously hostile. Not selfish, but hostile. And the findings verified the conventional wisdom that envy has a secretive quality about it that you wouldn't admit to the person you envy. And under the right circumstances it will lead to actual hostile behavior."

The reason people would be hostile in private seems obvious, given that envy is socially unacceptable. But why the hostility in the first place? Smith theorizes that the envied person's "superiority" emphasizes the envious person's low self-esteem in a specific area—the way the new real estate agent's communication skills affected Ann; Marcia's young date affected Dave; and Jim's happy family life affected Bill. Hostility is a way of putting down the envied person and devaluing his or her "superiority." Whether it takes the form of thought, word or deed, hostility pushes the envied person away, thereby allowing the envious person to restore his self-esteem.

Smith is looking at other implications of envy and hostility in society. For example, "We don't know much about why people are hostile to begin with, but since envy is related to hostility, maybe the way people respond to the way they compare themselves to others is at the root of hostility." And there are other subtle—and even more frightening—implications. For one thing, Smith notes, because envy is socially unacceptable, "It's often not conscious, and as a result people will arrange the details of their situation and their perception of the other person to make envy something they can label as 'resentment'—resentment in the sense of righteous indignation."

Smith goes as far as to propose that many intergroup conflicts in the world —between countries, races and religious groups, for example—may begin with envy. One group is better off economi-

RANKING JEALOUSY & ENVY

What situations are most likely to evoke feelings of jealousy and envy? In a study published in the *Journal of Personality and Social Psychology*, Peter Salovey and Judith Rodin asked subjects to rank 53 situations according to the degree of emotion each would evoke. Below are 25 of those situations, listed in decreasing order, that received the highest "jealousy/envy" ratings:

1 You find out your lover is having an affair.
2 Someone goes out with a person you like.
3 Someone gets a job that you want.
4 Someone seems to be getting closer to a person to whom you are attracted.
5 Your lover tells you how sexy his/her old girl/boyfriend was.
6 Your boyfriend or girlfriend visits the person he or she used to date.
7 You do the same work as someone else and get paid less than he or she.
8 Someone is more talented than you.
9 Your boyfriend or girlfriend would rather be with his/her friends than with you.
10 You are alone while others are having fun.
11 Your boyfriend or girlfriend wants to date other people.
12 Someone is able to express himself or herself better than you.
13 Someone else has something you wanted and could have had but don't.
14 Someone else gets credit for what you've done.
15 Someone is more intelligent than you.
16 Someone appears to have everything.
17 Your steady date has lunch with an attractive person of the opposite sex.
18 Someone is more outgoing and self-confident than you.
19 Someone buys something you wanted but couldn't afford.
20 You have to work while your roommate is out partying.
21 An opposite-sex friend gives another friend a compliment, but not you.
22 Someone has more free time than you.
23 You hear that an old lover of yours has found a new lover.
24 Someone seems more self-fulfilled than you.
25 You listen to someone tell a story about things he did without you. —J.Q.

cally, for example, than another, and the "inferior" group feels envy as a result. "But if they're just envious," he explains, "no one is going to give them any sympathy. So they tend to see their situation as unfair and unjust. In this way, envy becomes righteous resentment, which in turn gives them the 'right' to protest and conduct hostile—or even terrorist—activity."

This topic raises questions about what is fair or unfair in our society, how people cope with differences and whether, as a result, they feel envy, resentment or acceptance. Smith points out that coping with envy may depend on how well we learn to accept our inequalities. "I think as people mature, they learn to cope with differences by coming to terms with the fact that life *isn't* fair and that it's counterproductive to dwell on things you can't do anything about."

While some of Smith's ideas on coping with envy are speculative, Salovey has found that there are specific strategies—illustrated in the following anecdotes—that work in preventing and easing jealousy and envy. . . .

Rick and Liz are getting along much better these days even though Liz still works with her ex-boyfriend at the law firm. Rick isn't thrilled by this, but he overcame his jealousy by focusing instead on his relationship with Liz: spending time with her, planning vacations, discussing their future.

Ann no longer feels envy or hostility towards her co-worker at the real estate agency. She put an end to those negative feelings by simply ignoring her rival's superior communication skills and concentrating instead on her own achievements.

As illustrated here and isolated in a survey Salovey and Yale associate Judith Rodin conducted with *Psychology Today* readers, there are three major coping strategies: *Self-reliance*, in which a person does not give into the emotion, but continues to pursue the goals in the relationship; *Selective ignoring*, or simply ignoring the things that cause the jealousy or envy; and *Self-bolstering*, or concentrating on positive traits about yourself.

Surprisingly, "We found that the first two coping strategies are very effective in helping a person not feel jealousy," reports Salovey. "We thought that self-bolstering would also be good, since if something that's important to you is threatened, then maybe you should think of things in which you do well."

Although self-bolstering was not helpful in preventing jealousy. "Once you were *already* jealous, it was the only thing that kept you from becoming depressed and angry," notes Salovey. "So the first two keep jealousy in check, but if jealousy does emerge, self-bolstering keeps jealousy from its worst effects."

In the same study, Salovey and Rodin also uncovered some interesting data about how men and women experience jealousy and envy. "Men tend to be more envious in situations involving wealth and fame, and women more so in beauty and friendship," reports Salovey, but he emphasizes, "I should put that finding in context. We looked at a lot of variables and very rarely found differences. Men and women were very similar on nearly everything you could measure except that one difference."

Smith and Salovey do agree that while jealousy and envy can be devastating to those experiencing them, in milder forms they can actually be helpful. "I tend to think of jealousy and envy as normal," says Salovey. "In any relationship where you really care about the other person, when your relationship is threatened by someone, you're going to feel negative emotions. If you don't, maybe you don't care that much."

As for envy, Smith points out that, "I don't think it's a bad thing, necessarily. It's a motivator when it's in the form of admiration and hero worship." He does add, however, that at those levels the emotion may not be envy since envy, by definition is hostile. "It's hard to know where one stops and the other begins," he notes.

Nevertheless, Smith stresses that "Coping with differences is something we all do. Some of us do it in constructive ways and others in destructive, and that has implications for who is going to be happy or unhappy. Envy is a sign of not coping well—maybe." He adds that while "some people have a right to recognize that a situation is unfair, the next question is, what do you do? It may be best to recognize the unfairness and cope with it before it leads to more painful feelings."

Jealousy and envy can have an unpleasant knack of cropping up between the people who care most about each other. Our first reaction is often to blame the other person—my *wife* is the one who is lunching with her ex-boyfriend; my *co-worker* is undeservedly making more sales; why should my *brother* have a happy family *and* a big house? Part of our blame is understandable: life *is* unfair; society and circumstance *do* create differences between us beyond our control.

Nevertheless, the bottom line is not how we view other people, but how we view ourselves. When jealousy or envy become overwhelming, it is as much from passing judgment on ourselves as on others. That's when we owe it to everyone to talk it out, change what needs to be changed and—perhaps most importantly—accept ourselves for what we are.

Development During Middle and Late Adulthood

Developmentalists hold two extreme points of view about the latter part of the life span. Disengagement theory holds that the physical and intellectual deficits associated with aging are inevitable and should be accepted at face value by the aged. Activity theory acknowledges the decline in abilities associated with aging, but also notes that the aged can maintain satisfying and productive lives.

Extreme views in any guise risk stereotyping all individuals within a category or class as having the same needs and capabilities. Whether one's reference group is racial,

Unit 6

ethnic, cultural, or age-related, stereotyping usually leads to counterproductive, discriminatory social policy that alienates the reference group from mainstream society.

Evidence obtained during the past decade clearly illustrates the fallacy of extremist views of middle and late adulthood. Development during adulthood and aging is not a unitary phenomenon. Although there are common physical changes associated with aging, there are also wide individual differences in the rates of change and the degree to which changes are expressed. It is common to think of the changes associated with aging as solely physical and generally negative. The popular press devotes considerable space to discussions of the causes and treatment of such debilitating disorders such as Alzheimer's disease. However, there are also psychological changes associated with aging, and, as is the case at all age levels, some individuals cope well with change and others do not. New research on the aging process suggests that physical health and mental health changes do not correlate well. Although a variety of abuses can hasten physical and mental deterioration, proper diet and modest exercise can also slow the aging process. In addition, one cannot understate the importance of love, social interaction, and a sense of self-worth for combating the loneliness, despair, and futility often associated with aging. Leonard A. Sagan in his article, "Family Ties," offers evidence that family relationships and dynamics contribute to the good health, and thus the longevity, of people today.

Behavioral gerontology remains a specialization within human development that is absent even from most graduate programs in human development. Perhaps this is partly because of the natural tendency of people to avoid confronting the negative aspects of aging, such as loneliness and despair over the loss of one's spouse or over one's own impending death. Nevertheless, contemporary studies do provide fascinating information about the quality of life (see "The Vintage Years") and physical changes (as discussed in "Why Do We Age?" and "A Vital Long Life") during aging, and challenge many traditional views about interpersonal relationships and memory processes.

One physical side effect of aging affecting women is menopause. In "The Myths of Menopause," some of the changes that can occur with menopause are listed along with a "survival guide" for before, during, and after menopause.

Does the course of life really progress through an orderly series of universally experienced stages? Psychologist Bernice Neugarten in "The Prime of Our Lives," and Carol Tavris in "Don't Act Your Age!" do not think so and argue strongly against this view.

Because the proportion of the population represented by the aged is increasing rapidly, it is imperative that significant advances be made in our knowledge of the later years of development. Prolonging life and controlling physical illness should certainly be topics of interest in this regard. However, these biologically and medically oriented topics should be accompanied by studies on the psychological aspects of aging in pursuit of promotion and enhancement of the quality of life for the elderly.

Looking Ahead: Challenge Questions

Gerontologists suggest that significantly greater life expectancies are possible even without medical breakthroughs. Control of childhood diseases, better education, better physical fitness, and proper diet are factors that increase the life span. How would your life differ now if your expected life span was 150 years? How do you think your elderly years will differ from those of your parents?

Do you think that the life of most individuals can be described by a relatively constant sequence of life stages, or are life patterns too variable for general descriptions to be of use? What are some of the most common stereotypes about aging in our society?

What are the most important physical consequences of menopause? Do you think it's worth the risk for women to use estrogen therapy to counter the effects of menopause? Why or why not?

FAMILY TIES

The Real Reason People Are Living Longer

LEONARD A. SAGAN

LEONARD A. SAGAN *is an epidemiologist at the Electric Power Research Institute, in Palo Alto, California. His book* THE HEALTH OF NATIONS: TRUE CAUSES OF SICKNESS AND WELL-BEING *is published by Basic Books.*

WHEN MODERN MEDICINE made its debut at the Many Farms Navajo Indian community, in 1956, there was every reason to expect decisive results. The two thousand people who inhabited this impoverished and isolated Arizona settlement were living under extremely primitive conditions. Though nutrition was adequate, hygiene was poor, tuberculosis was widespread, and infant mortality rates were three times the national average. To a group of researchers from the Cornell University Medical College and the U.S. Public Health Service, the situation at Many Farms provided a perfect opportunity to introduce modern health care practices and measure the consequences. If the effort proved successful with this target population, they reasoned, it might become an example for underdeveloped communities worldwide.

Almost overnight, the Navajo settlement acquired an array of modern medical resources. The researchers set up a full-service clinic, staffed with physicians and nurses, as well as with public health consultants, a health teacher, and four Navajo health care workers. For medical emergencies, the community got a fleet of radio-equipped vehicles and a light airplane. Over the next six years, ninety percent of the Many Farms residents took advantage of the clinic. Two-thirds of them were seen at least once a year.

The result was a rapid decline in the transmission of tubercle bacillus (the agent that causes tuberculosis) and in the frequency of otitis media (an inflammation of the middle ear). Yet the population's overall health, as reflected in its mortality statistics, was virtually unchanged. Of the sixty-five deaths that occurred during the six-year study period, more than half involved infants, who made up less than four percent of the population. And, despite expert pediatric care, there was no reduction in the pneumonia–diarrhea complex that was the leading cause of childhood illness and death. In the end, the investigators were unsure whether the improved medical care had, on balance, produced any beneficial effect at all.

This outcome would be less unsettling if it were more unusual. Unfortunately, it is not unusual at all. Consider what happened in 1976, when the state legislature of North Carolina sponsored a study to determine the effects of improved maternal and perinatal health care on the state's poorer communities. Researchers identified a number of counties, similar in racial and socioeconomic characteristics, that had suffered high rates of infant mortality over the preceding decades. For the next five years, residents of some of those counties received state-of-the-art treatment at the medical centers of Duke University and the University of North Carolina while, for the purpose of comparison, similar counties were essentially left alone. As expected, infant mortality declined considerably in the areas that received the additional care, but it also declined in the areas that did not. In fact, the researchers found no significant differences between the two groups.

Similar stories can be told about much larger populations. When England established its National Health Service, in 1946, the country's lowest social classes had long suffered the poorest health and the shortest lives—presumably because of economic barriers to adequate health care. The new program effectively removed those barriers. Forty-two years later, however, the disparity in mortality rates remains undiminished; the life expectancy of the most affluent is almost twice that of the least affluent. The economists Lee and Alexandra Benham, of Washington University, in Saint Louis, have noted the similar failure of Medicare and Medicaid to affect mortality rates among the disadvantaged in the United States. This country's least educated classes now experience as much hospitalization and surgery as its most educated classes, yet overall health is still strongly associated with educational achievement.

What are we to make of all this? It is well known that life expectancy has risen dramatically in most societies over the past few centuries. As recently as 1900, the typical American lived only forty-nine years, and one in five children died during infancy. Today we live an average of seventy-five years, and infant mortality has declined to just ten deaths for every thousand births, or one percent. Both physicians and the public credit modern medicine for these bold achievements; we assume, almost reflexively, that people who lack expert medical attention die earlier, and that providing more care is the key to longer life.

Americans, therefore, have invested heavily in medicine. Our expenditures now total more than four hundred billion dollars a year, or eleven percent of the gross national product, the highest rate of any nation on Earth. Yet some measures of ill health, such as the rate of disability due to chronic illness among children, are on the rise. And though life expectancy continues to rise in the United States, it is rising more rapidly in countries that are spending at a lower rate. Many of those countries, including Greece, Spain, and Italy, now enjoy life expectancies greater than our own. And Japan, which leads the world in life expectancy, spends only a third of what the United States spends each year—about five hundred dollars per capita compared with fifteen hundred.

Clearly, we need to take a closer look at the relationship between our efforts at health care, on the one hand, and our actual health, on the other. If the United States is spending more on medicine than any other nation, while suffering poorer health than many, there may be something fundamentally wrong with the country's approach. The urgent questions are: What really makes people healthy? Why do we live so much longer than our ancestors and so much longer than the world's remaining premodern peoples? If medicine is not the source of this blessing, we would do well to find out what is—and to direct our medical and public health efforts accordingly.

THERE IS NO DENYING that modern medicine has accomplished much of value. It has done a great deal to alleviate suffering, and many treatments—including surgery for burns, bleeding, abdominal obstructions, and diabetic coma—undoubtedly save lives. Anything that saves lives would presumably contribute to overall life expectancy. But most therapy is not aimed directly at prolonging life. Rare is the patient for whom death would be the price of missing a doctor's appointment. Moreover, any medical procedure involves some risk; there is always a chance that the patient will have an adverse or fatal reaction to a given treatment—be it surgical, pharmaceutical, or even diagnostic. If treatments were administered only when patients stood to benefit, the net effect on mortality rates might be positive. But physicians have a well-documented tendency to overdo a good thing. And because there are no clear guidelines governing the use of most remedies, the cost of such zeal is that the benefits gained by those who require a particular treatment are often outweighed by the adverse effects on those who receive it unnecessarily. Thus, while such major medical advances as antibiotics, immunization, coronary bypass surgery, chemotherapy, and obstetric surgery all have saved lives, it is impossible to demonstrate that any of them has contributed significantly to overall life expectancy.

The introduction of antibiotics into clinical medicine is generally viewed as the turning point in mankind's war against infectious disease. Clearly, such illnesses as typhoid, cholera, measles, smallpox, and tuberculosis no longer claim lives at the rate they did during the nineteenth century. The decline began at different times in different nations, but it was under way in Scandinavia and the English-speaking countries by the mid-nineteenth century, roughly a hundred years before the first antibiotic drug, penicillin, became available, during the Second World War. By the time streptomycin, isoniazid, and other such agents came into wide use, during the 1940s and 1950s, death rates from the eleven most common infectious diseases had dwindled to a mere fraction of their nineteenth-century levels. Antibiotics did, for a time, hold tremendous therapeutic powers, and had they been used in moderation, they might have remained potent weapons against infection. But overuse has largely destroyed their effectiveness.

The indications that we rely too heavily on antibiotics are myriad. In 1973, scientists at the University of Wisconsin at Madison concluded, after reviewing the findings of other researchers, that enough antibiotics are manufactured and dispensed each year in the United States to treat two illnesses of average duration in every man, woman, and child in the country. The evidence suggests, however, that only once in five to ten years does the average individual experience an infection, such as meningitis or tuberculosis, that antibiotics might help control. The drugs are routinely prescribed for colds and flu, even though there is no evidence they have any effect on such viral ailments, and are given out like vitamins in many hospitals. In one recent survey of hospital patients, the internist Theodore C. Eickhoff, of the University of Colorado Medical Center, in Denver, found that thirty percent were receiving antibiotics—though only half of those receiving the drugs showed signs of infection. Other findings suggest that patients who might actually benefit from an antibiotic frequently receive the wrong one, or an incorrect dose.

One outcome of this overreliance on penicillin and the other so-called wonder drugs is that many bacteria have, through natural selection, become resistant to them, and infections that were easily controlled thirty years ago no longer respond well to treatment. Both gonococcus, the

6. MIDDLE AND LATE ADULTHOOD

pus-producing bacterium that causes the most common venereal disease, and pneumococcus, a bacterium frequently associated with lobar pneumonia, now show resistance to various antibiotics. And in hospitals, overall infection rates are on the rise. A 1985 study, published by Robert W. Haley and his colleagues at the Centers for Disease Control, concluded that hospital-acquired infections occur in almost six out of every one hundred patients, thereby producing a national toll of four million infections a year, and that this rate is increasing by two percent annually.

Immunization is another therapy widely believed to have reduced death rates from infectious disease. But studies indicate that the use of vaccines and their ostensible benefits are largely unrelated. There is no question that the smallpox vaccine, for one, is effective when properly administered. Historical records show, however, that the number of people dying of smallpox was already falling when the vaccine first became available in Europe, during the early nineteenth century. True, smallpox mortality continued to drop as the vaccine became more accessible, but so did the death rates associated with infectious diseases for which vaccines had *not* been developed. The parallel decline in mortality from typhoid and tuberculosis prompted speculation that the smallpox vaccine was somehow protecting people from those infections, too. But there was never any basis for such a conclusion. A more reasonable inference is that deaths from all three illnesses were declining on account of some other factor.

As in the case of smallpox, vaccines for polio, whooping cough, measles, and diphtheria are effective at protecting individuals from these diseases. As a result, they not only save lives but spare many people permanent disabilities. But the question, for our purposes, is whether such vaccines have caused a significant decline in overall mortality, and the evidence indicates they have not. The historical record shows that death rates for childhood diseases started falling before the vaccines became available, and there is no evidence that forgoing such vaccines shortens people's lives. When concern about the risks associated with the diphtheria vaccine led English physicians to stop administering it during the late 1970s, for example, there was a sharp increase in the incidence of the disease, yet diphtheria mortality barely changed.

LIKE INFECTIOUS DISEASE, cardiovascular illness seems to pose a less dire threat to most of us today than it has in the past. Coronary artery disease appears to be waning both in incidence and in deadliness. A twenty-six-year study of Du Pont Company employees found that the number of people afflicted with the disease declined by twenty-eight percent between 1957 and 1983. Other studies indicate that the rate at which Americans are killed by it fell from about three hundred and fifty for every hundred thousand in 1970 to just two hundred and fifty for every hundred thousand in 1985. If such outcomes could be attributed to medical intervention, they would indeed rank as major accomplishments. But here, as with infectious disease, the link between treatment and health is elusive—whether the treatment is directed at preventing the disease or curing it.

Consider the results of the Multiple Risk Factor Intervention Trial, or "Mr. Fit." In this study, a team of investigators from twenty-two health research centers randomly divided a sample population of nearly thirteen thousand men, aged thirty-five to fifty-seven, into two groups. For the next seven years, members of one group continued to receive routine care from their private physicians, while the other group participated in a therapeutic program to reduce the risk of coronary artery disease. Physicians supervised and monitored efforts to have them avoid smoking, reduce the amount of cholesterol in their diets, and control their blood pressure, using medication if necessary. At the end of the treatment period, the subjects who received the extra medical attention had indeed cut back on cigarettes and cholesterol, and they exhibited less hypertension. But they did not end up living any longer than the subjects who simply went about their business. In fact, their death rate from all causes (41.2 deaths for every thousand subjects) was slightly *higher* than that of the control group (40.4 deaths). The reasons for this failure were not readily evident; the researchers speculated that the ill effects of antihypertensive drugs may have outweighed any benefits derived from the program. Whatever the explanation, such results confirm that the recent decline in death from cardiovascular disease probably is not the fruit of preventive medicine.

If efforts at prevention have not caused the decline, might it reflect the advent of better therapeutic techniques, such as coronary bypass surgery? Saving lives was not the original intent of this operation when surgeons began performing it, during the late 1960s; bypassing portions of coronary arteries that had become partially clogged with fatty deposits was viewed as a way of alleviating the chest pain that accompanies such blockage. But when the operation was found to be effective for that purpose, physicians began touting it as a therapeutic measure—and even a preventive treatment for patients without symptoms—despite an utter lack of clinical evidence. Today coronary bypass is one of America's most commonly performed surgical procedures. Roughly two hundred thousand Americans undergo the operation each year, at a total cost of some five billion dollars. Yet only rarely does it contribute to anyone's survival. A study published in 1983 by the National Institutes of Health concluded that bypass surgery prolongs the life of roughly one bypass patient in ten but that it appears to add nothing to the life expectancies of the other nine.

IS CANCER TREATMENT, another major focus of modern medicine, perhaps the secret of our increased life expectancy? One might guess that it has made a contribution; after all, the average interval between the diagnosis of a malignancy and the death of the patient has increased considerably in recent years. Indeed, the percentage of cancer patients surviving at least five years rose from 38.5 percent in 1973 to 40.1 percent in 1978, an improvement of almost one percent a year. Regrettably, it does not necessarily follow that people with cancer are living longer, let alone that chemotherapy or surgical treatments are extending their lives. Many scientists speculate that earlier detection of the disease merely has created an illusion of increased survival.

There is evidence that physicians are diagnosing cancer at earlier stages of development, thanks largely to more frequent checkups and better diagnostic technology. But there is no indication that earlier treatment has improved patients' overall survival rates. In fact, for some forms of cancer, there is hard evidence that it has not. In one study, sponsored by the National Cancer Institute and published in 1984, a population of adult male smokers, all presumably at high risk of developing lung cancer, was divided into two groups. One group received only annual chest X rays; members of the second group underwent frequent X rays and had their sputum examined regularly for cancer cells. Not surprisingly, there were many more diagnoses of lung cancer among the closely monitored subjects. And because their malignancies were usually detected and treated at early stages, their survival rates from the time of diagnosis were impressive. Even so, the numbers of lung cancer deaths in the two groups were nearly identical. In short, the participants in the early-detection program gained no apparent advantage: they were no less likely to suffer recurrences of the disease, or to die of it, than were members of the control group.

If modern treatments were, on balance, helping cancer patients survive, those patients would be dying at a later average age, and this, in turn, would reduce the average person's chances of dying of cancer at any age. But age-adjusted cancer mortality has not declined at all in the United States during the past fifty years. The death rates have changed for particular forms of cancer (lung cancer mortality has increased, whereas deaths from stomach cancer have declined), and it is possible that treatment has played a role in some of the success stories. The relatively rare cancers of childhood, for example, seem to respond well to treatment. But such situations are the exception, not the rule. Most therapies are introduced without ever being thoroughly evaluated for effectiveness, and they are embraced by physicians and patients who are understandably eager to try anything.

Radical mastectomy, the standard treatment for breast cancer throughout most of this century, is a good example. Studies have shown that patients who have this operation —the mutilating removal of the breast and its underlying tissues—do not, as a group, live any longer than patients who undergo the less radical lumpectomy (removal of the cancerous mass only). In one study, published in 1985 in *The New England Journal of Medicine*, nearly two thousand breast cancer patients were randomly assigned to receive one treatment or the other. Those who had the traditional mastectomy died earlier. Still, most U.S. physicians continue to perform the more extensive operation, and some even recommend it as a preventive measure for women whose cystic (lumpy) breasts place them at a theoretical risk of developing the disease.

The point is not that cancer treatment is never justified, only that it has had no discernible effect on the overall survival rates of cancer patients, let alone the life expectancy of the general population. Indeed, no cancer treatment, however successful, could do much to increase life expectancy, for the disease does little to reduce it. Cancer strikes mostly among the aged. It has long been estimated that even if it were totally preventable or curable, the increase in U.S. life expectancy would be less than two years. Given that life expectancy has increased by twenty-five years during this century, it is impossible that the treatment of cancer has made much of a difference.

If modern medicine cannot be credited with taming infectious disease, cardiovascular illness, or cancer, one might expect to find that it has at least improved the odds that mothers and infants will survive the birth process. Cesarean section has undoubtedly contributed to the rapid decline in maternal mortality during this century. Like so many other medical procedures, however, it is now so grossly overused that it may be costing as many lives as it saves.

The maternal mortality rate had already dwindled to less than one death in ten thousand deliveries by the mid-1970s. Yet, since then, births by cesarean section have increased by three hundred percent in the United States. To confirm that this trend is not making childbirth any safer, one need only consider the survival statistics for societies that rely less heavily on surgical delivery. In 1965, the rate of cesarean births at Ireland's National Maternity Hospital, in Dublin, was equal to that in the United States—about five percent. Since then, the U.S. rate has climbed to twenty percent, but the rate in Dublin has remained stable, and perinatal mortality has fallen faster there than it has in the United States. The Netherlands, meanwhile, which enjoys one of the lowest perinatal and maternal mortality rates in Europe, also has one of the lowest rates of obstetric surgery.

IT SEEMS CLEAR that modern medicine, whatever it has done to save or improve individual lives, has had little effect on the overall health of large populations. Still, the fact is that life expectancy has increased spectacularly during the nineteenth and twentieth centuries. What else might explain such a change? There is no question that sanitation and nutrition, the other factors most often cited, have been beneficial. But neither of these developments accounts fully for the mystery at hand.

It is true that, toward the end of the nineteenth century, improvements in sanitation coincided with a decline in mortality from various infectious diseases in Europe and America. But there is no evidence of a cause-and-effect relationship. Sanitation worsened in many major cities during the Industrial Revolution, as the prospect of work drew hordes of immigrants from rural areas. Rotting meat, fish, and garbage were heaped in the streets of New York and London, and overflowing privies were still far more common than modern toilets in many crowded neighborhoods. Amazingly, though, mortality rates from infectious disease fell steadily over the same period.

Another problem with the sanitation argument is that the *incidence* of infection decreased little during the nineteenth century. What did decline was the frequency with which infections sickened or killed people. As recently as 1940, long after tuberculosis had ceased to be a major health threat, skin tests showed that ninety-five percent of all Americans were still being infected with tuberculosis bacteria by age forty-five. Yet the vast majority managed to fight it off. Even today, most of the micro-

6. MIDDLE AND LATE ADULTHOOD

organisms that caused so much disease and death in premodern times, particularly among children, are omnipresent in the environment. No amount of sanitation could eliminate them, for they are passed directly from one person to another. They exist harmlessly, for the most part, both in and on our bodies.

Could it be that improved nutrition has strengthened our resistance? This idea does not withstand scrutiny, either. If eating well were the key to long life, then the most privileged families of old Europe, who enjoyed better nutrition than their contemporaries, should also have enjoyed longer lives. But they died young (as did the first American settlers, for whom the threat of starvation was not a particular problem). Moreover, there is no evidence that the specific dietary changes that are associated with modernization have even been advantageous. Indeed, it is arguable that, on balance, those changes have been harmful.

In the United States—where modernization has been associated with less physical activity, and with increased consumption of white bread, cookies, doughnuts, alcohol, and red meat from fattened animals—an estimated twenty to twenty-five percent of adult men are overweight. Diet and inactivity are not the only factors that contribute to obesity, of course, but they clearly count. In one recent study, the University of Toronto anthropologists Andris Rode and Roy J. Shephard monitored body fat and physical fitness among members of an Eskimo community during a ten-year period of rapid modernization. They found that the community's adoption of a modern diet, along with its increased use of snowmobiles and snow-clearing equipment, accompanied a significant increase in body fat and decreases in several measures of fitness. If these Eskimo follow the usual pattern, modernization will bring about a net increase in life expectancy. But if their overall health improves, it will have improved *despite* the changes in diet and physical activity, not because of them.

IT IS, IN A WORD, impossible to trace the hardiness of modern people directly to improvements in medicine, sanitation, or diet. There is an alternative explanation for our increased life expectancy, however, one that has less to do with these developments than with changes in our psychological environment. We like to imagine that preindustrial peoples endured (and endure) less stress than we do—that, although they may have lacked physical amenities, they spent peaceful days weaving interesting fabrics and singing folk songs. But the psychic stresses of the simple life are, in fact, far greater than those experienced by the most harried modern executive. It is one thing to fret over a tax return or a real estate deal, and quite another to bury one's children, to wonder whether a fall's harvest will last the winter, or to watch one's home wash away in a flood.

To grow up surrounded by scarcity and ignorance and constant loss—whether in an African village or a twentieth-century urban slum—is to learn that misery is usually a consequence of forces beyond one's control and, by extension, that individual effort counts for naught. And there is ample evidence that such a sense of helplessness is often associated with apathy, depression, and death—whether in laboratory animals or in prisoners of war. The experimental psychologist Martin E. P. Seligman, of the University of Pennsylvania, has designed some remarkable studies to simulate in dogs the experience of helplessness in humans. His classic experiment involved placing dogs in a box in which they could avoid electric shocks by jumping over a barrier upon the dimming of a light. Naïve dogs quickly learned to avoid shocks entirely, leaping gracefully over the barrier whenever the light dimmed. But Seligman found that dogs responded differently if, before being placed in the box, they were confined and subjected to shocks they could not escape. Those dogs, having learned that effort is futile, just lay down and whined.

In many ways, the experiences and reactions of the second group resemble those of people raised in poverty, a shared feature of most premodern societies. Modernization, through such mechanisms as fire departments, building codes, social insurance, and emergency medical care, has cushioned most of us against physical, psychic, and economic disaster. But, more important, it has created circumstances in which few of us feel utterly powerless to control our lives. We now take for granted that we are, in large part, the masters of our own destinies, and that in itself leaves us better equipped to fight off disease.

How did this happen? What are the sources of this sense of personal efficacy and self-esteem? No institution has been so changed by modernization as the family. Until the late eighteenth century, it existed primarily as an economic unit; marriages were arranged for the purpose of preserving property, and children were viewed as a cheap source of labor or a hedge against poverty in old age. Beating and whipping were favored, even among royalty, as tools for teaching conformity and obedience. Then, during the Enlightenment, the standards and goals of child rearing began to change. If children were going to survive in a disorderly and unpredictable world, philosophers began to argue, they could not rely passively on traditional authority; they needed reasoned judgment. And if children were going to develop such judgment, they needed affection and guidance, not brute discipline. It was only gradually, as these ideas took root, that childhood came to be recognized as a special stage of life, and that affection and nurturing replaced obligation and duty as the cohesive forces among family members.

During the nineteenth century, as the upper classes came to view children as having needs of their own rather than serving the needs of the family—and, accordingly, started having fewer of them—their infant and childhood mortality rates began to fall. And as the trend toward smaller families spread to the lower social classes, theirs fell, too. It is unlikely that this was just coincidence, for family size is an excellent predictor of childhood survival even today. Young children of large families continue to suffer more infections, more accidents, and a higher overall mortality rate than the children of small families, regardless of social class. Indeed, as the Columbia University sociologist Joe D. Wray demonstrated in 1971, the effects of family size can outweigh those of social class: an only child in a poor family has about the same chance of surviving the first year of life as a child who is born into a professional-class family but who has four or more siblings.

Why should this be so? One explanation, supported by various lines of evidence, is that the children of small families are strengthened in every way by the extra nurture they receive from their parents. During the past forty years, studies have demonstrated that infants develop poorly, even die, when they are provided food and physical necessities but are denied intimate contact with care givers. In one experiment, orphans placed in an institution at an early age were separated into two groups. Members of one group stayed in the institution while the others were placed with foster parents. At the end of the first year, the children placed in foster homes were better developed, both mentally and physically, than those who received institutional care. And even after the institutionalized children were assigned to foster homes, they remained less developed than their counterparts for a number of years.

Other studies have produced even more arresting evidence. In 1966, Harold M. Skeels, of the National Institute of Mental Health, reported on an experiment that gauged the long-term effect of individual care on retarded institutionalized children. One group of children received routine institutional care, which is often physically adequate but emotionally sterile, while the other children were moved to a special ward to be cared for individually by retarded women. After three years, most of the children in the first group had lost an average of twenty-six IQ points, whereas those in the second group had *gained* an average of twenty-nine points. The differences were even more pronounced thirty years later. None of the children who received routine care had made it past the third grade, and most remained institutionalized. By contrast, many of those cared for by foster mothers had completed the twelfth grade and gone on to become self-supporting.

WE ARE ONLY BEGINNING to understand the mechanisms linking emotional and physical health (the endeavor has of late given rise to a new branch of medicine, known as psychoneuroimmunology). But whatever the connection, the fact stands that the affection and security associated with the modern family are the best available predictors of good health. In the end, it matters little whether sanitation, nutrition, and medical care are crude or sophisticated; children who receive consistent love and attention—who grow up in circumstances that foster self-reliance and optimism rather than submission and hopelessness—are better survivors. They are bigger, brighter, more resistant, and more resilient. And, as a result, they live longer.

It is ironic, in the light of this, that we continue to fret over the quality of our food and the purity of our environment, to spend billions of dollars on medical procedures of no proven value, and to pay so little attention to the recent deterioration of the American family. The divorce rate in the United States, though it appears to have leveled off during the past few years, has increased enormously since the 1950s, from less than ten percent to more than twenty percent today. The number of children being raised by single parents has doubled during the past decade alone, and divorce is not the only reason. Another ominous development is the rise in pregnancy among unwed teenagers. For whites, the rate increased from eight percent in 1940 to twenty percent in 1970, and to thirty percent in 1980. The problem is even worse among blacks, sixty percent of whom are now born out of wedlock. That this, in itself, constitutes a serious health problem is plain when one considers that fetal and infant death rates are twice as high for illegitimate children as for legitimate ones, and that a teenaged mother is at least seven times more likely than an older mother to abuse her child.

All of this suggests that good health is as much a social and psychological achievement as a physical one—and that the preservation of the family is not so much a moral issue as a medical one. Unless we recognize the medical importance of the family and find ways to stop its deterioration, we may continue to watch our health expenditures rise and our life-spans diminish. We will waste precious resources on unnecessary treatments, while ignoring a preventable tragedy.

The Vintage Years

THE GROWING NUMBER OF HEALTHY, VIGOROUS OLDER PEOPLE HAS HELPED OVERCOME SOME STEREOTYPES ABOUT AGING. FOR MANY, THE BEST IS YET TO COME.

Jack C. Horn and Jeff Meer

Jack C. Horn is a senior editor and Jeff Meer is an assistant editor at the magazine.

Our society is getting older, but the old are getting younger. As Sylvia Herz told an American Psychological Association (APA) symposium on aging last year, the activities and attitudes of a 70-year-old today "are equivalent to those of a 50-year-old's a decade or two ago."

Our notions of what it means to be old are beginning to catch up with this reality. During the past several decades, three major changes have altered the way we view the years after 65:

• The financial, physical and mental health of older people has improved, making the prospect of a long life something to treasure, not fear.

• The population of older people has grown dramatically, rising from 18 million in 1965 to 28 million today. People older than 65 compose 12 percent of the population, a percentage that is expected to rise to more than 20 percent by the year 2030.

• Researchers have gained a much better understanding of aging and the lives of older people, helping to sort out the inevitable results of biological aging from the effects of illness or social and environmental problems. No one has yet found the fountain of youth, or of immortality. But research has revealed that aging itself is not the thief we once thought it was; healthy older people can maintain and enjoy most of their physical and mental abilities, and even improve in some areas.

Because of better medical care, improved diet and increasing interest in physical fitness, more people are reaching the ages of 65, 75 and older in excellent health. Their functional age—a combination of physical, psychological and social factors that affect their attitudes toward life and the roles they play in the world—is much younger than their chronological age.

Their economic health is better, too, by almost every measure. Over the last three decades, for example, the number of men and women 65 and older who live below the poverty line has dropped steadily from 35 percent in 1959 to 12 percent in 1984, the last year for which figures are available.

On the upper end of the economic scale, many of our biggest companies are headed by what once would have been called senior citizens, and many more of them serve as directors of leading companies. Even on a more modest economic level, a good portion of the United States' retired older people form a new leisure class, one with money to spend and the time to enjoy it. Obviously not all of America's older people share this prosperity. Economic hardship is particularly prevalent among minorities. But as a group, our older people are doing better than ever.

In two other areas of power, politics and the law, people in their 60s and 70s have always played important roles. A higher percentage of people from 65 to 74 register and vote than in any other group. With today's increasing vigor and numbers, their power is likely to increase still further. It is perhaps no coincidence that our current President is the oldest ever.

Changing attitudes, personal and social, are a major reason for the increasing importance of older people in our society. As psychologist

45. Vintage Years

BY THE YEAR 2030 MORE THAN 20 PERCENT OF THE POPULATION IS EXPECTED TO BE 65 OR OLDER.

Bernice Neugarten points out, there is no longer a particular age at which someone starts to work or attends school, marries and has children, retires or starts a business. Increasing numbers of older men and women are enrolled in colleges, universities and other institutions of learning. According to the Center for Education Statistics, for example, the number of people 65 and older enrolled in adult education of all kinds increased from 765,000 to 866,000 from 1981 to 1984. Gerontologist Barbara Ober says that this growing interest in education is much more than a way to pass the time. "Older people make excellent students, maybe even better students than the majority of 19- and 20-year-olds. One advantage is that they have settled a lot of the social and sexual issues that preoccupy their younger classmates."

Older people today are not only healthier and more active; they are also increasingly more numerous. "Squaring the pyramid" is how some demographers describe this change in our population structure. It has always been thought of as a pyramid, a broad base of newborns supporting successively smaller tiers of older people as they died from disease, accidents, poor nutrition, war and other causes.

Today, the population structure is becoming more rectangular, as fewer people die during the earlier stages of life. The Census Bureau predicts that by 2030 the structure will be an almost perfect rectangle up to the age of 70.

The aging of America has been going on at least since 1800, when half the people in the country were younger than 16 years old, but two factors have accelerated the trend tremendously. First, the number of old people has increased rapidly. Since 1950 the number of Americans 65 and older has more than doubled to some 28 million—more than the entire current population of Canada. Within the same period, the number of individuals older than 85 has quadrupled to about 2.6 million (see "The Oldest Old," this article).

Second, the boom in old people has been paired with a bust in the proportion of youngsters due to a declining birth rate. Today, fewer than one American in four is younger than 16. This drop-off has been steady, with the single exception of the post-World War II baby boom, which added 76 million children to the country between 1945 and 1964. As these baby boomers reach the age of 65, starting in 2010, they are expected to increase the proportion of the population 65 and older from its current 12 percent to 21 percent by 2030.

The growing presence of healthy, vigorous older people has helped overcome some of the stereotypes about aging and the elderly. Research has also played a major part by replacing myths with facts. While there were some studies of aging before World War II, scientific interest increased dramatically during the 1950s and kept growing.

Important early studies of aging included three started in the mid or late 1950s: the Human Aging Study, conducted by the National Institute of Mental Health (NIMH); the Duke Longitudinal Studies, done by the Center for the Study of Aging and Human Development at Duke University; and the Baltimore Longitudinal Study of Aging, conducted by the Gerontological Institute in Baltimore, now part of the National Institute on Aging (NIA). All three took a multidisciplinary approach to the study of normal aging: what changes take place, how people adapt to them, how biological, genetic, social, psychological and environmental characteristics relate to longevity and what can be done to promote successful aging.

These pioneering studies and hundreds of later ones have benefited from growing federal support. White House Conferences on Aging in 1961 and 1971 helped focus attention on the subject. By 1965 Congress had enacted Medicare and the Older Americans Act. During the 1970s Congress authorized the establishment of the NIA as part of the National Institutes of Health and NIMH created a special center to support research on the mental health of older people.

All these efforts have produced a tremendous growth in our knowledge of aging. In the first (1971) edition of the *Handbook of the Psychology of Aging*, it was estimated that as much had been published on the subject in the previous 15 years as in all the years before then. In the second edition, published in 1985, psychologists James Birren and Walter Cunningham wrote that the "period for this rate of doubling has now decreased to 10 years...the volume of published research has increased to the almost unmanageable total of over a thousand articles a year."

Psychologist Clifford Swenson of Purdue

A man over 90 is a great comfort to all his elderly neighbours: he is a picket-guard at the extreme outpost; and the young folks of 60 and 70 feel that the enemy must get by him before he can come near their camp.
—Oliver Wendell Holmes,
The Guardian Angel.

6. MIDDLE AND LATE ADULTHOOD

University explained some of the powerful incentives for this tremendous increase: "I study the topic partly to discover more effective ways of helping old people cope with their problems, but also to load my own armamentarium against that inevitable day. For that is one aspect of aging and its problems that makes it different from the other problems psychologists study: We may not all be schizophrenic or neurotic or overweight, but there is only one alternative to old age and most of us try to avoid that alternative."

One popular misconception disputed by recent research is the idea that aging means inevitable physical and sexual failure. Some changes occur, of course. Reflexes slow, hearing and eyesight dim, stamina decreases. This *primary aging* is a gradual process that begins early in life and affects all body systems.

But many of the problems we associate with old age are *secondary aging*—the results not of age but of disease, abuse and disuse—factors often under our own control. More and more older people are healthy, vigorous men and women who lead enjoyable, active lives. National surveys by the Institute for Social Research and others show that life generally seems less troublesome and freer to older people than it does to younger adults.

In a review of what researchers have learned about subjective well-being—happiness, life satisfaction, positive emotions—University of Illinois psychologist Ed Diener reported that "Most results show a slow rise in satisfaction with age...young persons appear to experience higher levels of joy but older persons tend to judge their lives in more positive ways."

Money is often mentioned as the key to a happy retirement, but psychologist Daniel Ogilvie of Rutgers University has found another, much more important, factor. Once we have a certain minimum amount of money, his research shows, life satisfaction depends mainly on how much time we spend doing things we find meaningful. Ogilvie believes retirement-planning workshops and seminars should spend more time helping people decide how to use their skills and interests after they retire.

A thought that comes through clearly when researchers talk about physical and mental fitness is "use it or lose it." People rust out faster from disuse than they wear out from overuse. This advice applies equally to sexual activity. While every study from the time of Kinsey to the present shows that sexual interest and activity diminish with age, the drop varies greatly among individuals. Psychologist Marion Perlmutter and writer Elizabeth Hall have reported that one of the best predictors of continued sexual intercourse "is early sexual activity and past sexual enjoyment and frequency. People who have never had much pleasure from sexu-

WHILE THE OLD AND THE YOUNG MAY BE EQUALLY COMPETENT, THEY ARE DIFFERENTLY COMPETENT.

ality may regard their age as a good excuse for giving up sex."

They also point out that changing times affect sexual activity. As today's younger adults bring their more liberal sexual attitudes with them into old age, the level of sexual activity among older men and women may rise.

The idea that mental abilities decline steadily with age has also been challenged by many recent and not-so-recent findings. In brief, age doesn't damage abilities as much as was once believed, and in some areas we actually gain; we learn to compensate through experience for much of what we do lose; and we can restore some losses through training.

For years, older people didn't do as well as younger people on most tests used to measure mental ability. But psychologist Leonard Poon of the University of Georgia believes that researchers are now taking a new, more appropriate approach to measurement. "Instead of looking at older people's ability to do abstract tasks that have little or no relationship to what they do every day, today's researchers are examining real-life issues."

Psychologist Gisela Labouvie-Vief of Wayne State University has been measuring how people approach everyday problems in logic. She notes that older adults have usually done poorly on such tests, mostly because they fail to think logically all the time. But Labouvie-Vief argues that this is not because they have forgotten how to think logically but because they use a more complex approach unknown to younger thinkers. "The [older] thinker operates within a kind of double reality which is both formal and informal, both logical and psychological," she says.

In other studies, Labouvie-Vief has found that when older people were asked to give concise summaries of fables they read, they did so. But when they were simply asked to recall as much of the fable as possible, they concentrat-

*The pleasures that once were heaven
Look silly at sixty-seven.*
—Noel Coward,
"What's Going to Happen to the Tots?"

Old age consoles itself by giving good precepts for being unable to give bad examples.
—La Rochefoucauld,
The Maxims.

THE OLDEST OLD: THE YEARS AFTER 85

"Every man desires to live long, but no man would be old," or so Jonathan Swift believed. Some people get their wish to live long and become what are termed the "oldest old," those 85 and older. During the past 22 years, this group has increased by 165 percent to 2.5 million and now represents more than 1 percent of the population.

Who are these people and what are their lives like? One of the first to study them intensively is gerontologist Charles Longino of the University of Miami, who uses 1980 census data to examine their lives for the American Association of Retired People.

He found, not surprisingly, that nearly 70 percent are women. Of these, 82 percent are widowed, compared with 44 percent of the men. Because of the conditions that existed when they were growing up, the oldest old are poorly educated compared with young people today, most of whom finish high school. The average person now 85 years and older only completed the eighth grade.

Only one-quarter of these older citizens are in hospitals or institutions such as nursing homes, and more than half live in their own homes. Just 30 percent live by themselves. More than a third live with a spouse or with their children. There are certainly those who aren't doing well—one in six have incomes below the poverty level—but many more are relatively well-off. The mean household income for the group, Longino says, was more than $20,000 in 1985.

What of the quality of life? "In studying this group, we have to be aware of youth creep," he says. "The old are getting younger all the time." This feeling is confirmed by a report released late last year by the National Institute on Aging. The NIA report included three studies of people older than 65 conducted in two counties in Iowa, in East Boston, Massachusetts, and in New Haven, Connecticut. There are large regional differences between the groups, of course, and they aren't a cross-section of older people in the nation as a whole. But in all three places, most of those older than 85 seem to be leading fulfilling lives.

Most socialize in a variety of ways. In Iowa, more than half say they go to religious services at least once a week and the same percentage say they belong to some type of professional, social, church-related or recreational group. More than three-quarters see at least one or two children once a month and almost that many see other close relatives that often.

As you would expect, many of the oldest old suffer from disabilities and serious health problems. At least a quarter of those who responded have been in a hospital overnight in the past year and at least 8 percent have had heart attacks or have diabetes. In Iowa and New Haven, more than 13 percent of the oldest old had cancer, while in East Boston the rate was lower (between 7 percent and 8 percent). Significant numbers of the oldest old have suffered serious injury from falls. Other common health problems for this group are high blood pressure and urinary incontinence. However, epidemiologist Adrian Ostfeld, who directed the survey in New Haven, notes that "most of the disability was temporary."

Longino has found that almost 10 percent of the oldest old live alone with a disability that prevents them from using public transportation. This means that they are "isolated from the daily hands-on care of others," he says. "Even so, there are a surprising number of the oldest old who don't need much in the way of medical care. They're the survivors.

"I think we have to agree that the oldest old is, as a group, remarkably diverse," Longino says. "Just as it is unfair to say that those older than 85 are all miserable, it's not fair to say that they all lead wonderful lives, either." —*Jeff Meer*

---

ed on the metaphorical, moral or social meaning of the text. They didn't try to duplicate the fable's exact words, the way younger people did. As psychologists Nancy Datan, Dean Rodeheaver and Fergus Hughes of the University of Wisconsin have described their findings, "while [some people assume] that old and young are equally competent, we might better assume that they are differently competent."

John Horn, director of the Adult Development and Aging program at the University of Southern California, suggests that studies of Alzheimer's disease, a devastating progressive mental deterioration experienced by an estimated 5 percent to 15 percent of those older than 65, may eventually help explain some of the differences in thinking abilities of older people. "Alzheimer's, in some ways, may represent the normal process of aging, only speeded up," he says. (To see how your ideas about Alzheimer's square with the facts, see "Alzheimer's Quiz" and "Alzheimer's Answers," this article.)

6. MIDDLE AND LATE ADULTHOOD

Generalities are always suspect, but one generalization about old age seems solid: It is a different experience for men and women. Longevity is one important reason. Women in the United States live seven to eight years longer, on the average, than do men. This simple fact has many ramifications, as sociologist Gunhild Hagestad explained in *Our Aging Society*.

For one thing, since the world of the very old is disproportionately a world of women, men and women spend their later years differently. "Most older women are widows living alone; most older men live with their wives...among individuals over the age of 75, two-thirds of the men are living with a spouse, while less than one-fifth of the women are."

The difference in longevity also means that among older people, remarriage is a male prerogative. After 65, for example, men remarry at a rate eight times that of women. This is partly a matter of the scarcity of men and partly a matter of culture—even late in life, men tend to marry younger women. It is also a matter of education and finances, which Hagestad explains, "operate quite differently in shaping remarriage probabilities among men and women. The more resources the woman has available (measured in education and income), the less likely she is to remarry. For men, the trend is reversed."

The economic situations of elderly men and women also differ considerably. Lou Glasse, president of the Older Women's League in Washington, D.C., points out that most of these women were housewives who worked at paid jobs sporadically, if at all. "That means their Social Security benefits are lower than men's, they are not likely to have pensions and they are less likely to have been able to save the kind of money that would protect them from poverty during their older years."

Although we often think of elderly men and women as living in nursing homes or retirement communities, the facts are quite different. Only about 5 percent are in nursing homes and perhaps an equal number live in some kind of age-segregated housing. Most people older than 65 live in their own houses or apartments.

We also think of older people as living alone. According to the Census Bureau, this is true of 15 percent of the men and 41 percent of the women. Earlier this year, a survey done by Louis Harris & Associates revealed that 28 percent of elderly people living alone have annual incomes below $5,100, the federal poverty line. Despite this, they were four times as likely to give financial help to their children as to receive it from them.

In addition, fewer than 1 percent of the old people said they would prefer living with their children. Psychiatrist Robert N. Butler, chairman of the Commonwealth Fund's Commission on Elderly People Living Alone, which sponsored the report, noted that these findings dispute the "popular portrait of an elderly, dependent parent financially draining their middle-aged children."

There is often another kind of drain, however, one of time and effort. The Travelers Insurance Company recently surveyed more than 700 of its employees on this issue. Of those at least 30 years old, 28 percent said they directly care for an older relative in some way—taking that person to the doctor, making telephone calls, handling finances or running errands—for an average of 10 hours a week. Women, who are more often caregivers, spent an average of 16 hours, and men five hours, per week. One group, 8 percent of the sample, spent a heroic 35 hours per week, the equivalent of a second job, providing such care. "That adds up to an awful lot of time away from other things," psychologist Beal Lowe says, "and the stresses these people face are enormous."

Lowe, working with Sherman-Lank Communications in Kensington, Maryland, has formed "Caring for Caregivers," a group of professionals devoted to providing services, information and support to those who care for older relatives. "It can be a great shock to some people who have planned the perfect retirement," he says, "only to realize that your chronically ill mother suddenly needs daily attention."

Researchers who have studied the housing needs of older people predictably disagree on many things, but most agree on two points: We need a variety of individual and group living arrangements to meet the varying interests, income and abilities of people older than 65; and the arrangements should be flexible enough that the elderly can stay in the same locale as their needs and abilities change. Many studies have documented the fact that moving itself can be stressful and even fatal to old people, particularly if they have little or no influence over when and where they move.

> *AMONG OLDER PEOPLE TODAY, REMARRIAGE IS STILL LARGELY A MALE PREROGATIVE, DUE TO THE SEX DIFFERENCE IN LONGEVITY.*

This matter of control is important, but more complicated than it seemed at first. Psychologist Judith Rodin and others have demonstrated that people in nursing homes are happier, more alert and live longer if they are allowed to take responsibility for their lives in some way, even in something as simple as choosing a plant for their room, taking care of a bird feeder, selecting the night to attend a movie.

Rodin warns that while control is generally beneficial, the effect depends on the individuals involved. For some, personal control brings with it demands in the form of time, effort and the risk of failure. They may blame themselves if they get sick or something else goes wrong. The challenge, Rodin wrote, is to "provide but not impose opportunities. ... The need for self-determination, it must be remembered, also calls for the opportunity to choose not to exercise control. ..."

An ancient Greek myth tells how the Goddess of Dawn fell in love with a mortal and convinced Jupiter to grant him immortality. Unfortunately, she forgot to have youth included in the deal, so he gradually grew older and older. "At length," the story concludes, "he lost the power of using his limbs, and then she shut him up in his chamber, whence his feeble voice might at times be heard. Finally she turned him into a grasshopper."

The fears and misunderstandings of age expressed in this 3,000-year-old myth persist today, despite all the positive things we have learned in recent years about life after 65. We don't turn older people into grasshoppers or shut them out of sight, but too often we move them firmly out of the mainstream of life.

In a speech at the celebration of Harvard

> *If I had known when I was 21 that I should be as happy as I am now, I should have been sincerely shocked. They promised me wormwood and the funeral raven.*
> —Christopher Isherwood, letter at age 70.

University's 350th anniversary last September, political scientist Robert Binstock decried what he called The Spectre of the Aging Society: "the economic burdens of population aging; moral dilemmas posed by the allocation of health resources on the basis of age; labor market competition between older and younger workers within the contexts of age discrimination laws; seniority practices, rapid technologi-

ALZHEIMER'S QUIZ

Alzheimer's disease, named for German neurologist Alois Alzheimer, is much in the news these days. But how much do you really know about the disorder? Political scientist Neal B. Cutler of the Andrus Gerontology Center gave the following questions to a 1,500-person cross section of people older than 45 in the United States in November 1985. To compare your answers with theirs and with the correct answers, turn to the next page.

		True	False	Don't know
1.	Alzheimer's disease can be contagious.			
2.	A person will almost certainly get Alzheimer's if they just live long enough.			
3.	Alzheimer's disease is a form of insanity.			
4.	Alzheimer's disease is a normal part of getting older, like gray hair or wrinkles.			
5.	There is no cure for Alzheimer's disease at present.			
6.	A person who has Alzheimer's disease will experience both mental and physical decline.			
7.	The primary symptom of Alzheimer's disease is memory loss.			
8.	Among persons older than age 75, forgetfulness most likely indicates the beginning of Alzheimer's disease.			
9.	When the husband or wife of an older person dies, the surviving spouse may suffer from a kind of depression that looks like Alzheimer's disease.			
10.	Stuttering is an inevitable part of Alzheimer's disease.			
11.	An older man is more likely to develop Alzheimer's disease than an older woman.			
12.	Alzheimer's disease is usually fatal.			
13.	The vast majority of persons suffering from Alzheimer's disease live in nursing homes.			
14.	Aluminum has been identified as a significant cause of Alzheimer's disease.			
15.	Alzheimer's disease can be diagnosed by a blood test.			
16.	Nursing-home expenses for Alzheimer's disease patients are covered by Medicare.			
17.	Medicine taken for high blood pressure can cause symptoms that look like Alzheimer's disease.			

6. MIDDLE AND LATE ADULTHOOD

Alzheimer's Answers — National Sample

#	Answer	True	False	Don't know
1.	**False.** There is no evidence that Alzheimer's is contagious, but given the concern and confusion about AIDS, it is encouraging that nearly everyone knows this fact about Alzheimer's.	3%	83%	14%
2.	**False.** Alzheimer's is associated with old age, but it is a disease and not the inevitable consequence of aging.	9	80	11
3.	**False.** Alzheimer's is a disease of the brain, but it is not a form of insanity. The fact that most people understand the distinction contrasts with the results of public-opinion studies concerning epilepsy that were done 35 years ago. At that time, almost half of the public thought that epilepsy, another disease of the brain, was a form of insanity.	7	78	15
4.	**False.** Again, most of the public knows that Alzheimer's is not an inevitable part of aging.	10	77	13
5.	**True.** Despite announcements of "breakthroughs," biomedical research is in the early laboratory and experimental stages and there is no known cure for the disease.	75	8	17
6.	**True.** Memory and cognitive decline are characteristic of the earlier stages of Alzheimer's disease, but physical decline follows in the later stages.	74	10	16
7.	**True.** Most people know that this is the earliest sign of Alzheimer's disease.	62	19	19
8.	**False.** Most people also know that while Alzheimer's produces memory loss, memory loss may have some other cause.	16	61	23
9.	**True.** This question, like number 8, measures how well people recognize that other problems can mirror Alzheimer's symptoms. This is crucial because many of these other problems are treatable. In particular, depression can cause disorientation that looks like Alzheimer's.	49	20	30
10.	**False.** Stuttering has never been linked to Alzheimer's. The question was designed to measure how willing people were to attribute virtually anything to a devastating disease.	12	46	42
11.	**False.** Apart from age, research has not uncovered any reliable demographic or ethnic patterns. While there are more older women than men, both sexes are equally likely to get Alzheimer's.	15	45	40
12.	**True.** Alzheimer's produces mental and physical decline that is eventually fatal, although the progression varies greatly among individuals.	40	33	27
13.	**False.** The early and middle stages of the disease usually do not require institutional care. Only a small percentage of those with the disease live in nursing homes.	37	40	23
14.	**False.** There is no evidence that using aluminum cooking utensils, pots or foil causes Alzheimer's, although aluminum compounds have been found in the brain tissue of many Alzheimer's patients. They may simply be side effects of the disease.	8	25	66
15.	**False.** At present there is no definitive blood test that can determine with certainty that a patient has Alzheimer's disease. Accurate diagnosis is possible only upon autopsy. Recent studies suggest that genetic or blood testing may be able to identify Alzheimer's, but more research with humans is needed.	12	24	64
16.	**False.** Medicare generally pays only for short-term nursing-home care subsequent to hospitalization and not for long-term care. Medicaid can pay for long-term nursing-home care, but since it is a state-directed program for the medically indigent, coverage for Alzheimer's patients depends upon state regulations and on the income of the patient and family.	16	23	61
17.	**True.** As mentioned earlier, many medical problems have Alzheimer's-like symptoms and most of these other causes are treatable. Considering how much medicine older people take, it is unfortunate that so few people know that medications such as those used to treat high blood pressure can cause these symptoms.	20	19	61

cal change; and a politics of conflict between age groups."

Binstock, a professor at Case Western Reserve School of Medicine, pointed out that these inaccurate perceptions express an underlying ageism, "the attribution of these same characteristics and status to an artificially homogenized group labeled 'the aged.'"

Ironically, much ageism is based on compassion rather than ill will. To protect older workers from layoffs, for example, unions fought hard for job security based on seniority. To win it, they accepted mandatory retirement, a limitation that now penalizes older workers and deprives our society of their experience.

A few companies have taken special steps to utilize this valuable pool of older workers. The Travelers companies, for example, set up a job

GREAT EXPECTATIONS

If you were born in 1920 and are a . . .

	. . . white man	. . white woman
your life expectancy was . . .		
at birth	54.4 years	55.6 years
at age 40	71.7	77.1
at age 62	78.5	83.2

If you were born in 1940 and are a . . .

	. . . white man	. . . white woman
your life expectancy was . . .		
at birth	62.1 years	66.6 years
at age 20	70.3	76.3
at age 42	74.7	80.7

If you were born in 1960 and are a . . .

	. . . white man	. . . white woman
your life expectancy was . . .		
at birth	67.4 years	74.1 years
at age 22	73.2	80.0

SOURCE: U.S. NATIONAL CENTER FOR HEALTH STATISTICS

bank that is open to its own retired employees as well as those of other companies. According to Howard E. Johnson, a senior vice president, the company employs about 175 formerly retired men and women a week. He estimates that the program is saving Travelers $1 million a year in temporary-hire fees alone.

While mandatory retirement is only one example of ageism, it is particularly important because we usually think of contributions to society in economic terms. Malcolm H. Morrison, an authority on retirement and age discrimination in employment for the Social Security Administration, points out that once the idea of retirement at a certain fixed age was accepted, "the old became defined as a dependent group in society, a group whose members could not and should not work, and who needed economic and social assistance that the younger working population was obligated to provide."

We need to replace this stereotype with the more realistic understanding that older people are and should be productive members of society, capable of assuming greater responsibility for themselves and others. What researchers have learned about the strengths and abilities of older people should help us turn this ideal of an active, useful life after 65 into a working reality.

WHY DO WE AGE?

Ken Flieger
Ken Flieger is a free-lance writer in Washington, D.C.

In December 1987, Anna Williams died in a nursing home in Wales. At age 114, she was believed to be the oldest person on Earth. Scientists think that 115 to 120 years is probably the upper limit of human longevity. But why should that be? Why should the human body give out after 70, 80, or even 115 years? Why are older people more susceptible to disease, more inclined to have impaired vision and hearing, and likely to lose some of the physical and mental capacity they once enjoyed?

There are no fully satisfactory answers to these questions. No one knows exactly what the process of aging is or why it runs a different course in different people. Nor does anyone know how to increase human longevity, despite the often fraudulent and sometimes dangerous claims of the "life extension" hucksters and others who traffic in the fears and ills of the elderly.

What is known is that our bodies undergo more or less predictable changes with advancing age. Some—like hair turning gray or falling out, wrinkled skin, and reduced physical and mental vigor—are obvious. Others may not be visible, but they are happening nonetheless. Blood pressure usually rises with age; the ability to metabolize sugar decreases. There is a general decrease in lean body mass (primarily muscle and bone) as opposed to fat so that, although a person may weigh the same at age 70 as he or she did at 20, the body composition is considerably different. Medical tests will show that an older person's heart pumps less efficiently, the kidneys may not work as well, lung function is diminished, and bone density is reduced.

Gerontologists (scientists who study aging) stress, however, that these "common" age-related changes don't all happen to everyone or at predictable stages, and most definitely cannot provide a reliable picture of the individual's well-being. As one scientist observed, knowing that a 75-year-old man has diabetes and a history of heart disease doesn't provide a clue as to whether he's sitting in a nursing home or on the bench of the Supreme Court.

BY MISTAKE OR DESIGN?

Research is, however, beginning to piece together bits of the puzzle of aging. Scientists are exploring numerous theories that may help explain the changes that come with advancing age. "Error" theories speculate that with advancing age, we become less able to repair damage caused by internal malfunctions or external assaults to the body from, for example, pollutants, viruses, and cosmic and solar radiation. Much like an automobile, the human machine can sustain just so much damage until effective repair is no longer possible and it ceases to run.

"Program" theories of aging, on the other hand, suggest that an internal clock starts ticking at conception and is programmed to run just so long and no longer. These theories hold that genes carry specific instructions that facilitate not only growth and maturation, but decline and death as well.

Whichever broad concept turns out to offer a more satisfactory explanation of what aging is all about—and most authorities doubt that any single theory will account for all the complexities of the aging process—it seems clear that human beings age as a result of events that take place within cells. One "error" hypothesis, the "wear and tear" theory, suggests that cells gradually lose the ability to repair damaged DNA, the substance that passes genetic information from one cell to the next. As a consequence, cells become less efficient in carrying out vital functions, such as making proteins, and eventually they die. Other error theories relate aging to metabolic rate, the rate at which cells convert nutrients into the energy they need to live and reproduce. This view of aging implies that the faster an organism lives, the quicker it dies, which might explain why animals with short

46. Why Do We Age?

lifespans (compared to man's) usually have much higher metabolic rates than we do.

Studies done in the mid-1930s support this "rate of living" theory of aging. Investigators at Cornell University found that newly weaned rats fed a diet severely restricted in calories but nutritionally adequate lived extraordinarily long lives. Later studies demonstrated the same thing in mice and showed that undernutrition, as it is called, could lengthen lifespan even if it wasn't started until the animals had reached adulthood.

Caloric restriction is the only technique that has been repeatedly shown to alter the rate of aging of laboratory animals. Scientists speculate that reducing caloric intake may slow the animals' metabolism and thus reduce the rate of damage to cells or to DNA. There is little evidence to suggest that undernutrition will increase the human lifespan, although comprehensive studies have not been done.

Another theory attributes aging to damage caused by protein "cross-linking." With the passage of time, more and more protein molecules in cells and tissues become chemically bound to one another in ways that interfere with their normal functioning. When this happens to collagen—a protein that supports cells and tissues and is especially abundant in bone, cartilage, and tendons—it tends to become more rigid. The effect is all too apparent in the wrinkled skin that is virtually a cardinal sign of growing older. But whether "cross-linking" is at the root of the aging process is far from certain.

TICKING AWAY

Unlike error theories of aging, internal clock theories don't try to explain the process in terms of one particular biologic mechanism or defect. Instead, they view aging in terms of a genetically determined program. Support for this concept can be seen throughout our lives: the coming of adult teeth, and the onset of puberty and menopause all seem to confirm that human growth and development march to a built-in drummer. But an important scientific finding in the 1960s added even more powerful support to the evidence of simple observation.

A quarter century ago, Drs. Leonard Hayflick and Paul Moorehead at the Wistar Institute in Philadelphia found that embryonic human cells in tissue culture have an inherent capacity to divide only about 50 times. Once this limit is reached, cell division stops, and the cell line dies. If cultures are begun with cells taken from an adult rather than an embryo, the number of divisions they undergo is fewer than 50. The only exception seems to be abnormal cancer cells that can go on dividing indefinitely. (Human cells divide at widely different rates: tumor cells in tissue culture may divide as fast as once in every 18 hours; nerve cells, on the other hand, may never divide once nerve tissue has been formed.)

Why cells lose the ability to reproduce is not known, but apparently some element in the genetic code dictates how many times a cell can divide and then brings cell division to a halt. It's probably an oversimplification to say that the lifespan of an organism as complex as a human being is determined by how many times its cells can divide. Yet the "Hayflick phenomenon" rules out the view of earlier scientists working with tissue culture that, given the proper conditions, cells, tissues, organs and perhaps people could survive indefinitely.

If there is a genetically programmed clock at the heart of the aging process, how are its instructions communicated to cells and tissues? Both the endocrine (hormone) system and the immune system have been proposed as the medium through which those instructions are transmitted and carried out.

Most people have a fairly good notion of the role of hormones in sexual development and reproduction. Scientists, however, are learning a good deal more about hormones and aging. For example, we now understand that hormones in premenopausal women seem to protect against cardiovascular diseases and that the decline in estrogen levels after menopause is associated with bone loss and osteoporosis.

Another hormone, dehydroepiandrosterone (DHEA), produced by the adrenal glands, is present in extremely high levels in the blood of young adults and falls sharply with age. This has led to speculation that this hormone may play a role in aging. Laboratory studies seem to support this hunch. Mice given the hormone have lower breast cancer rates, increased survival, and a delayed loss of immune functions. They are more active and have a more youthful appearance (glossier coats) than untreated animals. But investigators have also noted that these mice don't gain weight as rapidly as untreated animals. Hence, as in other animal experiments, caloric restriction may be behind the apparent anti-aging effect of this hormone.

Products claiming to contain DHEA have been promoted for weight loss and life extension, but in 1985 FDA ordered a halt to their marketing because they have never been reviewed for safety and effectiveness. In fact, the agency warned at the time, risks from long-term use of such products were unknown. Furthermore, some of the products may have been made from human urine, and scientists had no idea what effect reintroducing into the body this bodily excess might have.

Such unproven products aside, the role of hormones is a busy and legitimate avenue of aging research.

Whether or not the immune system translates genetic information into programmed aging, scientists know that it is closely associated with the aging process. The thymus gland, located in the upper chest, is a major component of the immune system. It is present at birth, reaches its maximum size during adolescence, and is barely visible by age 50. The immune system itself runs along a somewhat parallel track. It is most efficient during childhood and young adulthood, but with advancing age, it is less able to recognize and counteract foreign substances, such as viruses, that enter the body. Nor is the aging immune system as efficient in recognizing the difference between normal and abnormal cellular components—in other words, between "self" and "non-self."

Some authorities believe that the gradual decline in the efficiency and effectiveness of the immune system explains not just the increased susceptibility to infection in older people, but also their increased rate of chronic diseases, such as cancer. They speculate that the body becomes progressively less able to recognize and destroy malignant cells, which then have the chance to become established and develop into tumors and other cancers. Similarly, the decline of the immune system may account for the so-called "autoimmune" diseases, such as rheumatoid arthritis, in which the body attacks its own cells. It is possible that a declining immune system may be associated with other factors related to the aging process, such as loss of the ability to repair damaged DNA.

DEFYING NATURE?

Living to a ripe old age has a certain appeal to most people, but it may in fact not be part of nature's plan. Wild animals don't survive much beyond the age of peak maturity. They tend to be killed

6. MIDDLE AND LATE ADULTHOOD

The Aging of America

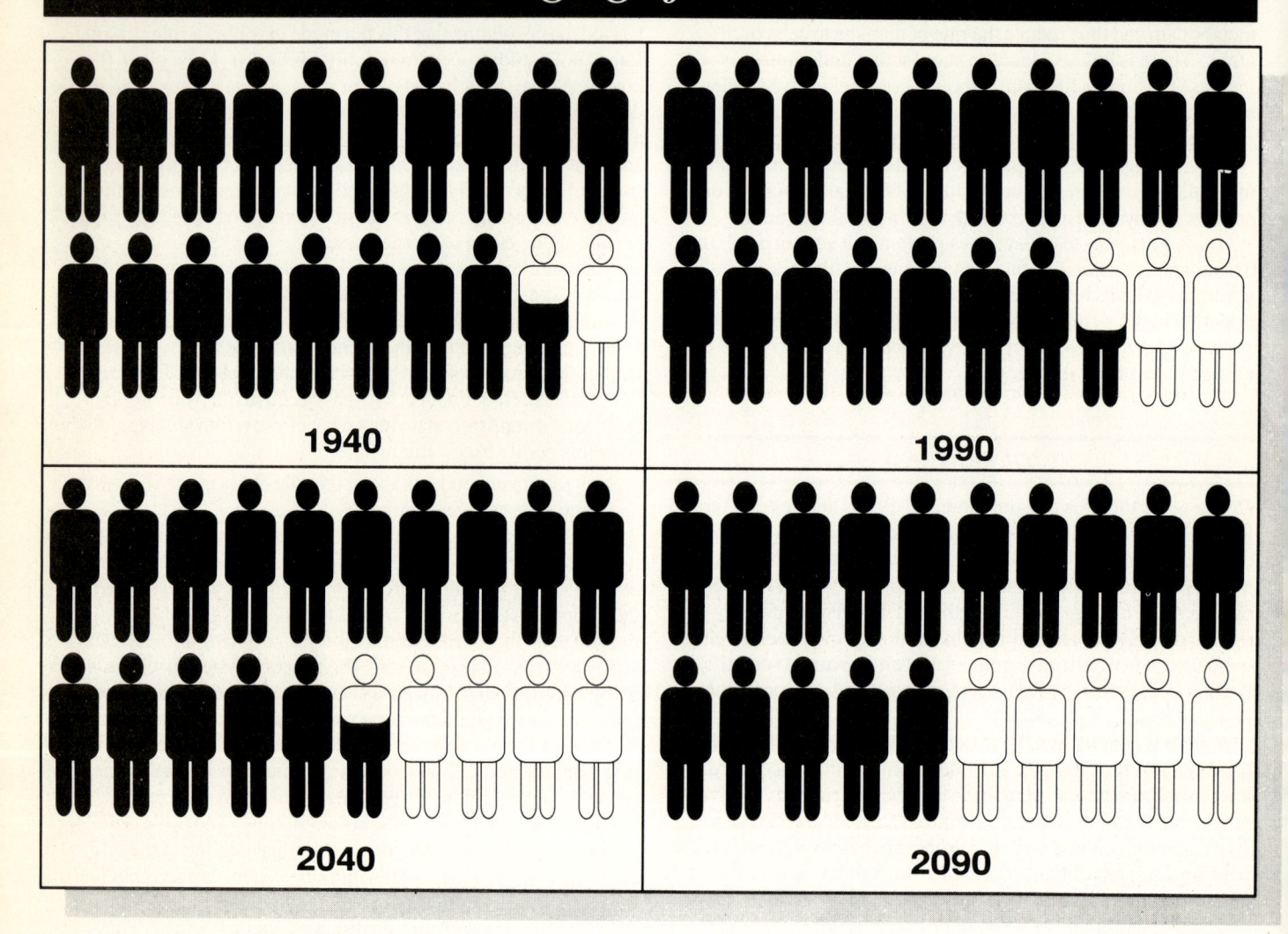

1940

1990

2040

2090

In 1940, 6.8 percent of the population was 65 or older. By 1990, 12.7 percent of us will have reached age 65; by 2040, 21.7 percent; and a century from now, nearly one out of four Americans will be 65 and older. An estimated 25,000 Americans now living are 100 years old or older.

Source: U.S. Bureau of the Census.

46. Why Do We Age?

off by predators once their physical abilities start to decline or they become sick. For humans living in primitive societies, that peak comes at around age 30, at about the age when childbearing and child-rearing come to an end.

Whether or not we are defying Mother Nature, Americans are undoubtedly living longer, thanks in part to improved health care, more responsive social and public assistance programs, and better understanding of the importance of nutrition, exercise, and personal interactions in sustaining vitality in advancing age.

At the same time, the burden of illness and disability among older people is tragically heavy, and as more and more of us live to advanced years, the care of the elderly will impose an increasing demand on the nation's resources.

Much of the puzzle of aging remains to be solved. But the field of aging research, one of the youngest branches of the health sciences, has added a wealth of new information in just the last few decades. Spearheaded by the National Institute on Aging, one of the National Institutes of Health in Bethesda, Md., aging studies are helping to dispel some myths—for example, that aging progresses rapidly as a result of an overwhelming event—and are enabling physicians to do a better job of caring for older patients. Knowing more precisely how older people absorb and excrete drugs can lead to better prescribing information for doctors who care for elderly patients. FDA is actively encouraging drug companies and others to make sure that, when appropriate, clinical studies of new investigational drugs do not arbitrarily exclude elderly persons, who may react to the drug very differently from younger patients.

Some people may be saddened by the realization that science doesn't hold the key to immortality (though the thought of a world in which no one ever died is rather alarming). But the goal of aging research is at once more realistic and more humane. It is to help people achieve the longest possible life with the least possible disability. Science just might be able to reach that goal.

Common Ailments in Older Americans

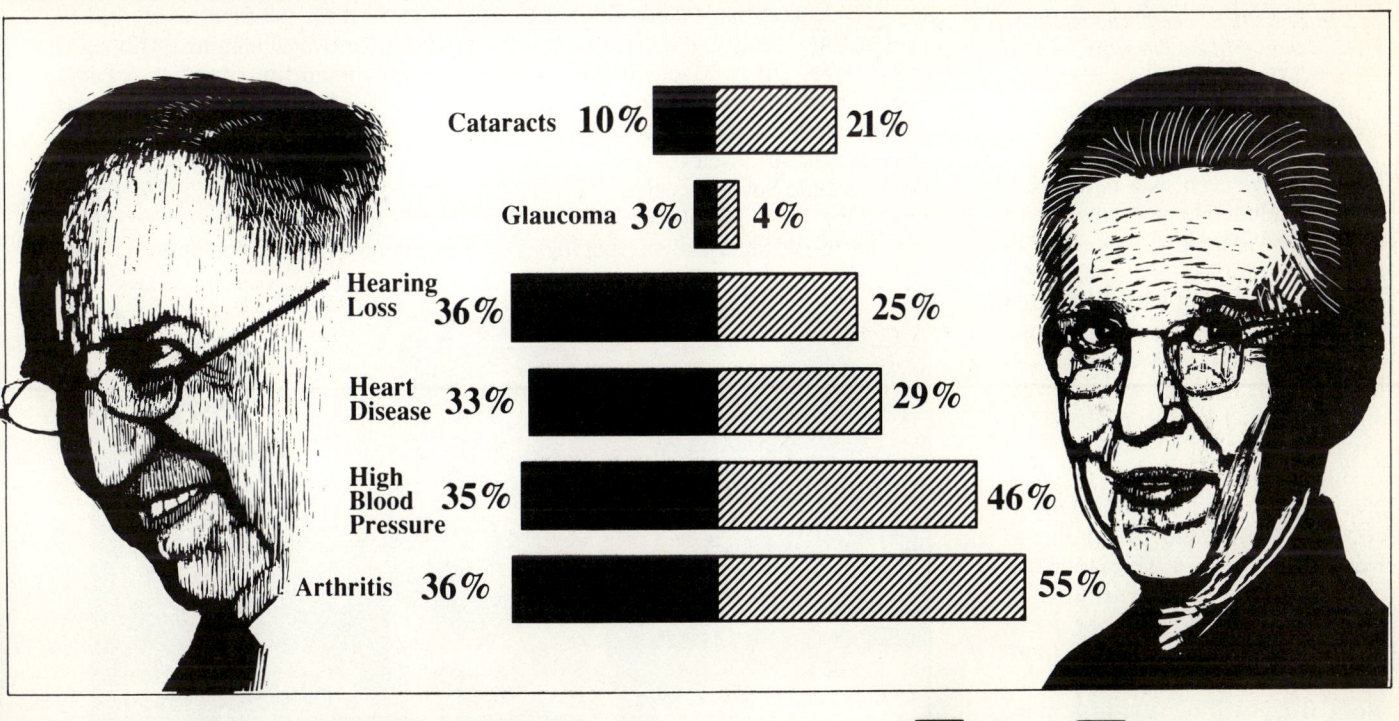

Ailment	Men	Women
Cataracts	10%	21%
Glaucoma	3%	4%
Hearing Loss	36%	25%
Heart Disease	33%	29%
High Blood Pressure	35%	46%
Arthritis	36%	55%

Percent of American men and women aged 65 or older who suffer from one or more of six common ailments.
Source: National Institute on Aging.

Article 47

A VITAL LONG LIFE

NEW TREATMENTS FOR COMMON AGING AILMENTS

Medical science enables us to live to a healthy ripe old age —and enjoy it.

Evelyn B. Kelly

Evelyn B. Kelly is vice president of the Florida chapter of the American Medical Writers' Association and a consultant on psychological and gerontological concerns.

Bill, seventy-seven, is a new model of older adult. He manages a vast network of athletic camps, flies to the World Series, and spends his spare time working with his church and the Gideons. He walks two miles a day and is very careful about his diet. Bill lives his life with zest and vigor and still contributes to society.

Bill is prototypical of the new elder culture predicted for the next cohort of older adults. According to Ken Dychtwald, a gerontological consultant, the next generation of older adults will be healthier, more mobile, better educated, and more politically astute than today's seniors. They will be part of a more powerful and energetic elder culture.

Like the old gray mare, aging is definitely not what it used to be. Time was when people believed they should eat, drink, and be merry—for tomorrow, they'd retire to their rockers. Invariably, it was held, the passing years meant steady mental and physical decline until one died from "old age." While many still cling to these myths, researchers marvel at the human potential to extend physical and intellectual capacities in later life.

Many factors have contributed to this greatly improved forecast for the aging. Medical advances of the last decade have made it possible for present and future generations to live longer lives. Demonstrably, healthier life-styles have played a major role in extending life span. The effects of disease, abuse, or disuse should never be called "normal aging." Put simply, people do not die of old age, but of specific conditions, over which they may have some control.

The sheer numbers of older adults are overwhelming. Although the maximum life span has not increased, the number of persons sixty-five and older has increased from 4 percent of the population in 1900 to 12 percent in 1985. Even more important, the fastest-growing segment of the population is the

Physical therapy, including exercise and weight reduction, is an important way to manage arthritis.

group eighty-five and older.

In 1900, only 25 percent of deaths occurred in the group over sixty-five. Advances in preventing childhood diseases had pushed the number to almost 70 percent by 1980. Impressive gains have been made in treating and preventing the four leading causes of death in the sixty-five plus population: heart disease, strokes, cancer, and pneumonia.

But this bounty of years has brought mixed blessings. As our aged population continues to grow, survivors may encounter nonfatal disorders, such as arthritis, diabetes, and dementia. Specified in the blueprint for a model healthy old age are education and healthy life-style choices.

Heart disease and strokes: silent stalkers

A diagnosis of high blood pressure surprises most people, who find the knowledge that they are potential candidates for stroke or heart failure traumatic. About 75 percent of hypertensives are over forty, but their condition is seldom diagnosed until they are fifty or sixty. Associated with excess weight, and found more commonly in black people than white and men than women, hypertension has been called the silent stalker.

According to Dr. Harvey Simon of Harvard Medical School, hearts may wear down a little, but they probably do not wear out. He adds that disease and disuse, rather than age, account for most of the deterioration. The heart's ability to pump blood declines about 8 percent each decade after adulthood, and blood pressure increases as fatty deposits clog the arteries. By middle age, the opening of the coronary arteries is 29 percent narrower than in the twenties.

Life-style changes, drugs, or a combination can reduce blood pressure. With more than fifty blood-pressure drugs on the market, if one produces side effects, another can be prescribed. A number of drugs are still being tested. In August 1987, the U.S. Food and Drug Administration (FDA) approved the use of lovastatin, a cholesterol-reducing drug, but recommended it only for patients who do not respond to diet and exercise.

Heart attacks are caused by a blood clot in one of the coronary arteries. Most deaths caused by heart attacks occur within minutes or a few hours. Until recently, doctors could do little to treat an attack and prevent damage to the heart muscle. Two recent developments, however, mean good news for older adults. A new class of drugs called "clotbusters" may be injected into the veins to dissolve clots, and the FDA has recently approved streptokinase, a blood thinner, and another "clotbuster," TPA—tissue plasminogen activator.

Other weapons against heart attacks are balloon angioplasty—a procedure that opens clogged arteries with small balloonlike structures—and laser angioplasty, an experimental method where light vaporizes plaque deposits on the arteries.

Diseases of the cerebral blood vessels causing stroke rank high among the disabilities of the elderly. Reducing blood pressure can cut the risk of stroke in half. In stroke, the blood flow to part of the brain stops. The effect is similar to that of a heart attack; without blood, the tissue has no oxygen and dies.

Although about half of the elderly are hypertensive, physicians do not agree on their treatment. In a recent American Medical Association report, Richard Davidson and George Caranosos of the University of Florida announced that treating elderly hypertensives appears to be effective in stroke prevention. Older adults are not more susceptible to the side effects of antihypertensive drugs than other groups.

Risk factors leading to heart disease include heredity, diabetes, high-fat diets, high blood pressure, and smoking. Changes in life-style dramatically reduce the risk of these silent stalkers.

Cancer and older adults

There's no question about it: Cancer occurs more frequently with increasing age. The passing of years may lengthen exposure to carcinogens, alter host immunity or chro-

The next generation of older adults will be healthier. Aging may occur from disuse, not from disease. This farmer and his wife stay active. He is 100 and she is 95 years old.

6. MIDDLE AND LATE ADULTHOOD

The Fountains of Youth

In 1513, Ponce de Leon combed Florida in search of mythical waters that would prevent aging. Today, Dr. J. Michael McGinnis, who launched the Healthy Older People program, has found that up to 50 percent of all premature mortality can be related to life-style habits, and that older adults are willing and able to make life-style changes to improve their health. Even in late age, behavioral changes, especially in the realms of exercise and nutrition, can produce health benefits.

Every cell in our bodies is dependent on oxygen. We can go weeks without food and days without water, but only a few minutes without oxygen. The heart pumps the blood that carries oxygen throughout the body, so anything that improves circulation is beneficial. Exercise fulfills this important function—and more!

Exercise

The symptoms of "old age"—no energy, stiff joints, poor circulation—may really be the signs of inactivity. You don't stop exercising because you are becoming old; you become old because you stop exercising. The same conditions attributed to aging can be induced in young people who sit around doing nothing. Aging may occur not from dis*ease* but from dis*use*.

● The relationship between inactivity and circulatory problems is immutable. The less active you are, the more chances you run of developing heart problems and high blood pressure.

● Exercise and the lowered risk of heart disease are connected with chemicals in blood cholesterol called lipoproteins. The so-called good lipoproteins—the HDLs (high-density lipoproteins)—are found in the blood of those on solid exercise programs. The bad lipoproteins—the LDLs (low-density lipoproteins)—outnumber the good in those who are inactive. Letter carriers—who walk all day—have fewer heart attacks than sedentary post office clerks.

● It is never too late to start exercising. Studies at the University of Toronto have shown that men and women over the age of sixty who began to exercise regularly were found to have achieved the fitness level of persons ten to twenty years younger in only seven weeks.

With exercise alone, cholesterol and triglyceride levels drop markedly. Appetite is suppressed and digestion improves. When people diet to lose weight, their metabolism slows down, and they reach a plateau of weight loss. With exercise, however, the rate of body metabolism increases, and body fat is absorbed. Because carbohydrate and complex-sugar metabolism is improved, diabetics who exercise can maintain lower insulin levels.

Nutrition

It sounds obvious, but a balanced diet is still the most important overall eating strategy. That means including something from each of the four food groups daily—grains; fruits and vegetables; meat, fish, and poultry; and dairy products. The emphasis today is on low-fat, high-fiber foods. By eating more fiber, you lower your cholesterol level and lessen the chances of getting heart disease.

The relationship between diet and many chronic conditions such as heart disease, stroke, hypertension, and cancer has been proven. However, Healthy Older People revealed that older adults are confused about what constitutes good nutrition. They seemed very knowledgeable about what *not* to eat, but were unable to describe the elements of a balanced diet.

A focal point of controversy in nutrition is the official U.S. recommended dietary allowances, or RDAs. One problem is that RDAs are established for only two adult groups—those under fifty and those over fifty. Certainly, the physiological differences between fifty-five- and eighty-five-year-olds reflect different dietary needs that must be addressed in education programs.

Also, nutrition and exercise act as natural tranquilizers that relieve tension and promote mental well-being. While exercise and nutrition can't reduce health risks to zero, through our behavior and life-style choices we can indeed create an inner fountain of youth.

---

mosomal linkage, or increase exposure to ontogenic viruses.

But the very group that needs screening is often disregarded because of preconceived notions about cancer in older adults. Some think older adults are too fragile to tolerate proper treatment. The American Cancer Society emphasizes that early detection is important at all ages.

In many parts of the body, such as the skin, the lungs, and digestive tract, the chances of cancer rise steadily with age. Prostate cancer is even more directly connected to age. Some cancers, theoretically preventable, are caused by habits or environmental factors, such as smoking or exposure to industrial chemicals. The chart on the next page shows the screening recommendations of the American Cancer Society.

Pneumonia: friend of the old?

Friend of the old: That's what Sir William Osler called pneumonia in 1912. He stated that a death from pneumonia is short, relatively painless, and peaceful.

Today, few would call pnuemonia a friend: We have little admiration for the No. 1 infectious killer of older adults.

And what's more—admissions to nursing homes and hospitals increase the chance of encounter with this "enemy." The yearly incidence of pneumonia, 20 to 40 cases per thousand, sharply rises with institutionalization to about 250 cases per thousand.

Hospitals are associated with an

increased risk for lower respiratory infections. Nosocomial pneumonia (the name given to infections acquired in a hospital) is involved in about 15 percent of pneumonia cases; the attack rate increases with age. In addition, nosocomial pneumonias are stubborn and hard to treat.

Pneumonia in older adults is a diagnostic and therapeutic nightmare. In many cases, pneumonia may not be suspected because the common signs of respiratory disease are absent. The ill person suspected of having pneumonia is often treated with a "broad-spectrum" antibiotic to kill different kinds of bacteria. This shotgun approach may or may not work. Despite the availability of potent antibiotics, mortality due to pneumonia in the aged remains substantial and comes with a high price tag. Over $550 million is spent each year in hospital care alone for elderly adults with pneumonia.

Education for prevention is imperative. Despite a well-recognized association between outbreaks of influenza and deaths from pneumonia, only about 20 percent of high-risk persons are vaccinated against the flu each year. Lederle, a pharmaceutical company, is marketing a vaccine that protects against twenty-three types of pneumococcal bacteria.

Future pharmaceutical development will help in the prevention and treatment of pneumonia in older people. Possible breakthroughs include vaccines against more bacterial strains, more potent oral antibiotics, and methods to keep people from breathing in microbes.

Arthritis: common and still confusing

Bury your body in horse manure up to your neck.... Sit in an abandoned uranium mine.... Take cobra and krait venom.... With such remedies, people have desperately sought relief from the pain of arthritis. Neanderthal cave paintings drawn more than forty thousand years ago show stooped beings with bent knees. Although medical science has conquered many old and exotic diseases, this condition continues to plague over thirty-one million Americans. In older people, its two most common forms are rheumatoid arthritis and osteoarthritis.

Rheumatoid arthritis (RA) is the most serious, painful, and disabling form. Affecting three times as many women as men, RA first affects the linings (synovial membranes) of small joints—such as those in the hands and feet—then moves to tissues and organs, such as the heart and lungs. Osteoarthritis (OA) is often called the "wear and tear" disease and is by far the commonest form of arthritis. The condition begins with the thinning, wearing down, or roughening of cartilage, which may lead to chem-

Cancer Screening Tests

Procedure	Sex	Age	Frequency
Sigmoidoscopy	M&F	over 50	every 3–5 years
Stool guaiac slide test	M&F	over 50	every year
Digital rectal examination	M&F	over 40	every year
Pap test	F	20–65	at least every 3 years
Pelvic examination	F	20–40 over 40	every 3 years every year
Endometrial tissue sample	F	at menopause	at menopause
Breast self-examination	F	over 20	every month
Breast exam by physician	F	20–40 over 40	every 3 years every year
Mammography	F	35–40 40–50 over 50	initial test 1–2 years every year
Health counseling and cancer checkup	M&F M&F	20–40 over 40	every 3 years every year

(Recommended by the American Cancer Society)

ical changes that cause the joint to become inflamed. Usually seen in older people, the condition can develop in anyone whose joints have taken a lot of punishment: the obese, those injured in accidents, or those subject to unusual stress in work or sports.

According to Dr. Paul Nickerson of the Geriatric Diagnosis Clinic in Cleveland, Ohio, "The association between age and OA is striking, but it should not be assumed that OA is caused by normal aging." He quickly adds that despite our ignorance of the etiology of arthritis, there is still much to offer the patient.

Dr. Nickerson outlines three areas of management: (1) physical therapy, including exercise and weight reduction; (2) medication with nonsteroid antiinflammation drugs (NSAID), which include aspirin and ibuprophen; and (3) surgery for those with severe pain. Advanced age is sometimes a positive factor in joint replacement, because highly active young people run a greater risk of loosening new joints.

Dr. Nickerson foresees research leading to more and safer NSAIDs and, ultimately, a drug that will inhibit cartilage-damaging enzymes.

In the next generation, research will bring relief from pain and help people learn to live with intractable discomfort. Pain clinics are using new techniques, such as nerve stimulation and biofeedback. Until arthritis is fully understood and conquered, the key to treatment lies in changing attitudes and developing a satisfying life-style despite arthritis.

Diabetes: dare to discipline

Not since the discovery of insulin in 1921 by Banting and Best has there been so much good news about diabetes for older Americans. Recent breakthroughs have led to better understanding of this maligned disease—and diabetes deserves its reputation. Heart attacks, blindness, limb loss, and kidney failure are only a few of its complications.

Diabetes mellitus is a disorder in which the body fails to convert the food we eat into the energy we need. Type I (insulin-dependent diabetes mellitus (IDDM)) may appear early in life and results from a failure of the pancreas to produce insulin. Type II (noninsulin-dependent diabetes mellitus (NIDDM)), found in 90 percent of diabetics, is more common with age. Some researchers believe diabetes is directly related to excess weight, because reducing body fat appears to make body cells more receptive to insulin. By daring to discipline themselves, type II diabetics can usually keep blood glucose levels normal by controlling weight, exercising, and taking oral medication.

Recent research at the National Institute on Aging revealed that the body's ability to handle glucose decreases with age. The finding has led to a revision of the official guidelines for diagnosing diabetes. As a result, fewer elderly people are at risk of being considered diabetic. The new guidelines mean that many persons once considered borderline have been freed from the emotional burden and consequent problems of a chronic degenerative disease.

Fear of the dementing Alzheimer's disease

Publicity about Alzheimer's disease has bred a widespread, morbid fear of the condition. Believing that memory decline is inevitable, middle-aged and older adults interpret any slip of memory as symptomatic of decline.

Minor forgetfulness is normal at any age. The key difference is that in Alzheimer's, dementia worsens until those affected cannot function normally. Fear of dementia will not prevent the condition: Remember, eighty percent of older people remain alert and active.

Alzheimer's disease is a real heartbreaker for the 2.5 million men and women plagued by it, and their families. Although no cure has been discovered, much can be done to make life more bearable for both patients and families.

Alzheimer's disease was discovered by German physician Alois Alzheimer in 1906. While symptoms may vary, patients can go from severe forgetfulness to personality changes with loss of all verbal skills and physical control. The course of the disease may run from less than three years to fifteen years or more. Death usually results from malnutrition or infection.

As some conditions produce symptoms similar to Alzheimer's, the disease is usually diagnosed by exclusion. For example, depression is closely correlated with memory loss. If the depression is successfully treated, the person may regain memory and dignity. Other conditions with similar symptoms include: malnutrition, stroke, drug reaction, metabolic change, or head injury.

An autopsy of the brain of an Alzheimer's patient reveals an abnormal disarray of litter. Abnormal neuron masses called "neurofibrillary tangles" appear in the outer cortex of the brain. Plaques of scarlike structures mark deteriorated nerve endings.

Some experts are looking for a genetic connection. Abnormal genes have been identified in most dementias, such as Huntington's chorea and Down's syndrome. Scientists thought they might be close to a cause when a protein amyloid present in the brains of Alzheimer's patients seemed to be leading them to the causative gene. However, the Winter 1987 issue of the Alzheimer's Disease and Related Disorders Association newsletters reported that Alzheimer's is probably not due to the replication of the gene that produces the amyloid. Another recent disappointment was the suspension of a test on tetrahydroaminoacridine (THA)—a drug that had alleviated some of the symptoms of memory loss. THA was found to cause liver damage in a significant number of patients. Scientists still look with hope for similar drugs to treat this disease associated with aging.

Adults of this and future generations can look forward to greater medical miracles and increased life expectancy. More knowledge and improved life-style choices will result in a new norm: living to a healthy old age. The nineteenth-century poet Robert Browning could have been addressing us today, when he wrote, "Grow old along with me! The best is yet to be."

The Myths of MENOPAUSE

As a catastrophe, most women say it's overrated. But even a trouble-free menopause can trigger problems down the line. Here's what you need to know, whatever your age.

Lisa Davis

Lisa Davis is a staff writer.

OUTSIDE, YOUNG WOMEN on billboards are showing off bottles of whiskey and getting into shiny cars. They're wearing tennis togs or slinky dresses and throwing back their hair; their frozen smiles are getting wet in the first rains of fall. But in this small clinic in suburban Cleveland it's warm and dry, and the woman sitting in one of the quiet rooms is no longer young. She's not really old, either; say mid-50s. That puts her at that awkward age, the one society holds in such disdain.

Too old to bear children, a visible reminder of the inevitability of decline, menopausal women are desexed and devalued by society. They're considered just, oh, déclassé; the change of life is such an embarrassing failure of the flesh. Perhaps it's out of delicacy, then, that menopause is such a rare subject for public discourse. No screenwriter has ever given Alexis Colby, *Dynasty*'s sexy dynamo of mature villainy, a hot flash or a night sweat. Ask a young woman what she knows about menopause, and you'll likely hear only that, in some indeterminate middle decade, monthly bleeding ends.

Ignorance obscures menopause; myth distorts it. Somewhere in the back of her mind, many a woman still harbors the fears that plagued her mother: that she will go crazy during menopause, or fall into a profound depression. Indeed, it wasn't until 1980 that the American Psychiatric Association struck "involutional melancholia"—a mental disorder that supposedly occurred around the time of menopause—from their official directory of psychiatric diagnoses.

Because of her years spent as a nurse and psychotherapist, Shirley Berner was spared these most frightening worries. She has the simultaneous grace and brusqueness that some women grow into, a directness that brooks no evasions. Even so, when she sought information to help her through bouts of anxiety and depression, she found it scarce. Sitting in the consultation room after her twice-a-year checkup at the Cleveland Menopause Clinic, she remembers, "Before I came here, the doctors I was seeing weren't particularly helpful. There was a patronizing, 'Oh my dear, this is all in your head. It'll go away if you just get serious; go to work, go have a hobby.' Or just, 'What do you expect? You're getting older.' I remember saying, 'It's not the getting older I mind. I'd like you to help me *get* older.' "

It's only in the last decade that researchers have attempted to trace the boundaries of the average menopause experience. For seven years, epidemiologist Sonja McKinlay has been following 2,500 randomly selected Massachusetts women with phone interviews as they've entered, and in many cases gone through, menopause. According to McKinlay, of Boston's New England Research Institute, somewhere between the ages of 45 and 55 women typically experience a year or two of irregular periods, fluctuations in body temperature, and some sleepless nights—and often, that is that.

"They have a lot of apprehension as they approach menopause," McKinlay says. "And then they come through it and say, 'Oh gee, it's not such a big deal after all.' "

If anything has become clear about menopause, however, it's that it can't be defined by averages. It's true for every woman that, eventually, her ovaries' production of the hormones estrogen and progesterone will take a dive. But that

By Lisa Davis. Reprinted from In Health, *formally* Hippocrates, *The Magazine of Health & Medicine, May/June 1989, pp. 52-59. Copyright © 1989 by Hippocrates Partners.*

6. MIDDLE AND LATE ADULTHOOD

may mean only a general sense of emotional fragility for one woman, and incontinence, itchy skin, and painful intercourse for another. "At my worst, I was having about fifty hot flashes a day," says one woman who attends a San Francisco menopause support group. Says Berner, "The hot flashes—it wasn't an incredible number, but when they came they were beauts. It was an overwhelming feeling of suddenly stepping into a blast furnace. I was shaky, depressed, and my energy was really low."

Doctors—even many gynecologists—can find the laundry list of symptoms confusingly various, irritatingly vague, and unprofitably time-consuming. So, like Berner, many a menopausal woman has been counseled to simply put up with any problems until her periods go away. Yet researchers say there *is* help available for the woman who is thrown off balance when her hormone levels begin to tilt and slide. They also say the most trouble-free menopause can trigger serious, even life-threatening, problems decades down the line—brittle bones, for instance, or an increased risk of heart attack. That didn't really matter at the turn of the century, when a typical woman was a matriarch at age 35, and dead at 48. But these days, a woman who is healthy at 52—the average age at which menopause occurs—will likely live into her 80s. What she does, or doesn't, do during menopause can have a profound effect on her health in the second half of her life.

Fortunately, attitudes about menopause have begun to change. It's no longer unheard of for gynecologists or internists to specialize in mid-life women, and there are about a dozen menopause clinics in the United States. When Shirley Berner first walked into the Cleveland Menopause Clinic five years ago, her experience was treated neither as a catastrophic end to meaningful existence nor as a tempest in a drying teapot, but as a major change with lifelong consequences.

That was more seriously than Berner herself took it at the time.

About twenty-five percent of my patients are desperate, twenty-five percent are bothered and concerned, and fifty percent just want to stay informed and take care of themselves," says clinic director Wulf Utian, in a South African accent that flattens vowels until they warp. "I find that once you ask the questions, it's like you've opened a valve. Women are

CHARTING THE CHANGES

Here are some of the major changes that can occur with menopause.* The symptoms—and their sheer number—can be intimidating, but most women experience only a few of them, often in a mild form.

WHEN	SYMPTOM	COMMENTS
BEFORE MENOPAUSE	irregular periods	Cycle may shorten or lengthen; flow may increase or decrease. Women may continue to be fertile until they have been without periods for one year.
DURING MENOPAUSE	periods stop	
	hot flashes	Skin temperature rises, then falls. Accompanied by sweating and sometimes heart palpitations, nausea, and anxiety. Frequency ranges from once a month to several an hour; most last about five minutes. Hot flashes may begin 12 to 18 months before menopause, and continue for some years after periods end.
	insomnia	Sometimes caused by nightly bouts of hot flashes. Dream-rich REM sleep may also decrease, disturbing sleep.
	psychological effects	Irritability, short-term memory loss, and problems with concentration are common. These symptoms may simply result from sleep deprivation. Some women experience decreased sexual desire.
AFTER MENOPAUSE	changes in nervous system	The perception of touch can become more or less sensitive.
	dry skin and hair	The skin can become thin, dry, and itchy. Hair may also thin out. Facial hair may increase.
	incontinence	Tissue shrinkage in the bladder and weakening pelvic muscles may lead to problems with bladder control. Susceptibility to urinary tract infections may increase.
	vaginal dryness	The mucosal membranes and walls of the vagina become thinner, which may lead to pain upon intercourse and susceptibility to infections.
	bone loss	Increases dramatically.
	cardiovascular changes	Blood vessels become less flexible. Cholesterol and triglyceride levels rise.

*Surgical menopause: It was once routine for surgeons to remove the ovaries along with the uterus when performing hysterectomies. That's now less common. If the ovaries are removed, a woman enters menopause instantly; some women have even reported waking from anesthesia with hot flashes. Surgical menopause may produce more severe symptoms, since in natural menopause the ovaries continue to produce small amounts of estrogen, and may also produce other hormones that are converted into estrogen by fat cells.

so happy that someone is finally willing to listen."

Utian's got a sort of one-stop shop for the mid-life woman here in the Cleveland Menopause Clinic. In a small room, a woman is hooked to an electrocardiogram machine for a stress test, pedaling away on a stationary bicycle as an exercise physiologist tells her to pick up the pace. Across the hall, there's an X-ray machine designed to scan the density of bones. There's a lab for testing hormone levels in blood, and a quiet office for women who want to talk with a counselor about growing older in a youth-obsessed society, about self-esteem or a changing sex life. And there's a large waiting room where first-time patients fill out a ten-page questionnaire covering everything from hot flashes to happiness levels. All the gadgets, specialists, and questions, because menopause can have such wide-ranging effects—from the physical to the psychological and back again.

There are 300 different tissues in a woman's body that respond to estrogen either directly or indirectly, so it makes sense that withdrawal from the hormone can produce a multitude of symptoms. Discomfort during sex, for instance: Estrogen stimulates the growth of cells lining the vagina, says Utian, and its lack can lead to thinner, drier, easily bruised tissue. Or incontinence: The hormone "switches on" collagen, the fibrous support tissue that's woven into the walls of the bladder and other organs. As the amount of collagen diminishes, the bladder shrinks and pelvic floor muscles weaken, creating problems with bladder control.

The classic symptom of menopause is the hot flash, striking 75 to 85 percent of women at some point. The cause of hot flashes is still unclear, says physiologist Fredi Kronenberg of New York's Columbia University. One hypothesis is that falling estrogen levels trigger an overreaction by the hypothalamus, the region in the brain that regulates body temperature.

Increasingly, researchers are also finding clues that the psyche, like physiology, responds to the loss of estrogen. Many women complain that they're irritable and fuzzy-headed, or that they have trouble coping with stress. "I feel sort of cobwebby," says Barbara Zierten, who attends the San Francisco support group. "What I really want to do is just sit and vegetate."

Menopause doesn't send otherwise stable women into full-blown clinical depression. But recent studies have shown that estrogen withdrawal does lead to a drop in endorphins, "feel-good" substances that the brain naturally produces. Estrogen derivatives act as chemical messengers in the brain, as well, so it's logical that there would be mental adjustments as supplies of the hormone sink.

And there's another, very simple reason that falling supplies of estrogen can cause irritability and vulnerability to stress: sleep deprivation. Night sweats

Most women could get relief with estrogen supplements, but only 10 percent take them. The reason: fear.

give many women wakeful nights for months on end. Even when a woman's sleep is unbroken, she may not feel rested. Work at Boston's Brigham and Women's Hospital suggests that there's a fall in dream-rich REM sleep as estrogen levels drop.

None of these short-term symptoms is life-threatening, but they certainly can do damage to the quality of life. Even so, few women are interested in the cure that medicine has to offer. Up to 70 percent of menopausal women can get relief with hormone supplements within a couple of months, says Utian, but only about 10 percent take them. The reason: fear. When Utian recommended hormones to Berner, for instance, she heard "estrogen," thought "cancer"—and turned the suggestion down.

ESTROGEN'S BAD REPUTATION was honestly earned. In the mid-sixties, a book called *Feminine Forever* warned women that failing ovaries doomed them to a sexless decrepitude. A woman "becomes the equivalent of a eunuch," author Robert A. Wilson wrote; "I have seen untreated women who had shriveled into caricatures of their former selves." Taking hormones was the only way to avoid this sure decline, Wilson said. The picture he painted was frightening enough that by 1975, estrogen had become the fifth most widely prescribed drug in the United States.

But the doctors who prescribed so eagerly didn't hear, or ignored, warnings of possible side effects. In a woman's normal menstrual cycle, both estrogen and progesterone ebb and flow as the uterus builds and sheds its lining. No attempt was made to mimic that cycle in treatment; doctors used a large dose of estrogen, and estrogen alone.

"I was seen as a wet blanket when I warned that we were going to see an increase in uterine cancer," Utian remembers. Sure enough, the first reports of estrogen-caused cancers were published in 1975. As the news spread, the feminine forever craze came to an abrupt end. The consensus at a 1979 conference held by the National Institutes of Health was that estrogen should be given at the lowest dose and for the shortest time possible—just long enough, say, to get a woman through the worst of her hot flashes—or not given at all. It's now clear that a woman who takes estrogen for more than 15 years increases her risk of uterine cancer at least fivefold.

These days, no well-informed doctor will recommend estrogen alone for a woman, unless she's had her uterus removed in a hysterectomy. Studies done in the early 1980s convinced researchers that they could successfully balance estrogen's effects by using a progestogen—any of several synthetic compounds that act like progesterone. Now the standard regimen is 12 to 14 days of progestogen every month and at least 25 days of estrogen—at about half the dose prescribed in the seventies. A woman on such a mix of hormones actually *lowers* her risk of uterine cancer.

Hormone replacement still has its costs, aside from the $100 or so that it takes to buy a year's worth of hormones. There's the inconvenience of daily pills or stick-on drug patches that must be changed twice a week. There's the loss of one of the few unadulterated blessings of menopause: In up to 90 percent of women, the estrogen-progestogen combination will prompt the return of monthly bleeding, sometimes along with bloating, irritability, and other PMS-like symptoms. There are more serious side effects as well. Hormone replacement brings a fourfold increase in the risk of developing gallstones, for instance, although the increased risk is still low. Very rarely, blood pressure can rise; equally rarely, blood clots can form.

Most worrisome for many women are the lingering questions about estrogen and breast cancer, especially with the recent spate of reports on a possible link between estrogen-based oral contracep-

6. MIDDLE AND LATE ADULTHOOD

In the decade after menopause, a woman might lose 6 percent of her bone mass every year. She can exercise and take calcium. But that won't be enough.

tives and breast cancer. Birth control pills, however, contain a much more potent form of estrogen than that given at menopause. And most studies on hormone replacement therapy have shown no increase in the risk of breast cancer; in the few that did, the risk occurred only after 15 to 20 years of use. Because the question remains open, women who have had breast cancer are advised against hormone replacement therapy.

Reputable doctors will counsel their patients about these risks. Still, epidemiologist Sonja McKinlay thinks we're once again jumping too quickly onto the hormone bandwagon. "There are a lot of changes going on around this time in a woman's life, and a doctor often doesn't ask the questions that would allow him to find out what's going on," she says. "For instance, sexual activity declines rapidly after age forty, and if you don't use it you lose it. But when a woman complains of vaginal dryness, many doctors don't even think of talking to her about sexual frequency. She's just told to try an estrogen-based cream."

Utian agrees that many of his colleagues are more comfortable dealing with questions of disease than quality of life. But, he argues, McKinlay misses an important point when she says that most women are able to get through menopause without drugs.

"McKinlay has gone out and asked, 'How has menopause affected your life, positively or negatively?'" says Utian. "But the effects of menopause occur over a number of years. After all, if a woman can predict that she's going to have a hip fracture or heart attack five years down the road, then she's doing better than the medical profession."

In the strongest and the weakest of us, bone is continually built and destroyed; the skeleton is the visible marker in a lifelong tug of war between those two processes. It's a losing battle. In both men and women, the destruction of bone begins to outstrip its replacement around age 35. The loss can be measured in infinitesimal increments at first, and increases only gradually through the years, but the rate drastically accelerates when a woman hits menopause. In the decade that follows, she can lose 6 percent or more of her bone mass every year. After ten or 12 years, for reasons still unclear, the rate of loss drops to 2 percent, about that experienced by men.

In women who lose bone quickly, or in those whose skeleton isn't particularly dense to start with—thin women or sedentary ones, smokers or women on thyroid medication—it may take only a decade for bone to become so brittle that it starts to fracture and crush. About a third of all women experience such fractures eventually; among Caucasian women, whose bones tend to be less dense, the rate is about half. Brittle bones *may* mean only a loss of height as the spine settles on itself, but hip fractures are a leading cause of disability and death in women over the age of 75.

Once bone is gone, it's nearly impossible to bring it back, says Delbert Booher, director of the menopause department at the Cleveland Clinic, across town from Utian. Booher puts all his patients on calcium supplements, since 55-year-old women need as much as 1,500 milligrams a day and generally get 500 milligrams in food. Utian pushes exercise as hard as calcium. But neither specialist thinks those steps are enough. Last year, researchers at the Mayo Clinic and the University of Arizona found estrogen receptors on the bone cells that produce skeletal material; it seems likely that the hormone must bond to the cell in order to turn on production. "You need *enough* calcium, but after that you can take a carload and it won't make a bit of difference," says Booher. "If you are at risk of osteoporosis, the only thing that will help is estrogen."

The same dose of estrogen may also protect against heart disease. Cardiovascular specialists have long been tantalized by the fact that 40-year-old men are nearly five times as likely as women of the same age to die of heart attacks—and that women lose their special protection as they age. Studies in recent years have come up with some explanations. Estrogen lowers the level of total cholesterol, while it raises the level of "good" cholesterol. Blood vessel walls remain elastic when there's enough of the hormone in circulation, and blood clots take longer to form. "All of those effects decrease the chance of sludge building up in blood vessels," says Booher.

Researchers at Harvard Medical School have been following 32,317 postmenopausal nurses for four years, comparing heart disease in those who have and haven't taken hormones. So far, those who use estrogen have been only half as likely to suffer a heart attack. It's unclear whether the estrogen-progestogen combination will be as protective.

Ultimately, every woman must weigh the risks and benefits of hormone replacement therapy for herself, factoring in her fears and the particularities of her situation. An obese woman, for instance, is at higher risk for heart problems, but she need not worry so much about osteoporosis—both because she's likely to have a denser skeleton and because fat cells can convert other hormones into estrogen. "The real question is, who's at risk for what?" says Utian. "It's the number one challenge for the clinician."

Because there are no simple answers to the questions raised by menopause, Utian does a lot of listening to the women who come to him. He was willing, for instance, to go along with Shirley Berner, who wanted to avoid hormones when she first came to the clinic. "It was plain old ignorance, really," she says now. "I was sure I needed an antidepressant. He questioned it, but he was willing to try it."

Six months of the drug, though, didn't touch her anxiety—or her hot flashes. Then Utian ran a bone scan on her, and found that Berner's spine had already begun to thin. She's an active woman who intends to stay that way, and that was enough to get her to try hormones—that, plus her family history of heart disease. Five years later, she's eating better, exercising more, and she's still on the regimen of hormone supplements.

"The most important thing for me is having vitality, confidence, zest for life," she says. "When your hormones are going nuts and society is saying you're through, it can really finish off a woman. God, it doesn't have to be that way."

DON'T ACT YOUR AGE!

Life stages no longer roll forward in a cruel numbers game

Carol Tavris, Ph.D.

Carol Tavris *is a social psychologist and writer, and author of* Anger: The Misunderstood Emotion.

A friend of mine has just had her first baby. Not news, exactly. It's not even news that she's 45 years old. The news is it's not news that a 45-year-old woman has just had her first baby.

Another friend, age 32, has decided to abandon the pursuit of matrimony and remodel her kitchen instead. The news is that her parents and friends don't think she's weird. They're giving her a Not-Wedding party.

It used to be that all of us knew what we were supposed to be doing at certain ages. The "feminine clock" dictated that women married in their early 20s, had a couple of kids by 30 (formerly the baby deadline), maybe went back to work in their 40s, came down with the empty nest blues in their 50s, and faded into grandmotherhood in their 60s. The "masculine clock" ticked along as men marched up the career ladder, registering their promotions and salaries with notches at each decade.

Nowadays, many women are following the masculine clock; many men are resetting their schedules; and huge numbers of both sexes have stopped telling time altogether. This development is both good news and bad.

The good news is that people are no longer expected to march in lock step through the decades of life, making changes on schedule. "No one is doing things on time anymore," says Dr. Nancy Schlossberg, an adult development expert at the University of Maryland, and the author of the forthcoming *Overwhelmed: Coping with Life's Ups and Downs* (Lexington Books). "Our lives are much too irregular and unpredictable. In my classes I've seen women who were first-time mothers at 43 and those who had their first baby at 17. I just met a woman who is newly married—at age 65. She quit her job, and with her husband is traveling around on their yacht writing articles. You can bet she won't be having an age-65 retirement crisis. She's having too much fun."

The bad news is that without timetables, many people are confused about what they're "supposed" to be doing in their 20s, 30s, 40s and beyond. They have confused *age* (a biological matter) with *stage* (a social matter). Women ask: "When is the best time to have children—before or after I've started working?" Men want to know: "Since I can't make up my mind about marriage, work, children and buying a dog, is it possible I'm having a midlife crisis even though I'm only 32?"

Although confusion can be unsettling, I prefer it to the imposed phoniness of the "life stage" theories of personal development. Actually, I date my dislike of stage theories to my childhood. My parents used to keep Gesell's *The First Five Years of Life* and *The Child From Five to Ten* on the highest shelf of their library (right next to Rabelais), and I *knew* they were consulting these volumes at regular intervals to check on my progress. I was

NO ONE IS DOING THINGS ON TIME ANYMORE.

indignant. For one thing, a 9¾-year-old person finds it humiliating to be lumped with six-year-olds. For another, I was sure I wasn't measuring up, though what I was supposed to be measuring up to I never knew.

I survived Gesell's stages only to find myself, as a college student in the '60s, assigned to read Erik Erikson's theory of the eight stages of man. Every few years in childhood, and then every 10 years or so after, Erikson said, people have a special psychological crisis to resolve and overcome.

The infant must learn to trust, or will forever mistrust the world. The toddler must develop a sense of autonomy and independence, without succumbing to shame and doubt. The school-aged child must acquire competence at schoolwork, or will risk lifelong feelings of inferiority. Teenagers, naturally, must overcome the famous "identity crisis," or they will wallow in "role confusion" and aimlessness. Once you have your identity, you must learn to share it; if you don't master this "intimacy crisis," you might become lonely and isolated. To Erikson, you're never home free. Older adults face the crises of stagnation versus generativity, and, in old age, "ego integrity" versus despair.

It turned out, of course, that Erikson meant the ages of "man" liter-

6. MIDDLE AND LATE ADULTHOOD

ally, but none of us knew that in those days. We female students all protested that our stages were out of order—but that was just further evidence, our instructor said, of how deviant, peculiar and irritating women are. Erikson's theory, he said, was a brilliant expansion of Freud's stage theory (which stopped at puberty). If women didn't fit, it was their own damned fault.

In the 1970s, stage theory struck again with an eruption of popular books. (Stage theories recur in predictable stages.) Journalist Gail Sheehy published *Passages: Predictable Crises of Adult Life* (no one asked how a crisis, by definition a "turning point" or "a condition of instability," could be predictable). Harvard psychiatrist George Vaillant, now at Dartmouth, studied privileged Harvard (male) students, and concluded that men go through orderly stages even if their lives differ. Yale psychiatrist Daniel J. Levinson, in *The Seasons of a Man's Life*, argued that the phases of life unfold in a natural sequence, like the four seasons. This book had nothing to say about women's seasons, possibly because women were continuing to irritate academics by doing things unseasonably.

By this time I was really annoyed. I wasn't having any of my crises in the right order. I hadn't married when I was supposed to, which put my intimacy and generativity crises on hold; leaving my job created an identity crisis at 32, far too late. My work-linked sense of competence, having reached a high of +9, now plunged to -2, and I was supposed to have resolved *that* one at around age seven.

I had only to look around to realize I was not alone. All sorts of social changes were detonating around me. Women who had been homebodies for 35 years were running off to start businesses, much to the annoyance of their husbands, who were quitting their businesses to take lute lessons. People who expected to marry didn't. People who expected to stay married didn't. Women who expected never to work were working. Men who expected never to care about babies were cooing over their own. Expectations were out the window altogether.

LIFE AS A FAN

Eventually, stages no longer mat-

EXPERTS CAN'T AGREE ON WHERE TO LOCATE "MIDLIFE."

tered, either. Psychological theories—which follow what people actually do—have had to change to keep up with the diversity of modern life. In recent years, researchers have discovered a few things that, once and for all, should drive a stake through the idea of fixed, universal life stages:

The psychology and biology of aging are not the same thing. Many of the problems of "old age" stem from psychological, not physical, losses. They would afflict most people at any age who were deprived of family, close friends, meaningful activity, intellectual stimulation and control over what happens to them. Today we've learned to distinguish the biology of normal aging from the decline caused by illness: Conditions once thought inevitable—osteoporosis, senility, excessive wrinkling, depression—can result from poor nutrition, overmedication, lack of exercise, cell damage or disease. For example, only 15% of people over 65 suffer serious mental impairment, and half of those cases are due to Alzheimer's disease.

These findings have played havoc with the basic definition of "old." It used to be 50. Then it was 60, then 70. Today there are so many vibrant octogenarians that "old" is getting even older. Researchers can't even agree on where to locate "midlife" (30 to 50? 40 to 60? 35 to 65?), let alone what problems constitute a midlife crisis.

Although children progress through biologically determined "stages," adults don't. Children go through a stage of babbling before they talk; they crawl before they walk; they wail before they can say, "Can we discuss this calmly, Mom?" These developments are governed by maturational and biological changes dictated by genes. But as children mature, genes become less of a driving force on their development, and the environment has greater impact.

Bernice Neugarten, a professor of behavioral science at the University of Chicago, observes that the better metaphor for life is a fan, rather than stages. When you open a fan, you can see all its diverse pieces linked at a common point of origin. As people age, their qualities and experiences likewise "fan out," which is why, she says, you find greater diversity in a group of 70-year-olds than in a passel of seven-year-olds.

The variety and richness of adult life can't be crammed into tidy "stages" anymore. Stage theorists such as Erikson assumed that growth is fixed (by some biological program or internal clock), progressive (you grow from a lower stage to a higher one), one-way (you grow up, not down; become more competent, not less), cumulative (reflecting your resolution of previous stages), and irreversible (once you gain a skill, there's no losing it).

Yet it has proven impossible to squash the great variety of adult experience into a fixed pattern, and there is no evidence to support the idea of neat stages that occur in five- or 10-year intervals. Why must you master an "identity crisis" before you learn to love? Don't issues of competence and inferiority recur throughout life? Why is the need for "generativity" relevant only to 30-year-olds?

EVENTS AND NONEVENTS

For all these reasons, new approaches to adult development emphasize not how old people are, but what they are doing. Likewise, new studies find that *having* a child has stronger psychological effects on mothers than the age at which they have the baby. (New mothers of any age feel more nurturing and less competent.) Entering the work force has a strong positive effect on your self-esteem and ambition, regardless of when you start working. Men facing retirement confront similar issues at 40, 50 or 60. Divorced people have certain

THE NEW APPROACH: NOT HOW OLD YOU ARE, BUT WHAT YOU ARE DOING.

common problems, whether they split at 30 or 50.

In their book *Lifeprints* (McGraw-Hill), Wellesley College psychologists Grace Baruch and Rosalind Barnett and writer Caryl Rivers surveyed 300 women, ages 35 to 55. They found that the differences among the women depended on what they were doing, not on their age. A career woman of 40, for example, has more in common with a career woman of 30 than with an unemployed woman her age.

At the heart of *Lifeprints* is the heretical notion that "there is no one lifeprint that insures all women a perpetual sense of well-being—nor one that guarantees misery, for that matter. American women today are finding satisfying lives in any number of role patterns. Most involve trade-offs at different points in the life cycle."

Instead of looking for the decade landmarks or the "crises" in life, the *transitions* approach emphasizes the importance of shifting from one role or situation to another. What matters are the events that happen (or fail to happen) and cause us to change in some way. Maryland's Schlossberg describes four kinds of transitions:

■ **Anticipated** transitions are the events you plan for, expect and rehearse: going to school, getting married, starting a job, getting promoted, having a child, retiring at 65. These are the (previously) common milestones of adult life, and because they're predictable, they cause the least difficulty.

■ **Unanticipated** transitions are the things that happen when you aren't prepared: flunking out of school, being fired, having a baby after being told you can't, being forced to retire early. Because these events are bolts from the blue, they can leave you reeling.

■ **Nonevent** transitions are the changes you expect to happen that don't: You don't get married; you can't have children; you aren't promoted; you planned to retire but need to keep working for the income. The challenge here is knowing when to accept these ongoing events as specific transitions and learning to live with them.

■ **Chronic Hassle** transitions are the situations that may eventually require you to change or take action, but rumble along uncomfortably for a long stretch: You aren't getting along with your spouse; your mother gets a chronic illness and needs constant care; you have to deal with discrimination at work; your child keeps getting into trouble.

There are no rules: An anticipated change for one person (having a baby) might be unanticipated for another. An upsetting "nonevent" transition for one person (not getting married) can be a planned decision for another and thus not a transition at all. And even unexpected good news—you recover despite a hopeless diagnosis—can require adjustment. This approach acknowledges that nonevents and chronic situations cause us to change just as surely as dramatic events do, though perhaps less consciously.

Seeing our lives in terms of these transitions, says Schlossberg, frees us from the old stereotypes that say we "should" be doing one thing or another at a certain time in our lives. But it also helps us understand why we can sail through changes we thought would be traumatic—only to be torpedoed by transitions expected to be a breeze. Our reactions have little to do with an internal clock, and everything to do with expectations, goals and, most of all, what else is happening to us.

For example, says Schlossberg, people have very different reactions to "significant" birthdays. For some, 30 is the killer; for others it's 50. For an aunt of mine, who breezed through decade markers without a snivel, 70 was traumatic. "To determine why a birthday marker creates a crisis," says Schlossberg, "I'd ask what was going on in the person's life, not their age. How old were they when their last parent died? How is their work going? Have they lost a loved one?" If they see a birthday as closing down options, then the event can feel negative, she adds.

"All of us carry along a set of psychological needs that are important throughout our lives, not just at one particular age or stage," says Schlossberg. "We need to feel we *belong* to a family, group or community, for example. Changing jobs, marriages or cities often leaves people feeling temporarily left out.

"We also need to feel we matter to others, that we count. At some phases of life, people are burdened by mattering too much to too many people. Many women in their 30s must care for children, husbands and parents, to say nothing of working at their paid jobs. At other phases, people suffer from a sense of mattering too little."

In addition to belonging and "mattering," says Schlossberg, people need to feel they have a reasonable amount of control over their lives; they need to feel competent at what they do; they need identity—a strong sense of who they are; and they need close attachments and commitments that give their lives meaning.

These themes, says Schlossberg, reflect our common humanity, uniting men and women, old and young. A freshman in college and a newly retired man may both temporarily feel marginal, "out of things." A teenager and her grandmother may both feel they don't "matter" to enough other people, and be lonely as a result. A man may feel he has control over his life until he's injured in a car accident. A woman's identity changes when she goes back to school in midlife. A newly

6. MIDDLE AND LATE ADULTHOOD

divorced woman of 30 and a recently laid-off auto worker of 40 may both feel inadequate and incompetent. When people lose the commitments that give their lives meaning, they feel adrift.

"By understanding that these emotional feelings are a normal response to what is going on in your life, and not an inevitable crisis that occurs at 23 or 34, or whatever," says Schlossberg, "people can diagnose their problems more accurately—and more important, take steps to fix them." If you say, "No wonder I'm miserable; I'm having my Age 30 Decade Panic," there's nothing to do but live through it—getting more panicked when you're still miserable at age 36½.

But if you can say, "No wonder I'm miserable—I don't feel competent at work, I don't feel I matter to enough people, I feel like a stranger in this neighborhood," then more constructive possibilities present themselves. You can learn new skills, join new groups, start a neighborhood cleanup committee, and quit whining about being 30.

None of this means that age doesn't matter, as my 83-year-old mother would be the first to tell you. She mutters a lot about irritating pains, wrinkles, forgetfulness and getting shorter. But mostly my mother is too busy to complain, what with her paralegal counseling, fund-raising, organizing programs for shut-in older women, traveling around the world, and socializing. She knows she belongs; she matters to many; she has countless commitments; she knows who she is.

And yet the transitions approach reminds us that adult concerns aren't settled, once and for all, at some critical stage or age. It would be nice if we could acquire a sense of competence in grammar school and keep it forever, if we had only one identity crisis per lifetime, if we always belonged. But adult development is more complicated than that, and also more interesting. As developmental psychologist Leonard Pearlin once said, "There is not one process of aging, but many; there is not one life course, but many courses; there is no one sequence of stages, but many." The variety is as rich as the diversity of human experience.

Let's celebrate the variety—and leave stages to children, geologists, rocket launches and actors.

A WOMAN'S IDENTITY CHANGES WHEN SHE GOES BACK TO SCHOOL.

The Prime of Our Lives

WHAT SEEMS TO MARK OUR ADULT YEARS MOST IS OUR SHIFTING PERSPECTIVE ON OURSELVES AND OUR WORLD. IS THERE A COMMON PATTERN TO OUR LIVES?

Anne Rosenfeld and Elizabeth Stark

Anne Rosenfeld and Elizabeth Stark, both members of Psychology Today's *editorial staff, collaborated across cohorts to write this article.*

"My parents had given me everything they could possibly owe a child and more. Now it was my turn to decide and nobody ... could help me very far...." That's how Graham Greene described his feelings upon graduation from Oxford. And he was right. Starting on your own down the long road of adulthood can be scary.

But the journey can also be exciting, with dreams and hopes to guide us. Maybe they're conventional dreams: getting a decent job, settling down and starting to raise a family before we've left our 20s. Or maybe they're more grandiose: making a million dollars by age 30, becoming a movie star, discovering a cure for cancer, becoming President, starting a social revolution.

Our youthful dreams reflect our unique personalities, but are shaped by the values and expectations of those around us—and they shift as we and our times change. Twenty years ago, college graduates entered adulthood with expectations that in many cases had been radically altered by the major upheavals transforming American society. The times were "a-changin'," and almost no one was untouched. Within a few years many of the scrubbed, obedient, wholesome teenagers of the early '60s had turned into scruffy, alienated campus rebels, experimenting with drugs and sex and deeply dissatisfied with their materialistic middle-class heritage.

Instead of moving right on to the career track, marrying and beginning families, as their fathers had done, many men dropped out, postponing the obligations of adult life. Others traveled a middle road, combining "straight" jobs with public service rather than pursuing conventional careers. And for the first time in recent memory, large numbers of young men refused to serve their country in the military. In the early 1940s, entire fraternities went together to enlist in World War II. In the Age of Aquarius, many college men sought refuge from war in Canada, graduate school, premature marriages or newly discovered medical ailments.

Women were even more dramatically affected by the social changes of the 1960s. Many left college in 1967 with a traditional agenda—work for a few years, then get married and settle down to the real business of raising a family and being a good wife—but ended up following a different and largely unexpected path. The women's movement and changing economics created a whole new set of opportunities. For example, between 1967 and 1980, women's share of medical degrees in the United States rocketed from 5 percent to 26 percent, and their share of law degrees leaped from 4 percent to 22 percent.

6. MIDDLE AND LATE ADULTHOOD

A group of women from the University of Michigan class of 1967 who were interviewed before graduation and again in 1981 described lives very different from their original plans. Psychologists Sandra Tangri of Howard University and Sharon Jenkins of the University of California found that far more of these women were working in 1981 than had expected to, and far more had gotten advanced degrees and were in "male" professions. Their home lives, too, were different from their collegiate fantasies: Fewer married, and those who did had much smaller families.

Liberation brought problems as well as opportunities. By 1981, about 15 percent of the women were divorced (although some had remarried), and many of the women who "had it all" told Tangri and Jenkins that they felt torn between their careers and their families.

Living out our dreams in a rapidly changing society demands extreme flexibility in adjusting to shifting social realities. Our hopes and plans, combined with the traditional rhythms of the life course, give some structure, impetus and predictability to our lives. But each of us must also cope repeatedly with the unplanned and unexpected. And in the process, we are gradually transformed.

For centuries, philosophers have been trying to capture the essence of how people change over the life course by focusing on universally experienced stages of life, often linked to specific ages. Research on child development, begun earlier in this century, had shown that children generally pass through an orderly succession of stages that correspond to fairly specific ages. But recent studies have challenged some of the apparent orderliness of child development, and the pattern of development among adults seems to be even less clear-cut.

When we think about what happens as we grow older, physical changes leap to mind—the lessening of physical prowess, the arrival of sags, spreads and lines. But these take a back seat to psychological changes, according to psychologist Bernice Neugarten of Northwestern University, a pioneer in the field of human development. She points out that although biological maturation heavily influences childhood development, people in young and middle adulthood are most affected by their own experiences and the timing of those experiences, not by biological factors. Even menopause, that quintessentially biological event, she says, is of relatively little psychological importance in the lives of most adult women.

In other words, chronological age is an increasingly unreliable indicator of what people will be like at various points. A group of newborns, or even 5-year-olds, shows less variation than a group of 35-year-olds, or 50-year-olds.

What seems to mark our adult years most is our shifting perspective on ourselves and our

> *STAGE THEORIES ARE A LITTLE LIKE HOROSCOPES—VAGUE ENOUGH TO LET EVERYONE SEE SOMETHING OF THEMSELVES IN THEM. THAT'S WHY THEY'RE SO POPULAR.*

world—who we think we are, what we expect to get done, our timetable for doing it and our satisfactions with what we have accomplished. The scenarios and schedules of our lives are so varied that some researchers believe it is virtually impossible to talk about a single timetable for adult development. However, many people probably believe there is one, and are likely to cite Gail Sheehy's 1976 best-seller *Passages* to back them up.

Sheehy's book, which helped make "midlife crisis" a household word, was based on a body of research suggesting that adults go through progressive, predictable, age-linked stages, each offering challenges that must be met before moving on to the next stage. The most traumatic of these transitions, Sheehy claimed, is the one between young and middle adulthood—the midlife crisis.

Sheehy's ideas were based, in part, on the work of researchers Daniel Levinson, George Vaillant and Roger Gould, whose separate studies supported the stages of adult development Erik Erikson had earlier proposed in his highly influential model (see "Erikson's Eight Stages," next page).

Levinson, a psychologist, had started his study in 1969, when he was 49 and intrigued with his own recent midlife strains. He and his Yale colleagues intensively interviewed 40 men between the ages of 35 and 45 from four occupational groups. Using these interviews, bolstered by the biographies of great men and the development of memorable characters in literature, they described how men develop from 17 to 65 years of age (see "Levinson's Ladder," this article).

At the threshold of each major period of adulthood, they found, men pass through predictably unstable transitional periods, including a particularly wrenching time very close to age 40. At each transition a man must confront issues that may involve his career, his marriage, his family and the realization of his dreams if he is to progress successfully to the

50. Prime of Our Lives

next period. Seventy percent to 80 percent of the men Levinson interviewed found the midlife transition (ages 40 to 45) tumultuous and psychologically painful, as most aspects of their lives came into question. The presumably universal timetable Levinson offered was very rigid, allowing no more than four years' leeway for each transition.

Vaillant's study, although less age-bound than Levinson's, also revealed that at midlife men go through a period of pain and preparation—"a time for reassessing and reordering the truth about adolescence and young adulthood." Vaillant, a psychiatrist, when he conducted his study at Harvard interviewed a group of men who were part of the Grant Study of Adult Development. The study had tracked almost 270 unusually accomplished, self-reliant and healthy Harvard freshmen (drawn mostly from the classes of 1942 to 1944) from their college days until their late 40s. In 1967 and 1977 Vaillant and his team interviewed and evaluated 94 members of this select group.

They found that, despite inner turmoil, the men judged to have the best outcomes in their late 40s "regarded the period from 35 to 49 as the happiest in their lives, and the seemingly calmer period from 21 to 35 as the unhappiest." But the men least well adapted at midlife "longed for the relative calm of their young adulthood and regarded the storms of later life as too painful."

While Levinson and Vaillant were completing their studies, psychiatrist Roger Gould and his colleagues at the University of California, Los Angeles, were looking at how the lives of both men and women change during young and middle adulthood. Unlike the Yale and Harvard studies, Gould's was a one-time examination of more than 500 white, middle-class people from ages 16 to 60. Gould's study, like those of Levinson and Vaillant, found that the time around age 40 was a tough one for many people, both personally and maritally. He stressed that people need to change their early expectations as they develop. "Childhood delivers most people into adulthood with a view of adults that few could ever live up to," he wrote. Adults must confront this impossible image, he said, or be frustrated and dissatisfied.

The runaway success of *Passages* indicated the broad appeal of the stage theorists' message with its emphasis on orderly and clearly defined transitions. According to Cornell historian Michael Kammen, "We want predictability, and we desperately want definitions of 'normality.'" And almost everyone could find some relationship to their own lives in the stages Sheehy described. Stage theories, explains sociologist Orville Brim Jr., former president of the Russell Sage Foundation, are "a little like horoscopes. They are vague enough so that everyone can see something of themselves in them. That's why they're so popular."

But popularity does not always mean validity. Even at the time there were studies contradicting the stage theorists' findings. When sociologist Michael Farrell of the State University of New York at Buffalo and social psychologist Stanley Rosenberg of Dartmouth Medical School looked for a crisis among middle-aged men in 1971 it proved elusive. Instead of finding a "universal midlife crisis," they discovered several different developmental paths. "Some men do appear to reach a state of crisis," they found, "but others seem to thrive. More typical than either of these responses is the tendency for men to bury their heads and deny and avoid all the pressures closing in on them."

Another decade of research has made the picture of adult development even more complex. Many observations and theories accepted earlier as fact, especially by the general public, are now being debated. Researchers have espe-

Erikson's Eight Stages

According to Erik Erikson, people must grapple with the conflicts of one stage before they can move on to a higher one.

BONNIE SCHIFFMAN

	1	2	3	4	5	6	7	8
Old Age								Integrity vs. Despair, Disgust
Maturity							Generativity vs. Self-absorption	
Young Adulthood						Intimacy vs. Isolation		
Adolescence					Identity vs. Identity Confusion			
School Age				Industry vs. Inferiority				
Play Age			Initiative vs. Guilt					
Early Childhood		Autonomy vs. Shame, Doubt						
Infancy	Trust vs. Mistrust							

SOURCE: ADAPTED FROM "REFLECTION ON DR. BORG'S LIFE CYCLE", ERIK H. ERIKSON, DAEDALUS, SPRING 1976.

Oh, God, I'm only twenty and I'll have to go on living and living and living.
—Jean Rhys, *Diary*

At thirty a man should know himself like the palm of his hand, know the exact number of his defects and qualities, know how far he can go, foretell his failures—be what he is. And above all accept these things.
—Albert Camus
Carnets.

6. MIDDLE AND LATE ADULTHOOD

cially challenged Levinson's assertion that stages are predictable, tightly linked to specific ages and built upon one another.

In fact, Gould, described as a stage theorist in most textbooks, has since changed his tune, based upon his clinical observations. He now disagrees that people go through "formal" developmental stages in adulthood, although he says that people "do change their ways of looking at and experiencing the world over time." But the idea that one must resolve one stage before going on to the next, he says, is "hogwash."

Levinson, however, has stuck by his conceptual guns over the years, claiming that no one has evidence to refute his results. "The only way for my theory to be tested is to study life structure as it develops over adulthood," he says. "And by and large psychologists and sociologists don't study lives, they study variables."

Many researchers have found that changing times and different social expectations affect how various "cohorts"—groups of people born in the same year or time period—move through the life course. Neugarten has been emphasizing the importance of this age-group, or cohort, effect since the early 1960s. Our values and expectations are shaped by the period in which we live. People born during the trying times of the Depression have a different outlook on life from those born during the optimistic 1950s, according to Neugarten.

The social environment of a particular age group, Neugarten argues, can influence its so-

WHAT WAS TRUE FOR PEOPLE BORN IN THE DEPRESSION ERA MAY NOT HOLD FOR TODAY'S 40-YEAR-OLDS, BORN IN THE UPBEAT POSTWAR YEARS.

cial clock—the timetable for when people expect and are expected to accomplish some of the major tasks of adult life, such as getting married, having children or establishing themselves in a work role. Social clocks guide our lives, and people who are "out of sync" with them are likely to find life more stressful than those who are on schedule, she says.

Since the 1960s, when Neugarten first measured what people consider to be the "right" time for major life events, social clocks have changed (see "What's the Right Time?" this article), further altering the lives of those now approaching middle age, and possibly upsetting the timetable Levinson found in an earlier generation.

As sociologist Alice Rossi of the University of Massachusetts observes, researchers trying to tease out universal truths and patterns from

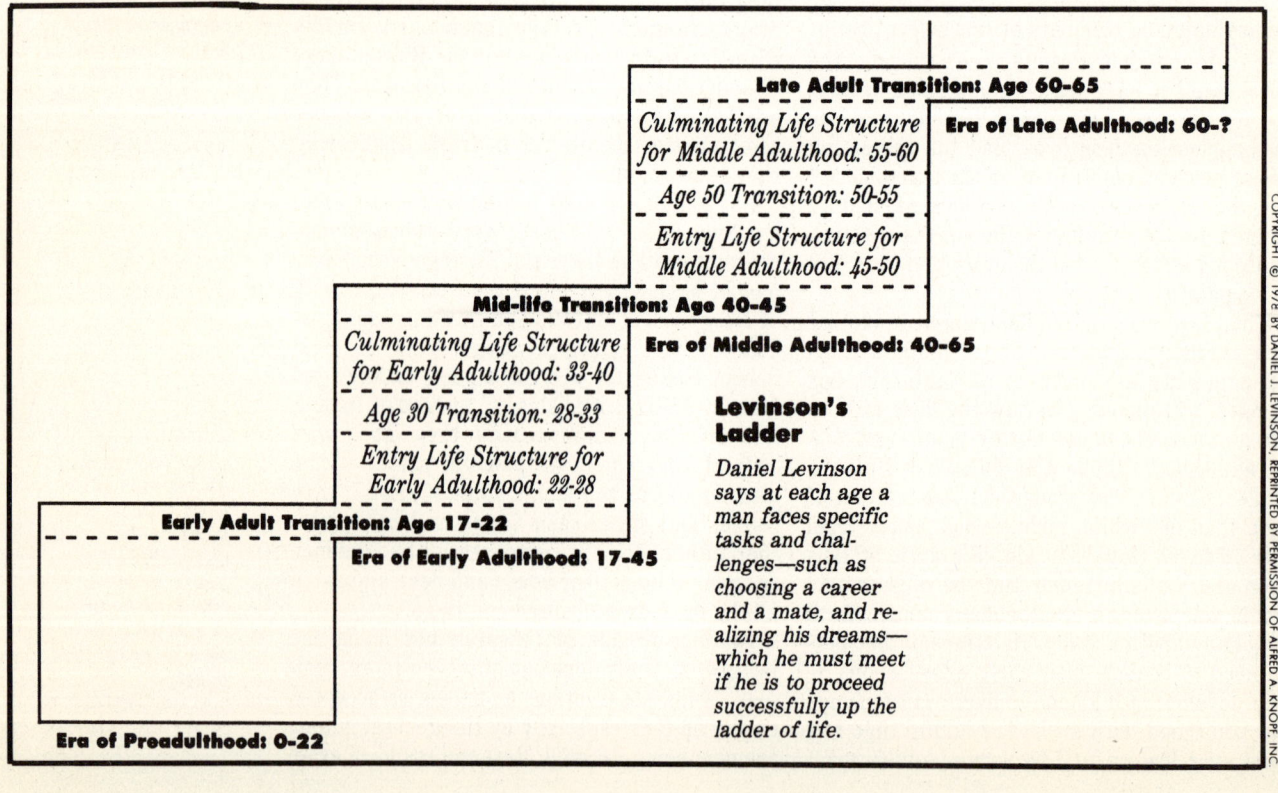

Levinson's Ladder

Daniel Levinson says at each age a man faces specific tasks and challenges—such as choosing a career and a mate, and realizing his dreams—which he must meet if he is to proceed successfully up the ladder of life.

Late Adult Transition: Age 60-65
Culminating Life Structure for Middle Adulthood: 55-60
Era of Late Adulthood: 60-?
Age 50 Transition: 50-55
Entry Life Structure for Middle Adulthood: 45-50
Mid-life Transition: Age 40-45
Culminating Life Structure for Early Adulthood: 33-40
Era of Middle Adulthood: 40-65
Age 30 Transition: 28-33
Entry Life Structure for Early Adulthood: 22-28
Early Adult Transition: Age 17-22
Era of Early Adulthood: 17-45
Era of Preadulthood: 0-22

the lives of one birth cohort must consider the vexing possibility that their findings may not apply to any other group. Most of the people studied by Levinson, Vaillant and Gould were born before and during the Depression (and were predominantly male, white and upper middle class). What was true for these people may not hold for today's 40-year-olds, born in the optimistic aftermath of World War II, or the post baby-boom generation just approaching adulthood. In Rossi's view, "The profile of the midlife men in Levinson's and Vaillant's studies may strike a future developmental researcher as burned out at a premature age, rather than reflecting a normal developmental process all men go through so early in life."

Based on her studies of women at midlife, Nancy Schlossberg, a counselor educator at the University of Maryland, also disagrees that there is a single, universal timetable for adult development—or that one can predict the crises in people's lives by knowing their age. "Give me a roomful of 40-year-old women and you have told me nothing. Give me a case story about what each has experienced and then I can tell if one is going to have a crisis and another a tranquil period." Says Schlossberg: "What matters is what transitions she has experienced. Has she been 'dumped' by a husband, fired from her job, had a breast removed, gone back to school, remarried, had her first book published. It is what has happened or not happened to her, not how old she is, that counts.... There are as many patterns as people."

Psychologist Albert Bandura of Stanford University adds more fuel to the anti-stage fire by pointing out that chance events play a big role in shaping our adult lives. Careers and marriages are often made from the happenstance of meeting the right—or wrong—person at the right—or wrong—time. But, says Bandura, while the events may be random, their effects are not. They depend on what people do with the chance opportunities fate deals them.

The ages-and-stages approach to adult development has been further criticized because it does not appear to apply to women. Levinson claims to have confirmed that women do follow the same age-transition timetable that men do. But his recent study of women has yet to be published, and there is little other evidence that might settle the case one way or the other.

Psychologists Rosalind Barnett and Grace Baruch of the Wellesley Center for Research on Women say, "It is hard to know how to think of women within this [stage] theory—a woman may not enter the world of work until her late 30s, she seldom has a mentor, and even women with lifelong career commitments rarely are in a position to reassess their commitment pattern by age 40."

But University of Wisconsin-Madison psychologist Carol Ryff, who has directly compared the views of men and women from different age groups, has found that the big psychological issues of adulthood follow a similar developmental pattern for both sexes.

Recently she studied two characteristics highlighted as hallmarks of middle age: Erikson's "generativity" and Neugarten's "complexity." Those who have achieved generativity, according to Ryff, see themselves as leaders and decision makers and are interested in helping and guiding younger people. The men and women Ryff studied agreed that generativity is at its peak in middle age.

Complexity, which describes people's feeling that they are in control of their lives and are actively involved in the world, followed a somewhat different pattern. It was high in young adulthood and stayed prominent as people matured. But it was most obvious in those who are now middle-aged—the first generation of middle-class people to combine family and work in dual-career families. This juggling of roles, although stressful, may make some men and women feel actively involved in life.

Psychologist Ravenna Helson and her colleagues Valory Mitchell and Geraldine Moane at the University of California, Berkeley, have recently completed a long-term study of the lives of 132 women that hints at some of the forces propelling people to change psychologically during adulthood. The women were studied as seniors at Mills College in California in the late 1950s, five years later and again in 1981, when they were between the ages of 42 and 45.

Helson and her colleagues distinguished three main groups among the Mills women: family-oriented, career-oriented (whether or not they also wanted families) and those who followed neither path (women with no children who pursued only low-level work). Despite their different profiles in college, and their diverging life paths, the women in all three groups underwent similar broad psychological changes over time, although those in the third group changed less than those committed to career or family.

Personality tests given through the years revealed that from age 21 to their mid-40s, the Mills women became more self-disciplined and committed to duties, as well as more independent and confident. And between age 27 and the early 40s, there was a shift toward less traditionally "feminine" attitudes, including greater dominance, higher achievement motivation, greater interest in events outside the family and more emotional stability.

To the Berkeley researchers, familiar with the work of psychologist David Gutmann of Northwestern University, these changes were not surprising in women whose children were mostly grown. Gutmann, after working with Neugarten and conducting his own research, had theorized that women and men, largely

SUDDENLY I'M THE ADULT?
BY RICHARD COHEN

Several years ago, my family gathered on Cape Cod for a weekend. My parents were there, my sister and her daughter, too, two cousins and, of course, my wife, my son and me. We ate at one of those restaurants where the menu is scrawled on a blackboard held by a chummy waiter and had a wonderful time. With dinner concluded, the waiter set the check down in the middle of the table. That's when it happened. My father did not reach for the check.

In fact, my father did nothing. Conversation continued. Finally, it dawned on me. Me! I was supposed to pick up the check. After all these years, after hundreds of restaurant meals with my parents, after a lifetime of thinking of my father as the one with the bucks, it had all changed. I reached for the check and whipped out my American Express card. My view of myself was suddenly altered. With a stroke of the pen, I was suddenly an adult.

Some people mark off their life in years, others in events. I am one of the latter, and I think of some events as rites of passage. I did not become a young man at a particular year, like 13, but when a kid strolled into the store where I worked and called me "mister." I turned around to see whom he was calling. He repeated it several times—"Mister, mister"—looking straight at me. The realization hit like a punch: Me! He was talking to me. I was suddenly a mister.

There have been other milestones. The cops of my youth always seemed to be big, even huge, and of course they were older than I was. Then one day they were neither. In fact, some of them were kids—short kids at that. Another milestone.

The day comes when you suddenly realize that all the football players in the game you're watching are younger than you. Instead of being big men, they are merely big kids. With that milestone goes the fantasy that someday, maybe, you too could be a player—maybe not a football player but certainly a baseball player. I had a good eye as a kid—not much power, but a keen eye—and I always thought I could play the game. One day I realized that I couldn't. Without having ever reached the hill, I was over it.

For some people, the most momentous milestone is the death of a parent. This happened recently to a friend of mine. With the burial of his father came the realization that he had moved up a notch. Of course, he had known all along that this would happen, but until the funeral, the knowledge seemed theoretical at best. As long as one of your parents is alive, you stay in some way a kid. At the very least, there remains at least one person whose love is unconditional.

For women, a milestone is reached when they can no longer have children. The loss of a life, the inability to create one—they are variations on the same theme. For a childless woman who could control everything in life but the clock, this milestone is a cruel one indeed.

I count other, less serious milestones—like being audited by the Internal Revenue Service. As the auditor caught mistake after mistake, I sat there pretending that really knowing about taxes was for adults. I, of course, was still a kid. The auditor was buying none of it. I was a taxpayer, an adult. She all but said, Go to jail.

There have been others. I remember the day when I had a ferocious argument with my son and realized that I could no longer bully him. He was too big and the days when I could just pick him up and take him to his room/isolation cell were over. I needed to persuade, reason. He was suddenly, rapidly,

Richard Cohen is a syndicated columnist for The Washington Post.

locked into traditional sex roles by parenthood, become less rigidly bound by these roles once the major duties of parenting decline; both are then freer to become more like the opposite sex—and do. Men, for example, often become more willing to share their feelings. These changes in both men and women can help older couples communicate and get along better.

During their early 40s, many of the women Helson and Moane studied shared the same midlife concerns the stage theorists had found in men: "concern for young and old, introspectiveness, interest in roots and awareness of limitation and death." But the Berkeley team described the period as one of midlife "consciousness," not "crisis."

In summing up their findings, Helson and Moane stress that commitment to the tasks of young adulthood—whether to a career or family (or both)—helped women learn to control impulses, develop skills with people, become independent and work hard to achieve goals.

50. Prime of Our Lives

older. The conclusion was inescapable: So was I.

One day you go to your friends' weddings. One day you celebrate the birth of their kids. One day you see one of their kids driving, and one day those kids have kids of their own. One day you meet at parties and then at weddings and then at funerals. It all happens in one day. Take my word for it.

I never thought I would fall asleep in front of the television set as my father did, and as my friends' fathers did, too. I remember my parents and their friends talking about insomnia and they sounded like members of a different species. Not able to sleep? How ridiculous. Once it was all I did. Once it was what I did best.

I never thought that I would eat a food that did not agree with me. Now I meet them all the time. I thought I would never go to the beach and not swim. I spent all of August at the beach and never once went into the ocean. I never thought I would appreciate opera, but now the pathos, the schmaltz and, especially, the combination of voice and music appeal to me. The deaths of Mimi and Tosca move me, and they die in my home as often as I can manage it.

I never thought I would prefer to stay home instead of going to a party, but now I find myself passing parties up. I used to think that people who watched birds were weird, but this summer I found myself watching them, and maybe I'll get a book on the subject. I yearn for a religious conviction I never thought I'd want, exult in my heritage anyway, feel close to ancestors long gone and echo my father in arguments with my son. I still lose.

One day I made a good toast. One day I handled a headwaiter. One day I bought a house. One day—what a day!—I became a father, and not too long after that I picked up the check for my own. I thought then and there it was a rite of passage for me. Not until I got older did I realize that it was one for him, too. Another milestone.

COPYRIGHT 1986, WASHINGTON POST WRITERS GROUP. REPRINTED WITH PERMISSION.

According to Helson and Moane, those women who did not commit themselves to one of the main life-style patterns faced fewer challenges and therefore did not develop as fully as the other women did.

The dizzying tug and pull of data and theories about how adults change over time may frustrate people looking for universal principles or certainty in their lives. But it leaves room for many scenarios for people now in young and middle adulthood and those to come.

People now between 20 and 60 are the best-educated and among the healthiest and most fit of all who have passed through the adult years. No one knows for sure what their lives will be like in the years to come, but the experts have some fascinating speculations.

For example, Rossi suspects that the quality of midlife for baby boomers will contrast sharply with that of the Depression-born generation the stage theorists studied. Baby boomers, she notes, have different dreams, values and opportunities than the preceding generation. And they are much more numerous.

Many crucial aspects of their past and future lives may best be seen in an economic rather than a strictly psychological light, Rossi says. From their days in overcrowded grade schools, through their struggles to gain entry into college, to their fight for the most desirable jobs, the baby boomers have had to compete with one another. And, she predicts, their competitive struggles are far from over. She foresees that many may find themselves squeezed out of the workplace as they enter their 50s—experiencing a crisis at a time when it will be difficult to redirect their careers.

But other factors may help to make life easier for those now approaching midlife. People are on a looser, less compressed timetable, and no longer feel obliged to marry, establish their careers and start their families almost simulta-

> *The first forty years of life furnish the text, while the remaining thirty supply the commentary.*
> —Schopenhauer, *Parerga and Paralipomena.*

neously. Thus, major life events may not pile up in quite the same way they did for the older generation.

Today's 20-year-olds—the first wave of what some have labeled "the baby busters"—have a more optimistic future than the baby boomers who preceded them, according to economist Richard Easterlin of the University of Southern California. Easterlin has been studying the life patterns of various cohorts, beginning with the low-birthrate group born in the 1930s—roughly a decade before the birthrate exploded.

The size of a birth cohort, Easterlin argues, affects that group's quality of life. In its simplest terms, his theory says that the smaller the cohort the less competition among its members and the more fortunate they are; the larger the cohort the more competition and the less fortunate.

Compared with the baby boomers, the smaller cohort just approaching adulthood "will have much more favorable experiences as they grow

6. MIDDLE AND LATE ADULTHOOD

WHAT'S THE RIGHT TIME?

Two surveys asking the same questions 20 years apart (late 1950s and late 1970s) have shown a dramatic decline in the consensus among middle-class, middle-aged people about what's the right age for various major events and achievements of adult life.

Activity/Event	Appropriate Age Range	Late '50s Study % Who Agree Men	Late '50s Study % Who Agree Women	Late '70s Study % Who Agree Men	Late '70s Study % Who Agree Women
Best age for a man to marry	20-25	80%	90%	42%	42%
Best age for a woman to marry	19-24	85	90	44	36
When most people should become grandparents	45-50	84	79	64	57
Best age for most people to finish school and go to work	20-22	86	82	36	38
When most men should be settled on a career	24-26	74	64	24	26
When most men hold their top jobs	45-50	71	58	38	31
When most people should be ready to retire	60-65	83	86	66	41
When a man has the most responsibilities	35-50	79	75	49	50
When a man accomplishes most	40-50	82	71	46	41
The prime of life for a man	35-50	86	80	59	66
When a woman has the most responsibilities	25-40	93	91	59	53
When a woman accomplishes most	30-45	94	92	57	48

SOURCE: ADAPTED FROM "AGE NORMS AND AGE CONSTRAINTS TWENTY YEARS LATER," P. PASSUTH, D. MAINES AND B.L. NEUGARTEN, PAPER PRESENTED AT THE MIDWEST SOCIOLOGICAL SOCIETY MEETING, CHICAGO, APRIL 1984.

up—in their families, in school and finally in the labor market," he says. As a result, they will "develop a more positive psychological outlook."

The baby busters' optimism will encourage them to marry young and have large families—producing another baby boom. During this period there will be less stress in the family and therefore, Easterlin predicts, divorce and suicide rates will stabilize.

Psychologist Elizabeth Douvan of the University of Michigan's Institute for Social Research shares Easterlin's optimistic view about the future of these young adults. Surprisingly, she sees as one of their strengths the fact that, due to divorce and remarriage, many grew up in reconstituted families. Douvan believes that the experience of growing up close to people who are not blood relatives can help to blur the distinction between kinship and friendship, making people more open in their relationships with others.

Like many groups before them, they are likely to yearn for a sense of community and ritual, which they will strive to fulfill in many ways, Douvan says. For some this may mean a turn toward involvement in politics, neighborhood or religion, although not necessarily the religion of their parents.

In summing up the future quality of life for today's young adults and those following them, Douvan says: "Life is more open for people now. They are judging things internally and therefore are more willing to make changes in the external aspects. That's pretty exciting. It opens up a tremendous number of possibilities for people who can look at life as an adventure."

Index

abortion: drug use during pregnancy and, 34; prenatal testing and, 17–18
activism, twentysomething generation and, 193
adolescents: 131, 158, 175–179, 180–183, 189–190, 235; new research regarding, 172–174
adults: 189–190, 208–215; abused children as, 123–126, 127; stages of, 233–240; see also, aging; men; parents; twentysomething generation; women
African Americans, see blacks
"age segregation," 182–183
Agent Orange, effect of, on sperm, 20
aggression: 76, 177; gender differences in, 44–46, 151–152, 153, 154
aging: of adults, 208–215, 220–224; sperm production and, 21; theories of, 216–219
alcohol: fetal alcohol syndrome and, 25–26, 33; effect of, on sperm, 21–22
alienation, four worlds of childhood and, 156–160
Alzheimer's disease, 13, 211, 213–214, 224
antibiotics: 223; life expectancy and, 203, 204
approach/withdrawal, in difficult children, 76
art: of children, 63–69, 92; therapy, and effect of violence on children, 137
arthritis, 223
asthma, maternal smoking and, 25
"attachment theory," 39
attention-deficit hyperactivity disorder, 98–104
"authoritarian" parents, 115
"authoritative" parents, 115
autonomy: in adolescents, 182, 189–190; in children, 87

"back-to-basics" movement, in education, 90, 92, 93, 165
Baumrind, Diana, 115
Belsky, Jay, 61, 119–120
benzene, effect of, on sperm, 19
birthweight, drug use during pregnancy and, 24–25
blacks: hidden obstacles to success of, 146–150; teenage pregnancy and, 182
Bloom, Benjamin: 162; mastery learning and, 163, 165–167
bonding, parent/infant, 75
Bowlby, John, 39
brain function, 109
Brazelton, T. Berry, 75, 140

caloric restriction, aging and, 217
cancer: 221–222, 223; genetic testing for, 13, 15; maternal smoking and, 25; treatment for, and life expectancy, 203, 204, 205, 228
careers, twentysomething generation and, 192–193
child abuse: effects of, 123–127; fetal abuse and, 31–35; mental disorders and, 10–11
child care, 60–62, 119–122, 141, 157
child development: 38–39; influence of child care on, 60–62; language acquisition and, 94–97
child-directed speech, 96
children: abuse of, 10–11, 31–35, 123–127; art of, 63–69; child care and, 60–62, 119–122, 141, 157; difficult, 75–77; discipline and, 114–115; of divorce, 128–133, 143; effect of early childhood education on, 49–52; historical view of, 206; indulged, 116–118; learning of, 88–93; motivation and, 72–74; effect of parent's careers on, 119–122; psychological problems of, 10–11; resilient, 72–74, 79–81; spending on, 6–9; stress and, 50, 76, 77, 79–81, 90, 109, 140–145, 157, 178; effect of violence on, 134–139; see also, adolescents; infants
China, preschool in, 53, 54, 56–57, 58, 59
chorionic villus sampling, 17
chromosomes, gene mapping and, 12–15
cocaine: effect of, on sperm, 22; see also, crack-exposed children
competition: blacks and, 148, 150; in children, 73
componential theory of intelligence, Sternberg's, 83, 84, 85
computers, learning disabled children and, 105–107
conformity, in adolescents, 186
contextual intelligence, Sternberg's theory of, 84, 85, 86, 87
cornucopia kids, 116–118
coronary bypass surgery, life expectancy and, 203, 204
corrective approaches, to treat dyslexia, 110
corrective instruction, in mastery learning, 166
courts, drug abuse during pregnancy and, 31, 33–35
crack, see cocaine; crack-exposed children
crack-exposed children, 26, 27, 29–30, 31–35
creative thinking, 83–84
creativity, 86
criminal behavior: abused children and, 124–125; gender differences in, 151, 152, 154
"criterion-referenced tests," 162
critical thinking, 83
"cross-linking," of protein, and aging, 217
cultural conservation, preschool in United States, China, and Japan as, 53
cycle of maltreatment, of child abuse, 124
cystic fibrosis, prenatal testing for, 12, 14, 16

dating, twentysomething generation and, 192
day care, see child care
dehydroepiandrosterone (DHEA), aging and, 217
delayed gratification, in children, 117
depression, 129, 130, 178, 186, 224, 227
developmentally appropriate practice, for children's learning, 88–93
diary, of baby, 40–43
difficult children, 75–77
dioxin, effect of, on sperm, 20
discipline, for children, 114–115, 117, 130
discrimination: against blacks, 34, 146; in fetal-abuse prosecution, 34; sex, 155
divorce: 158, 207, 234, 239; children of, 128–133, 143
Down's syndrome: prenatal genetic testing for, 14, 17; age of father and, 21

drug therapy, for hyperactive children, 101–103, 104
drug use: 120, 159, 185, 187; children's learning and, 23–28; pregnancy and, 23–28, 29–30, 31–35; effect of, on sperm, 22
dyslexia, 108–111

early childhood education, 49–52
education: for blacks, 146–150; in China, 53, 54, 56–57, 58, 59; in Japan, 51, 53–56, 58, 59; learning by children and, 88–93; for older people, 209; testing and, 161–164; twentysomething generation and, 193
effort-reward relationship, children's need to learn, 118
emotional development, 39
emotionality, gender differences in, 152
envy, 195–199
Erikson, Erik, 182, 229–230, 234, 235, 236
"error" theories, of aging, 216–217
eugenics, 18
exercise, for older adults, 222, 224
expectancy communications, 148, 150
experiential intelligence, Sternberg's theory of, 84, 85
expressive language learners, 95

family, twentysomething generation and, 192
family ties, life expectancy and, 202–207
feminists: fetal-abuse prosecution and, 34; gender differences and, 233, 234
fetal abuse, 31–35
fetal alcohol syndrome, 25–26, 33
"fetal protection policies," 19–21
fetuses, genetic testing of, 14, 16–18
formal operations, in children's thinking, 177

Gardner, Howard, 63
gender differences, 44–48, 108, 151–155, 177, 178, 199, 212
gene mapping, 12–15
genetics, diagnostic tests and, 16–18
genome, 12, 13, 15
gestural, children's art as, 63, 64
gifted people, 85, 86, 152
"good-citizens topics," 173
Gould, Roger, 234, 235, 236, 237
gratification, delayed, in children, 117

hands-on learning, 88–93
Head Start, 8, 167
health care, life expectancy and, 202–207
heart disease, genetic testing for, 15; in elderly, 221
hereditary disorders, 12–15, 16–18
hormone replacement therapy, 227–228
hormones: sex, and gender differences, 44–46; aging and, 217
hot flashes, 225, 226, 227
Human Genome Initiative, 13, 14, 16
Huntington's disease, prenatal genetic testing for, 13, 17
hyperactivity, 98–104
hysterectomies, 226, 227

idiot savants, see Savant syndrome
immune system, aging and, 217
immunization, life expectancy and, 203, 204
independence, in children, 76

infant development: 38–39; see also, child development
infants: 89; development of, 38–39; diary of, 40–43; speech and, 94–97
inferiority, rumors of black Americans, 146–150
insight, in gifted people, 85–86
intelligence: of blacks, 146–150; Sternberg's triarchic theory of, 82–87
intelligence quotient, see IQ
internalization, of inner controls, by children, 115
Intervention Program, for learning disabled children, 105–106
intuitive thinking, in women, 151, 152, 153
IQ: 67, 68, 82, 83, 84, 86, 109, 110, 148, 149; effect of fetal drug exposure on, 27–28

Japan: education in, 51, 91; preschool in, 53–56, 58, 59
jealousy, 195–199
job interests, gender differences in, 153

kindergarten, children under stress and, 141–142

language, development of, in children, 89, 91, 94–97
language learners, types of, 95
language receptive organization(LRO), effect of crack on, 30
latchkey children, 120
learning disability: dyslexia as, 108–111; computer programs for children with, 105–107
letter reversals, dyslexia and, 108
Levin, Henry M., 162–163
Levinson's ladder, of adult developmental stages, 234–235, 236, 237
life expectancy, influence of family ties on, 202–207; new treatments to increase, 220–224
life stage theories, 229–232, 233–240
life-span developmental theory, 123
limbic system, effect of crack exposure on child, 29, 30
limits, parental enforced, on children, 114, 115
linear perspective, in children's drawings, 64–65, 67
linguists, infants ability to learn language and, 94–97
look-say method, of teaching reading, 109

manic depression, as inherited disorder, 13, 15
marijuana: effect of, on child, 26, 27; effect of, on sperm, 22; use of, by child, 143
marriage, twentysomething generation and, 192
masculine role model, father as, 121, 123
mastery learning, Bloom's theory of, 163, 165–167
materialism, in children, 116–118
math ability, gender differences in, 151, 152
memory, 64, 68, 89, 213, 214, 224, 226
men: gender differences in, 151–155; effect of outside influence on sperm and, 19–22
menopause, 234, 235, 237
mental disorders, in children, 10–11
miseducation, 50, 90

mood: in adolescents, 176, 178, 186, 188; in difficult children, 76–77
morality, see ethics
mothers: drug use during pregnancy and, 31–35; infants and, 39; teenagers as 181; see also, adults; parents
motivation, to learn, 72–74, 116
multiple intelligence, 50
multisensory approach, to treat dyslexia, 110

negative expectancy communication, 148
negotiation, adolescents and, 173
neurofibrillary tangles, in Alzheimer's disease, 224
nicotine, see smoking
"norm-referenced tests," 162
nosocomical pneumonia, 223
nucleotides, 12, 14
nutrition: 222; life expectancy and, 205–206

Oakes, Jeannie, 162
Offer, Daniel, 172–173
oldest old, 211
open education, 90, 91
osteoarthritis (OA), 223–224
osteoporosis, 228
overburdened child, 130, 131

parents: 114–115, 167; effect of careers of, on children, 119–122; child's education and, 164; infant's speech and, 95, 96, 97; puberty and, 175, 176, 177, 184–188, 189–190; types of, 115
performance gap, for blacks, 147, 150
"permissive" parents, 115
personality, 15
perspective, linear, in children's drawings, 64–66
Piaget, Jean, 38, 64, 91, 177
plaques, in Alzheimer's disease, 224
pneumonia, in elderly, 221, 222–223
positive expectancy communication, 148, 150
post-traumatic stress disorder, effect of inner-city violence on children and, 134–139
poverty, 146, 206; children and, 142, 158, 159; elderly and, 208, 212
pregnancy: drug use and, 31–35; teen, 130, 181
prenatal testing, 14, 16–18
prerequisite training, in mastery learning, 166–167
preschool, in China, Japan, and United States, 53–59
primary aging, 210
primary effects, of memory, 64
problem solving, 165–166
prodigies, child, 13
progesterone, 225, 228
"program" theories, of aging, 216–217
progressive education, 89, 93
prospective studies, of child abuse cases, 124
protective factors, resilient children and, 79–81
protein "cross-linking," aging and, 217
puberty: parents and, 184–188; see also, adolescents

racism, 146, 147
radiation, effect of, on sperm, 22
"rate of living" theory, of aging, 217

rational thinking, 152
rebellion, in adolescents, 188
recency effects, memory and, 64
reference group expectancies, 149
referential language learners, 95
remedial approaches, to treat dyslexia, 109–110
representation, stages of, in children's art, 63–69
resilient children, 72–74, 79–81
restriction fragment-length polymorphisms (RFLPs), 14
retirement, 208, 209, 215
retrospective studies, of child abuse, 124
rheumatoid arthritis (RA), 223
Ritalin, 101, 102, 103, 104

sanitation, improvement in, and life expectancy, 205–206
savant syndrome, 66
school(s): 75, 158; adolescents and, 176, 177, 178–179; children's learning in, 88–93; as cultural communication, 156–160; elite, and early childhood education, 49–52; hyperactive children and, 99–100; preschool in China, United States, and Japan, 53–59; see also, child care; education; teachers
science, importance of learning, 168–169
secondary aging, 210
selective ignoring, to cope with jealousy and envy, 199
self-confidence: 50, 147; gender differences in, 151, 152, 153
self-esteem: 206, 227, child's, 50, 72, 74, 92, 117, 118, 167; envy and jealousy and, 196, 198, 199
sex: elderly and, 209; teenage, 143, 158, 177, 187
sex differences, see gender differences
sex hormones, gender differences and, 44–46
sex selection, prenatal testing and, 18
sexual abuse, 123, 125, 126
Sheehy, Gail, 230, 234, 235
sickle cell anemia, 14, 16
single-parent homes, 121, 143
"situational depression," hyperactive children and, 100
sleeper effect, in children of divorce, 129, 130, 131
"slippery slope" argument, fetal-abuse prosecution and, 34
slow-to-warm-up children, 75
Smetana, Judith G., 173
smoking: effect of maternal, on child, 25; effect of, on sperm, 21
social comparisons, in envy and jealousy, 196, 197
speech, of infants, 94–97
speech pathologists, 95
spelling errors, dyslexia and, 108
sperm, effect of outside influences on, and infant's health, 19–22
Spock, Benjamin, 114–115, 144
"spoiled-child syndrome," 114
stage theories, of adult development, 229–232, 233–240
stereotypes: of aging, 209; racial, 147, 149; sex role, 151, 155, 177, 238
Sternberg, Robert, triarchic intelligence theory of, 82–87
Strauss, Murray, 124

stress: 187, 206; adolescents and, 180–183; in children, 50, 76, 77, 79–81, 90, 109, 140–145, 157, 178; violence and, 134–139
strokes, 221, 224
suicide, children under stress and, 143
surgical menopause, 226

"tadpole," human, in children's art, 64, 65, 66
Tay-Sachs, prenatal testing and, 16, 17, 18
teachers: children's learning and, 88–93, 110; influence of, 150; strategies for, in dealing with difficult young children, 75–77; see also, education
teenage pregnancy, 130, 181
teenagers, see adolescents

television, effect of, on children, 143, 144, 157
temperment: children's, 75, 76, 77; see also, personality
testing, effect of, on education, 161–164
testosterone: 154, 188; gender differences and, 44–46
thalassemia, 16, 18
transitions, life stages and, 231
triarchic theory of intelligence, Sternberg's, 82–87
twentysomething generation, 191–194

UCLA Intervention Program, 105–106
United States, preschool in, 53, 54, 57–59
"use it or lose it," and older people, 210, 228

Vaillant, George, 234, 235, 237
values: 160, 184, 188; shift in paternal, 116, 117
violence: effect of inner-city, on children, 134–139; on television, and children, 143, 144
Visual Expressive Organization (VEO), effect of crack on, 30
Visual Receptive Organization (VRO), effect of crack on, 30

"wear and tear" theory, of aging, 216–217
women: "fetal protection policies" and, 19–21; gender differences and, 44–48, 151–155; life stages of, 229–232, 237–239; menopause and, 225–228

Zigler, Edward, 121–122

Credits/Acknowledgments

Cover design by Charles Vitelli

1. Genetic and Prenatal Influences
Facing overview—WHO photo.

2. Infancy and Early Childhood
Facing overview—United Nations photo by John Isaac.

3. Childhood
Facing overview—United Nations photo by John Isaac.

4. Family, School, and Cultural Influences
Facing overview—United Nations photo by L. Barnes.

5. Adolescence and Early Adulthood
Facing overview—United Nations photo by Jeffrey J. Foxx.
194—United Nations photo by L. Barnes.

6. Middle and Late Adulthood
Facing overview—United Nations photo by Jeffrey J. Foxx.

PHOTOCOPY THIS PAGE!!!*

ANNUAL EDITIONS ARTICLE REVIEW FORM

■ NAME: _____ DATE: _____

■ TITLE AND NUMBER OF ARTICLE: _____

■ BRIEFLY STATE THE MAIN IDEA OF THIS ARTICLE: _____
_____
_____
_____
_____

■ LIST THREE IMPORTANT FACTS THAT THE AUTHOR USES TO SUPPORT THE MAIN IDEA:
_____
_____
_____
_____
_____
_____
_____

■ WHAT INFORMATION OR IDEAS DISCUSSED IN THIS ARTICLE ARE ALSO DISCUSSED IN YOUR TEXTBOOK OR OTHER READING YOU HAVE DONE? LIST THE TEXTBOOK CHAPTERS AND PAGE NUMBERS:
_____
_____
_____
_____
_____
_____
_____

■ LIST ANY EXAMPLES OF BIAS OR FAULTY REASONING THAT YOU FOUND IN THE ARTICLE:
_____
_____
_____
_____

■ LIST ANY NEW TERMS/CONCEPTS THAT WERE DISCUSSED IN THE ARTICLE AND WRITE A SHORT DEFINITION:
_____
_____
_____
_____

*Your instructor may require you to use this Annual Editions Article Review Form in any number of ways: for articles that are assigned, for extra credit, as a tool to assist in developing assigned papers, or simply for your own reference. Even if it is not required, we encourage you to photocopy and use this page; you'll find that reflecting on the articles will greatly enhance the information from your text.

ANNUAL EDITIONS: HUMAN DEVELOPMENT 92/93
Article Rating Form

Here is an opportunity for you to have direct input into the next revision of this volume. We would like you to rate each of the 50 articles listed below, using the following scale:

1. **Excellent: should definitely be retained**
2. **Above average: should probably be retained**
3. **Below average: should probably be deleted**
4. **Poor: should definitely be deleted**

Your ratings will play a vital part in the next revision. So please mail this prepaid form to us just as soon as you complete it.
Thanks for your help!

We Want Your Advice

Annual Editions revisions depend on two major opinion sources: one is our Advisory Board, listed in the front of this volume, which works with us in scanning the thousands of articles published in the public press each year; the other is you—the person actually using the book. Please help us and the users of the next edition by completing the prepaid article rating form on this page and returning it to us. Thank you.

Rating	Article	Rating	Article
	1. Suffer the Little Children: Shameful Bequests to the Next Generation		28. Children After Divorce
	2. The Gene Dream		29. Children of Violence
	3. Made to Order Babies		30. Children Under Stress
	4. Sperm Under Siege		31. Rumors of Inferiority
	5. Clipped Wings		32. Biology, Destiny, and All That
	6. What Crack Does to Babies		33. Alienation and the Four Worlds of Childhood
	7. Motherhood on Trial		34. Tracked To Fail
	8. How Infants See the World		35. Master of Mastery
	9. Diary of a Baby		36. Not Just for Nerds
	10. Guns and Dolls		37. The Myth About Teen-Agers
	11. Preschool: Head Start or Hard Push?		38. Those Gangly Years
	12. How Three Key Countries Shape Their Children		39. A Much Riskier Passage
	13. The Day Care Generation		40. Puberty and Parents: Understanding Your Early Adolescent
	14. Where Pelicans Kiss Seals		41. Therapists Find Last Outpost of Adolescence in Adulthood
	15. Building Confidence		
	16. Dealing With Difficult Young Children		42. Proceeding With Caution
	17. The Miracle of Resiliency		43. Jealousy and Envy: The Demons Within Us
	18. Three Heads Are Better Than One		
	19. How Kids Learn		44. Family Ties: The Real Reason People Are Living Longer
	20. Now We're Talking!		
	21. Suffer the Restless Children		45. The Vintage Years
	22. Tykes and Bytes		46. Why Do We Age?
	23. Facts About Dyslexia		47. A Vital Long Life: New Treatments for Common Aging Ailments
	24. Dr. Spock Had It Right		
	25. Positive Parenting		48. The Myths of Menopause
	26. Can Your Career Hurt Your Kids?		49. Don't Act Your Age!
	27. The Lasting Effects of Child Maltreatment		50. The Prime of Our Lives

(Continued on next page)

ABOUT YOU

Name_____ Date_____

Are you a teacher? ☐ Or student? ☐
Your School Name _____
Department _____
Address _____
City _____ State _____ Zip _____
School Telephone # _____

YOUR COMMENTS ARE IMPORTANT TO US!

Please fill in the following information:

For which course did you use this book? _____
Did you use a text with this Annual Edition? ☐ yes ☐ no
The title of the text? _____
What are your general reactions to the Annual Editions concept?

Have you read any particular articles recently that you think should be included in the next edition?

Are there any articles you feel should be replaced in the next edition? Why?

Are there other areas that you feel would utilize an Annual Edition?

May we contact you for editorial input?

May we quote you from above?

ANNUAL EDITIONS: HUMAN DEVELOPMENT 92/93

BUSINESS REPLY MAIL
First Class   Permit No. 84   Guilford, CT

Postage will be paid by addressee

The Dushkin Publishing Group, Inc.
Sluice Dock
DPG **Guilford, Connecticut 06437**

No Postage
Necessary
if Mailed
in the
United States